Q 172.5 .C45 M66 1992
Moon, F. C., 1939-
Chaotic and fractal dynamics

5/26/93

D1541952

CHAOTIC AND
FRACTAL DYNAMICS

D. HIDEN RAMSEY LIBRARY
U.N.C. AT ASHEVILLE
ASHEVILLE, N. C. 28804

CHAOTIC AND FRACTAL DYNAMICS

An Introduction for Applied Scientists and Engineers

FRANCIS C. MOON

Mechanical and Aerospace Engineering
Cornell University
Ithaca, New York

A Wiley-Interscience Publication

JOHN WILEY & SONS, INC.

New York · Chichester · Brisbane · Toronto · Singapore

This text is printed on acid-free paper.

Copyright © 1992 by John Wiley & Sons, Inc.

All rights reserved. Published simultaneously in Canada.

Reproduction or translation of any part of this work beyond
that permitted by Section 107 or 108 of the 1976 United
States Copyright Act without the permission of the copyright
owner is unlawful. Requests for permission or further
information should be addressed to the Permissions Department,
John Wiley & Sons, Inc., 605 Third Avenue, New York, NY
10158-0012.

Library of Congress Cataloging in Publication Data:
Moon, F. C., 1939–
   Chaotic and fractal dynamics : an introduction for applied
scientists and engineers / Francis C. Moon.
       p.   cm.
   "A Wiley-Interscience publication."
   Includes bibliographical references and index.
   ISBN 0-471-54571-6
   1. Chaotic behavior in systems.   2. Dynamics.   3. Fractals.
I. Title.
Q172.5.C45M66   1992
003'.7—dc20   92-25350

Printed in the United States of America

10 9 8 7 6 5 4 3 2 1

*To my grandchildren
Anne Leise and Calum
Sources of new creative chaos in my world*

D. HIDEN RAMSEY LIBRARY
U.N.C. AT ASHEVILLE
ASHEVILLE, N. C. 28804

CONTENTS

PREFACE

Had anyone predicted that new discoveries could be made in dynamics 300 years after publication of Newton's *Principia*, they would have been thought naive or foolish. Yet, in the decade 1977–1987, new phenomena in nonlinear dynamics were discovered, principal among these being chaotic and unpredictable behavior from apparently deterministic systems. Since publication in 1987 of *Chaotic Vibrations*, the first edition of this book, new discoveries in dynamics have been made in many of the sciences, including biology. And, what should be of special interest for the applied scientist or engineer is the emergence of applications of the new ideas in chaotic dynamics and fractals. Chaotic dynamics has been known to be a common occurrence in fluid mechanics, and turbulence remains one of the unsolved problems of classical physics. However, it is now generally accepted that unpredictable dynamics can be found quite easily in simple electrical and mechanical systems as well as in other physical systems.

The purpose of this book is to help translate the new mathematical ideas in nonlinear dynamics into language that engineers and scientists can use and apply to physical systems. Many fine books have been written on chaos, fractals, and nonlinear dynamics (see e.g., Appendix D), but most have focused on the mathematical principles. Many readers of the first edition cited the inclusion of many physical examples as an important feature of the book, and they have urged me to keep the physical nature of chaos as a hallmark of any new edition. The decision to make a substantial rewrite of *Chaotic Vibrations* was based on feedback from a number of readers. They asked for more tutorial material on maps or difference equations and fractals, and they wanted some problems so that the book could be used as a basis of a course.

In this book I have tried to start from a background that a B.S. engineering or science graduate would have; namely, ordinary differential equations and some intermediate-level dynamics and vibrations or system dynamics courses. I have also taken the view of an experimentalist, namely that the book should provide some tools to measure, predict, and quantify chaotic dynamics in physical systems.

Chapter 1 includes an introduction to classical nonlinear dynamics; however, if the book is used as a text, additional supplemental material is recommended. Chapter 2 presents an experimentalist's view of chaotic dynamics along with some simple tools such as the Poincaré map. Chapter 3 introduces maps and is entirely new. It is an attempt to summarize the basic concepts of coupled iterative difference equations as they relate to chaotic dynamics. Chapter 4 is a much expanded litany of physical applications with lots of references to experimental observations of chaos along with the appropriate mathematical models. Many readers have found the discussion of experimental methods (Chapter 5) to be useful, and this too has been expanded. If Chapter 2 asks the question, "How do we recognize chaos?," then in Chapter 6 we ask, "How do we predict when chaos will occur?" Topics such as period doubling, homoclinic bifurcations, Shilni'kov chaos, and Lyaponov exponents are discussed here. The treatment of fractals has been much expanded in the new Chapter 7, including an introduction to multifractals. One of the new directions in chaos research has been in spatiotemporal dynamics. An introduction to some of the simple models of spatially extended systems including dynamics of chain systems and Lagrangian chaos are discussed in Chapter 8. Finally, in Appendix C, an expanded list of chaotic toys and experiments is presented; a guide to some of the more popular books on chaos and fractals is given in Appendix D.

Although over 100 new references have been included in this new edition, it became clear that the tremendous growth in papers on chaos and fractals in the last few years would make it impossible to cover all the significant papers. I apologize to those researchers whose fine contributions have not been cited, especially those who took the time to send me papers, photos, and software. The inclusion of more of the papers from my own Cornell research laboratory must be interpreted as an author's vanity and not any measure of their relative importance to the field.

I have written this new edition not only because of the success of the first, but because I believe the new ideas of chaos and fractals are important to the fields of applied and engineering dynamics. It is already evident that these new geometric and topological concepts have become part of the laboratory tools in dynamics analysis in the same way that Fourier analysis became an important part of engineering systems dynamics decades ago. Already, these tools have found application in areas such as machine noise, impact printer dynamics, nonlinear circuit design, laser instabilities, mixing of chemicals, and even in understanding the dynamics of the human heart. This book is only an introduction to the subject, and it is hoped that interested students would be inspired to explore the more advanced aspects of chaos and fractals, not only for its potential application, but for the fascination and beauty of the basic mathematical ideas which underlie this subject.

ACKNOWLEDGMENTS

Many of the examples presented in this book reflect one and a half decades of research on nonlinear dynamics at Cornell University, especially in the Nonlinear Dynamics and Magneto-Mechanics Laboratory. This unique facility has had many contributors and sponsors. First, I must thank my colleagues at Cornell in Theoretical and Applied Mechanics, John Guckenheimer, Phillip Holmes, Subrata Mukherjee, and Richard Rand, who have always been a source of advice and criticism to myself and other students of the Laboratory. Any deliberate lack of mathematical rigor in this book, however, must be blamed on me.

With regard to the sponsors, I acknowledge generous support from the Air Force Office of Scientific Research, the Army Research Office, the Department of Energy, IBM Corporation, the National Science Foundation, the National Aeronautics and Space Administration, and the Office of Naval Research. I am especially grateful for the long-time support of the Air Force Office of Scientific Research.

Among the present and former graduate students who have contributed much to the life of this laboratory are G. S. Copeland (David Taylor Naval Lab), J. Cusumano (Penn State University), M. Davies, B. F. Feeny (Institut für Robotik, ETH Zürich), M. Golnaraghi (University of Waterloo), P.-Y. Chen (Chung-Shan Institute of Science and Technology, Taiwan), C.-K. Lee (IBM San José), G.-X. Li (Heroux, Inc., Montreal), G. Muntean, O. O'Reilly, and P. Schubring (U. California, Berkley). Many undergraduate students have made contributions, including P. Khoury (University of California at Berkeley).

Among the many visiting faculty and postdoctoral researchers who have worked with me are F. Benedettinni (Universitá Degli Studi di L'Aquila, Italy), M. El Naschie (Cambridge University), M. Païdoussis (McGill University), C. H. Pak (Inha University, Korea), K. Sato (Hachinohe National College of Technology, Japan), and T. Valkering (University of Twente, The Netherlands).

Finally, I want to acknowledge the contribution of two staff members of this Laboratory: Cora Lee Jackson, who deciphered the scribblings of me and my students in typing our many reports and papers and who typed the final manuscript, and William Holmes, who kept computers and research equipment in top shape and who helped design many of the experiments.

1

INTRODUCTION: A NEW AGE OF DYNAMICS

In the beginning, how the heav'ns and earth rose out of chaos.
J. Milton
Paradise Lost, 1665

1.1 WHAT IS CHAOTIC DYNAMICS?

For some, the study of dynamics began and ended with Newton's Law of $F = mA$. We were told that if the forces between particles and their initial positions and velocities were given, one could predict the motion or history of a system forever into the future, given a big enough computer. However, the arrival of large and fast computers has not fulfilled the promise of infinite predictability in dynamics. We now know that the motion of very simple dynamical systems cannot always be predicted far into the future. Such motions have been labeled *chaotic*, and their study has promoted a discussion of some exciting new mathematical ideas in dynamics. Three centuries after the publication of Newton's *Principia* (1687), it is appropriate that new phenomena have been discovered in dynamics and that new mathematical concepts from topology and geometry have entered this venerable science.

1

The nonscientific concept of chaos[1] is very old and is often associated with a physical state or human behavior without pattern and out of control. The term *chaos* often stirs fear in humankind because it implies that governing laws or traditions no longer have control over events such as prison riots, civil wars, or a world war. Yet there is always the hope that some underlying force or reason is behind the chaos or can explain why seemingly random events appear unpredictable.

In the physical sciences, the paragon of chaotic phenomena is turbulence. Thus, a rising column of smoke or the eddies behind a boat or aircraft wing[2] provide graphic examples of chaotic motion. For example, the flow pattern behind a cylinder (Figure 1-1) and the mixing of drops of color in paint (Color Plate 10) illustrate the basic nature of chaotic dynamics. The fluid mechanician, however, believes that these events are not random because the governing equations of physics for each fluid element can be written down. Also, at low velocities, the fluid patterns are quite regular and predictable from these equations. Beyond a critical velocity, however, the flow becomes turbulent. A great deal of the excitement in nonlinear dynamics today is centered around the hope that this transition from ordered to disordered flow may be explained or modeled with relatively simple mathematical equations. What we hope to show in this book is that these new ideas about turbulence extend to other problems in physics as well. It is the recognition that chaotic dynamics are inherent in all of nonlinear physical phenomena that has created a sense of revolution in physics today.

We must distinguish here between so-called random and chaotic motions. The former is reserved for problems in which we truly do not know the input forces or we only know some statistical measures of the parameters. The term *chaotic* is reserved for those *deterministic* problems for which there are no random or unpredictable inputs or

[1] The origin of the word *chaos* is a Greek verb which means *to gape open* and which was often used to refer to the primeval emptiness of the universe before things came into being (*Encyclopaedia Britannica*, Vol. 5, p. 276). To the stoics, chaos was identified with water and the watery state which follows the periodic destruction of the earth by fire. In *Metamorphoses*, Ovid used the term to denote the raw and formless mass in which all is disordered and from which the ordered universe is created. A modern dictionary definition of chaos (Funk and Wagnalls) provides two meanings: (i) utter disorder and confusion and (ii) the unformed original state of the universe.

[2] The reader should look at the beautiful collection of photos of fluid turbulent phenomena compiled by Van Dyke (1982).

Figure 1-1 Turbulent eddies in the flow of fluid behind a cylinder. (Courtesy of C. Williamson, Cornell University.)

parameters. The existence of chaotic or unpredictable motions from the classical equations of physics was known by Poincaré.[3] Consider the following excerpt from his essay on *Science and Method*:

> It may happen that small differences in the initial conditions produce very great ones in the final phenomena. A small error in the former will produce an enormous error in the latter. Prediction becomes impossible.

[3] Henri Poincaré (1854–1912) was a French mathematician, physicist, and philosopher whose career spanned the grand age of classical mechanics and the revolutionary ideas of relativity and quantum mechanics. His work on problems of celestial mechanics led him to questions of dynamic stability and the problem of finding precise mathematical formulas for the dynamic history of a complex system. In the course of this research he invented the "the method of sections," now known as the *Poincaré section* or *Poincaré map*. See Holmes (1990b) for a modern discussion of Poincaré's work.

An excellent discussion of uncertainties and determinism and Poincaré's ideas on these subjects may be found in the very readable book by L. Brillouin (1964, Chapter IX).

In the current literature, *chaotic* is a term assigned to that class of motions in deterministic physical and mathematical systems whose time history has a *sensitive dependence on initial conditions*.

Two examples of mechanical systems that exhibit chaotic dynamics are shown in Figure 1-2. The first is a thought experiment of an idealized billiard ball (rigid body rotation is neglected) which bounces off the sides of an elliptically shaped billiard table. When elastic impact is assumed, the energy remains conserved, but the ball may wander around the table without exactly repeating a previous motion for certain elliptically shaped tables.

Another example, which the reader with access to a laboratory can see for oneself (see Appendix C), is the ball in a two-well potential shown in Figure 1-2*b*. Here the ball has two equilibrium states when the table or base does not vibrate. However, when the table vibrates with periodic motion of large enough amplitude, the ball will jump from one well to the other in an apparently random manner; that is, periodic input of one frequency leads to a randomlike output with a

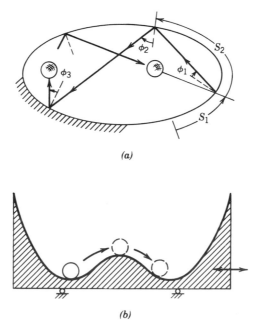

(a)

(b)

Figure 1-2   (*a*) The motion of a ball after several impacts with an eliptically shaped billiard table. The motion can be described by a set of discrete numbers (s_i, ϕ_i) called a *map*. (*b*) The motion of a particle in a two-well potential under periodic excitation. Under certain conditions, the particle jumps back and forth in a periodic way—that is, LRLR ... , or LLRLLR ... , and so on. For other conditions the jumping is chaotic—that is, it shows no pattern in the sequence of symbols L and R.

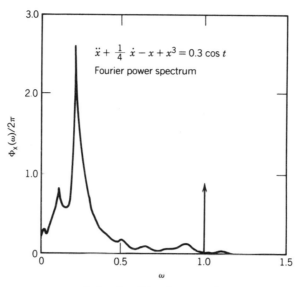

Figure 1-3   The power spectral density (Fourier transform) of chaotic motion in a two-well potential. (After Y. Ueda, Kyoto University.)

broad spectrum of frequencies. The generation of a continuous spectrum of frequencies below the single input frequency is one of the characteristics of chaotic vibrations (Figure 1-3).

Loss of information about initial conditions is another property of a chaotic system. Suppose one has the ability to measure a position with accuracy Δx and a velocity with accuracy Δv. Then in the position–velocity plane (known as the *phase plane*) we can divide up the space into areas of size $\Delta x \, \Delta v$ as shown in Figure 1-4. If we are given initial conditions to the stated accuracy, we know the system is somewhere in the shaded box in the phase plane. But if the system is chaotic, this uncertainty grows in time to $N(t)$ boxes as shown in Figure 1-4b. The growth in uncertainty given by

$$N \approx N_0 e^{ht} \tag{1-1.1}$$

is another property of chaotic systems. The constant h is related to the concept of *entropy* in information theory (e.g., see Shaw, 1981, 1984) and will also be related to another concept called the *Lyapunov exponent* (see Chapter 6), which measures the rate at which nearby trajectories of a system in phase space diverge. A *positive* value of this Lyapunov exponent for a particular dynamical system is a quantitative measure of chaos.

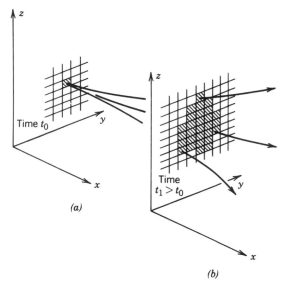

Figure 1-4 An illustration of the growth of uncertainty or loss of information in a dynamical system. The black box at time $t = t_0$ represents the uncertainty in initial conditions.

Why Fractal Dynamics?

The reader may ask: With predictability lost in chaotic systems, is there any order left in the system? For dissipative systems the answer is yes; there is an underlying structure to chaotic dynamics. This structure is not apparent by looking at the dynamics in the conventional way, that is, the output versus time or from frequency spectra. One must search for this order in phase space (position versus velocity). There one will find that chaotic motion exhibits a new geometric property called *fractal* structure. Examples of fractal patterns are illustrated in the color plates. Fractals are geometric structures that appear at many scales. One of the goals of this book is to teach how to discover the fractal structure in chaotic vibrations as well as to measure the loss of information in these randomlike motions.

Why Study Chaotic Dynamics?

The subject of chaos has certainly become newsworthy over the past few years—the study of mathematical chaos that is. Many popular magazines have carried articles on the new studies into mathematics of chaotic dynamics. But engineers have always known about chaos—it was called *noise* or *turbulence,* and "fudge" factors or factors of safety were used to design around these apparent random unknowns that seem to crop up in every technical device. So what is new about chaos?

First, the recognition that chaotic vibrations can arise in low-order, nonlinear deterministic systems raises the hope of understanding the source of randomlike noise and doing something about it. Second, the new discoveries in nonlinear dynamics bring with them new concepts and tools for detecting chaotic vibrations in physical systems and for quantifying this "deterministic noise" with new measures such as fractal dimensions and Lyapunov exponents.

Since the turn of the century, mathematicians have also known that certain dynamical systems possessed irregular solutions. Poincaré, as noted in the above quote, was aware of chaotic solutions, as was Birkhoff in the early part of this century. Van der Pol and Van der Mark (1927) reported "irregular noise" in experiments with an electronic oscillator in the magazine *Nature.* So what is new about chaos?

What is new about chaotic dynamics is the discovery of a seemingly underlying order which holds out the promise of being able to predict certain properties of noisy behavior. Perhaps the greatest hope lies in the possibility of understanding turbulence in fluid, thermofluid, and thermochemical systems. Turbulence is one of the few remaining unsolved problems of classical physics, and the recent discovery of deterministic systems which exhibit chaotic oscillations has created much optimism about solving the mysteries of turbulence. But already this optimism has been tempered by the complexities of chaotic dynamics in thermofluid systems, especially the spatial aspects of fluid flow as illustrated in Figure 1-1. However, there may be more immediate payoffs in the study of chaotic phenomena in systems with fewer degrees of freedom, such as low-order nonlinear mechanical devices and nonlinear circuits.

Sources of Chaos

Chaotic vibrations occur when some strong nonlinearity exists. Examples of nonlinearities in mechanical and electromagnetic systems include the following:

Gravitational forces in the solar system
Nonlinear elastic or spring elements
Nonlinear damping such as friction

Backlash, play, or limiters or bilinear springs
Fluid-related forces
Nonlinear boundary conditions
Nonlinear feedback control forces in servosystems
Nonlinear resistive, inductive, or capacitative circuit elements
Diodes, transistors, and other active devices
Electric and magnetic forces
Nonlinear optical properties, lasers

In mechanical continua, nonlinear effects arise from a number of different sources which include the following:

1. Kinematics—for example, convective acceleration, Coriolis and centripetal accelerations
2. Constitutive relations—for example, stress versus strain
3. Boundary conditions—for example, free surfaces in fluids, deformation-dependent constraints
4. Nonlinear body forces—for example, magnetic or electric forces
5. Geometric nonlinearities associated with large deformations in structural solids such as beams, plates and shells

How such nonlinearities enter the laws of mechanics can be seen by looking at the equation of momentum balance in continuum mechanics,

$$\nabla \cdot \mathbf{t} + \mathbf{f} = \frac{\partial \mathbf{v}}{\partial t} + \mathbf{v} \cdot \nabla \mathbf{v} \tag{1-1.2}$$

where $\mathbf{t}$ is the stress tensor, $\mathbf{f}$ is the body force, and the right-hand side represents the acceleration. Nonlinearities can enter this equation through the stress–strain or stress–strain rate relations in the first left-hand term. Nonlinear body forces such as occur in magnetohydrodynamics or plasma physics can enter the body force term $\mathbf{f}$. Finally, on the right-hand side of Eq. (1-1.2), we see an explicit nonlinear term in the convective acceleration. This term appears in many fluid flow problems and is one of the sources of turbulence in fluids.

In the classic Navier–Stokes equations of fluid mechanics, derived from the momentum balance Eq. (1-1.2), one can see that the nonlinearity resides in the convective acceleration or kinematic term:

$$\nu \nabla^2 \mathbf{v} - \nabla P = \frac{\partial \mathbf{v}}{\partial t} + \mathbf{v} \cdot \nabla \mathbf{v} \qquad (1\text{-}1.3)$$

where ν is the kinematic viscosity and P is the pressure. The viscous term on the left-hand side is linear and is based on the assumption of a Newtonian fluid.

One can imagine that if one goes beyond the study of the Navier–Stokes equation to include nonlinear viscous fluids (non-Newtonian fluids) or elastoplastic materials, there is a vast array of nonlinear and chaotic phenomena to be discovered in mechanics, electromagnetics, and acoustics.

Where Have Chaotic Vibrations Been Observed?

From the previous discussion, one can see that chaotic phenomena can be observed in many physical systems. Since the writing of the first edition of this book, many new phenomena have been reported in the scientific and engineering literature. A partial list of the physical systems known to exhibit chaotic vibrations includes the following:

Selected closed- and open-flow fluid problems
Selected chemical reactors
Vibrations of buckled elastic structures
Mechanical systems with play or backlash such as gears
Flow-induced or aeroelastic problems
Magnetomechanical actuators
Large, three-dimensional vibrations of structures such as beams and shells
Systems with sliding friction
Rotating or gyroscopic systems
Nonlinear acoustic systems
Simple forced circuits with diodes or p–n transistor elements
Harmonically forced circuits with nonlinear capacitance and inductance elements
Feedback control devices
Laser and nonlinear optical systems
Video feedback

Systems which are suspected of behaving in a chaotically dynamic way include:

A few objects in the solar system (e.g., Hyperion, Halley's comet)
Cardiac oscillations
Earthquake dynamics
Extreme maneuver aircraft and ship dynamics
Iterative optimal design algorithms
Econometric models

These are but a few of the many phenomena in which chaos has been uncovered. Descriptions of specific examples are given in Chapter 4. A question asked by most novices to the field of chaotic dynamics is: If chaos is so pervasive, why was it not seen earlier in experiments? Two responses to this question come to mind. First, if one goes back and reads earlier papers on experiments in nonlinear vibrations, one often finds a brief mention of nonperiodic phenomena buried in a discussion of more classical nonlinear vibrations (see Chapter 4 for examples). Second, Joseph Keller, an applied mathematician at Stanford University, in responding to this question in a lecture, speculates that earlier scientists and engineers were taught almost exclusively in linear mathematical ideas, including linear algebra and differential equations. Hence, it was natural, Keller summarizes, that when approaching dynamic experiments in the laboratory, they looked only for phenomena that fit the linear mathematical models.

As to why theorists had not come upon those ideas earlier, there is evidence that some did, like Poincaré and Birkhoff. And, those dynamicists working in energy-conserving systems (Hamiltonian dynamics), especially theorists in the former Soviet Union, knew about stochastic behavior in certain theoretical models (see, e.g., Sagdeev et al., 1988). However, specific manifestations of chaotic solutions had to wait for the arrival of powerful computers with which to calculate the long time histories necessary to observe and measure chaotic behavior. Some day in the future an interesting history will be written on the interdependence between computer technology and the mathematics of fractal and nonlinear processes in the late 20th century.

1.2   CLASSICAL NONLINEAR VIBRATION THEORY: A BRIEF REVIEW

In this section, we present a short review of classical vibration theory, both linear and nonlinear. This is meant simply to define and review a few ideas in nonlinear dynamics concerning periodic vibration so we may later be able to contrast these with chaotic vibration. Readers desiring more detailed discussion in classical nonlinear vibration should consult books such as Stoker (1950), Minorsky (1962), Nayfeh and Mook (1979), or Hagedorn (1988). We begin with a brief review of linear vibration concepts.

Linear Vibration Theory

The classic paradigm of linear vibrations is the spring–mass system shown in Figure 1-5 along with its electric circuit analog. When there is no disturbing force, the undamped system vibrates with a frequency that is independent of the amplitude of vibration:

$$\omega_0 = \left(\frac{k}{m}\right)^{1/2} = \left(\frac{1}{LC}\right)^{1/2} \tag{1-2.1}$$

In this state, energy flows alternately between elastic energy in the spring (electric energy in the capacitor C) and kinetic energy in the mass (magnetic energy in the inductor L). The addition of damping $c \neq 0$, $R \neq 0$) introduces decay in the free vibrations so that the amplitude of the mass (or charge in the circuit) exhibits the following time dependence:

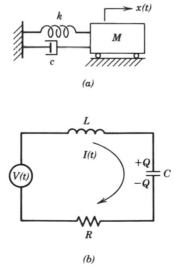

(a)

(b)

Figure 1-5 (*a*) The classic, mechanical spring–mass–dashpot oscillator. (*b*) The electrical circuit analog of a damped oscillator.

$$x(t) = A_0 e^{-\gamma t} \cos[(\omega_0^2 - \gamma^2)^{1/2} t + \varphi_0] \qquad (1\text{-}2.2)$$

where

$$\gamma = \frac{c}{2m} \quad \text{or} \quad \gamma = \frac{R}{2L}$$

The system is said to be *underdamped* when $\gamma^2 < \omega_0^2$, *critically damped* when $\gamma^2 = \omega_0^2$, and *overdamped* when $\gamma^2 > \omega_0^2$.

One of the classic phenomena of linear vibratory systems is that of *resonance* under harmonic excitation. For this problem, the differential equation that models the system is of the form (e.g., see Thompson, 1965)

$$\ddot{x} + 2\gamma\dot{x} + \omega_0^2 x = f_0 \cos \Omega t \qquad (1\text{-}2.3)$$

If one fixes f_0 and varies the driving frequency Ω, the absolute magnitude of the steady-state displacement of the mass (after transients have damped out) reaches a maximum close to the natural frequency ω_0, or more precisely at $\Omega = (\omega_0^2 - \gamma^2)^{1/2}$. This phenomenon is sketched in Figure 1-6. The effect is more pronounced when the damping is small. This is indeed the case in structural systems, and engineers are familiar with the problem of fatigue failures in structures and machines owing to large, resonance-excited vibrations.

When a linear mechanical system has many degrees of freedom,

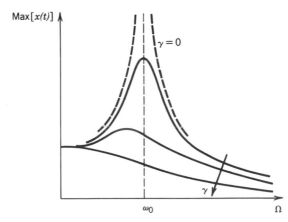

Figure 1-6   Classical resonance curves (response amplitude versus frequency) for the forced motion of a damped *linear* oscillator for different values of the damping γ.

one often models it as a coupled set of spring–mass oscillators leading to the phenomena of multiple resonant frequencies when the system is harmonically forced. This behavior has often led vibration analysts to assume that every peak in a vibration frequency spectrum is associated with at least one mode of degree of freedom. In nonlinear vibrations, this is not the case. A one-degree-of-freedom nonlinear system can generate many frequency spectra in contrast to its linear counterpart, as was shown in Figure 1-3. In any event, the mathematical theory of linear systems is well understood and has been codified in sophisticated computer software packages. Nonlinear problems are another story.

Nonlinear Vibration Theory

Nonlinear effects can enter the problem in many ways. A classic example is a nonlinear spring where the restoring force is not linearly proportional to the displacement. For the case of a symmetric nonlinearity (equal effects for compression or tension), the equation of motion takes the following form:

$$\ddot{x} + 2\gamma\dot{x} + \alpha x + \beta x^3 = f(t) \qquad (1\text{-}2.4)$$

When the system is undamped and $f(t) = 0$, periodic solutions exist where the natural frequency increases with amplitude for $\beta > 0$. This model is often called a *Duffing equation,* after the mathematician who studied it.

If the system is acted on by a periodic force, in the classical theory one assumes that the output will also be periodic. When the output has the same frequency as the force, the resonance phenomena for the nonlinear spring is shown in Figure 1-7. If the amplitude of the forcing function is held constant, there exists a range of forcing frequencies for which three possible output amplitudes are possible as shown in Figure 1-7. One can show that the dashed curve in Figure 1-7 is unstable so that a *hysteretic effect* occurs for increasing and decreasing frequencies. This is called a *jump phenomenon* and can be observed experimentally in many mechanical and electrical systems.

Other periodic solutions can also be found such as *subharmonic* and *superharmonic* vibrations. If the driving force has the form $f_0 \cos \omega t$, then a subharmonic oscillation may take the form $x_0 \cos(\omega t/n + \varphi)$ plus higher harmonics (n is an integer). Subharmonics play an important role in prechaotic vibrations, as we shall see later.

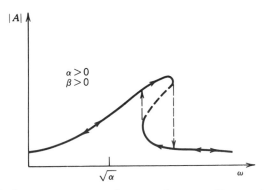

Figure 1-7   Classical resonance curve for a *nonlinear* oscillator with a hard spring when the response is periodic with the same period as the driving force. [α and β refer to Eq. (1-2.4).]

Nonlinear resonance theory depends on the assumption that periodic input yields periodic output. However, it is this postulate that has been challenged in the new theory of chaotic vibrations.

Self-excited oscillations are another important class of nonlinear phenomena. These are oscillatory motions which occur in systems that have no periodic inputs or periodic forces. Several examples are shown in Figure 1-8. In the first, the friction created by relative motion between a mass and moving belt leads to oscillations. In the second example there exists the whole array of aeroelastic vibrations in which the steady flow of fluid past a solid object on elastic restraints produces steady-state oscillations. A classic electrical example is the vacuum tube circuit studied by Van der Pol and shown in Figure 1-9.

In each case, there is a steady source of energy, a source of dissipation, and a nonlinear restraining mechanism. In the case of the Van der Pol oscillator, the source of energy is a dc voltage. It manifests itself in the mathematical model of the circuit as a negative damping:

$$\ddot{x} - \gamma \dot{x}(1 - \beta x^2) + \omega_0^2 x = 0 \qquad (1\text{-}2.5)$$

For low amplitudes, energy can flow into the system, but at higher amplitudes the nonlinear damping limits the amplitude.

In the case of the Froude pendulum (e.g., see Minorsky, 1962, Chap. 28), the constant rotation of the motor provides an energy input. For small vibrations the nonlinear friction is modeled as negative damping, whereas for large vibrations the amplitude of the vibration is limited by the nonlinear term $\beta \dot{\theta}^3$:

$$\ddot{\theta} + \alpha \sin \theta = T_0 + \gamma \dot{\theta}(1 - \beta \dot{\theta}^2) \qquad (1\text{-}2.6)$$

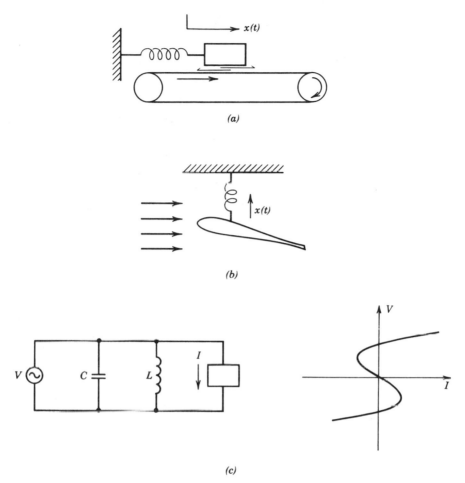

Figure 1-8 Example of self excited oscillations: (*a*) dry friction between a mass and moving belt; (*b*) aeroelastic forces on a vibrating airfoil; and (*c*) negative resistance in an active circuit element.

The oscillatory motions of such systems are often called *limit cycles*. The phase plane trajectories for the Van der Pol equation is shown in Figure 1-10. Small motions spiral out to the closed asymptotic trajectory, whereas large motions spiral onto the limit cycle. (In Figures 1-10 and 1-11, $y = \dot{x}$.)

Two questions are often asked when studying problems of this kind: (1) What is the amplitude and frequency of the limit cycle vibrations? (2) For what parameters will stable limit cycles exist?

In the case of the Van der Pol equation, it is convenient to normalize the dependent variable by $\sqrt{\beta}$ and the time by ω_0^{-1} so the equation assumes the form

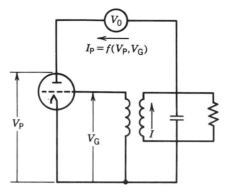

Figure 1-9   Sketch of a vacuum tube circuit with limit cycle oscillation of the type studied by Van der Pol.

$$\ddot{x} - \varepsilon \dot{x}(1 - x^2) + x = 0 \qquad (1\text{-}2.7)$$

where $\varepsilon = \gamma/\omega_0$. For small ε, the limit cycle solution is a circle of radius 2 in the phase plane; that is,

$$x \simeq 2 \cos t + \cdots \qquad (1\text{-}2.8)$$

where the $+ \cdots$ indicates third-order harmonics and higher. When ε is larger, the motion takes the form of *relaxation oscillations* shown

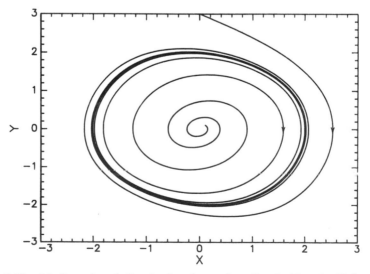

Figure 1-10   Limit cycle solution in the phase plane for the Van der Pol oscillator.

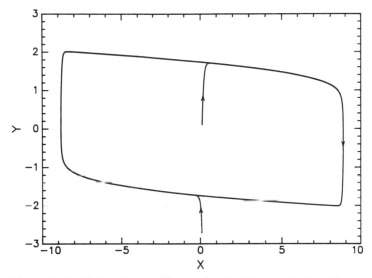

Figure 1-11 Relaxation oscillation for the Van der Pol oscillator.

in Figure 1-11, with a nondimensional period of around 1.61 when $\varepsilon > 10$.

Quasiperiodic Oscillators

A more complicated problem is the case when a periodic force is added to the Van der Pol system:

$$\ddot{x} - \gamma\dot{x}(1 - \beta x^2) + \omega_0^2 x = f_0 \cos \omega_1 t \qquad (1\text{-}2.9)$$

Because the system is nonlinear, *superposition of free and forced oscillations is not valid.* Instead, if the driving frequency is close to the limit cycle frequency, the resulting periodic motion will become *entrained* at the driving frequency. Frequency locking is a well-known classical phenomenon in nonlinear oscillations.

When the difference between driving and free oscillation frequencies is large, a new phenomenon is possible in the Van der Pol system—*combination oscillations*—sometimes called *almost periodic* or *quasiperiodic* solutions. Combination oscillation solutions take the form

$$x = b_1 \cos \omega_1 t + b_2 \cos \omega_2 t \qquad (1\text{-}2.10)$$

When ω_1 and ω_2 are incommensurate, that is, ω_1/ω_2 is an irrational

number, the solution is said to be *quasiperiodic*. In the case of the
Van der Pol equation [Eq. (1-2.9)], $\omega_2 \equiv \omega_0$; this is the free oscillation
limit cycle frequency (e.g., see Stoker, 1950, p. 166).

More will be said about quasiperiodic vibrations later, but because
they are not periodic, they may be mistaken for chaotic solutions,
which they are not. [For one, the Fourier spectrum of Eq. (1-2.10) is
just two spikes at $\omega = \omega_1$, ω_2, whereas for chaotic solutions the
spectrum is broad and continuous.]

The phase plane portrait of (1-2.10) is not closed when ω_1 and
ω_2 are incommensurate, so another method is used to portray the
quasiperiodic function graphically. To do this we stroboscopically
sample $x(t)$ with a period equal to $2\pi/\omega_1$; that is, let

$$t_n = \frac{n2\pi}{\omega_1} \tag{1-2.11}$$

and denote $x(t_n) = x_n$, $\dot{x}(t_n) = v_n$. Then Eq. (1-2.10) becomes

$$x_n = b_1 + b_2\cos\frac{2\pi n\omega_0}{\omega_1}, \qquad v_n = -\omega_0 b_2\sin\frac{2\pi n\omega_0}{\omega_1} \tag{1-2.12}$$

As n increases, the points x_n, v_n move around an ellipse in the
stroboscopic phase plane (called a *Poincaré map*), as shown in Figure
1-12. When ω_0/ω_1 is incommensurate, the set of points $\{x_n, v_n\}$ for
$n \to \infty$ fill in a closed curve given by

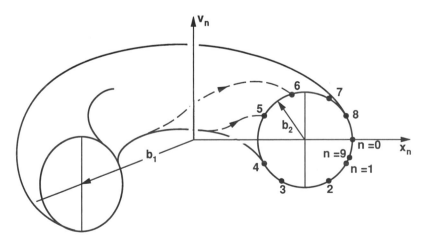

Figure 1-12   Stroboscopic plot of quasiperiodic solutions of the Van der Pol equation
[Eq. (1-2.9)] in the phase plane.

$$(x_n - b_1)^2 + \left(\frac{v_n}{\omega_0}\right)^2 = b_2^2 \qquad (1\text{-}2.13)$$

Quasiperiodic oscillations also occur in systems with more than one degree of freedom.

Dynamics of Lossless or Conservative Systems

In some areas of physics, one can assume that there are no energy dissipation mechanisms. Furthermore, in many mechanical and electromagnetic systems the forces and voltages can be derived from a potential energy function. Sometimes these systems are called *conservative* or *Hamiltonian* dynamics after the great Irish dynamicist, William R. Hamilton (1805–1865), whose mathematical formulation helped clarify the analysis of such systems. These systems exhibit a set of dynamic phenomena that are discussed in many of the classical and modern books on dynamics such as Goldstein (1980), Arnold (1978), and Sagdeev, Usikov, and Zaslavsky (1988). Among the interesting phenomena that differ from linear dynamics are nonlinear resonance, stochastic chaos, and diffusion in phase space. What further distinguishes conservative systems from dissipative oscillators is that there are no transient motions in the dynamics followed by limiting motions. In other words, there are no attractors in conservative dynamics. Each initial condition results is a unique orbit, which may be periodic, quasiperiodic, or chaotic. However, the chaotic motion does not have the kind of fractal structure that we find in dissipative systems.

Conservative or Hamiltonian dynamics is often used in applications to orbital dynamics in astronomy or the motions of charged particles in plasma devices or high-energy accelerators. Also, the mathematics of such lossless systems is sometimes used as the starting point in the analysis of systems with small dissipation.

Nonlinear Resonance in Conservative Systems

This is a phenomenon that is central to the study of conservative dynamics yet is not easily accessible to the novitiate. This is because some of the discoveries were only made in the second half of the 20th century and also because the phenomena is still coded by names of the theoreticians who made these discoveries such as Kolmogorov,

Arnold, and Moser (KAM theory). However, we will make an attempt at a brief description without the mathematical rigor.

Resonance is a phenomenon that occurs between two or more coupled oscillating systems. Two models are the following:

$$\ddot{x} + \frac{\partial V(x, y)}{\partial x} = 0$$

$$\ddot{y} + \frac{\partial V(x, y)}{\partial y} = 0$$

(1-2.14)

and

$$\ddot{x} + \frac{\partial V(x)}{\partial x} = A \cos \Omega t$$

(1-2.15)

By rewriting (1-2.15) we obtain a form similar to (1-2.14)

$$\ddot{x} + \frac{\partial V(x)}{\partial x} = y$$

$$\ddot{y} + \Omega^2 y = 0$$

(1-2.16)

For example, for the rotation of a pendulum under gravity, we obtain $V(x) = \omega_0^2(1 - \cos x)$, using nondimensional variables. The second model (1-2.15) is a special case of the first (1-2.14) where the second oscillator is uncoupled from the first.

An example of a conservative system with periodic, quasiperiodic, and chaotic or stochastic motions can be found in the periodically forced pendulum. Different initial conditions can lead to all three types of motion as illustrated in Figure 1-13 for a fixed forcing amplitude and frequency. In this figure the continuous motion is replaced by a set of points $(\theta_n, \dot{\theta}_n)$ which represent the angular position and velocity at times synchronous with the phase of the driving force. Using this so-called Poincaré section, periodic orbits show up as a finite set of points, quasiperiodic orbits show up as closed curves, and stochastic orbits show up as the diffuse set of points shown in Figure 1-13. These chaotic or stochastic orbits seem to be close to the saddle points of the unforced motions of the pendulum shown in Figure 1-13. As we shall see in Chapters 3 and 6, the existence of saddle points gives one a clue to the possibilities for chaos.

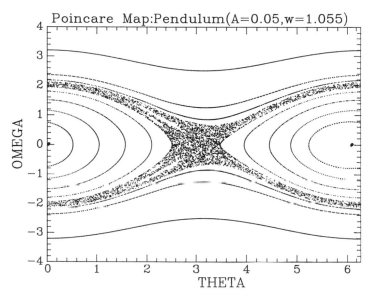

Figure 1-13 Stroboscopic pictures $[t_n = n(2\pi/\Omega)]$ of the dynamics of a periodically forced pendulum with no damping for different initial conditions. Isolated dots indicate periodic motion, continuous lines represent quasiperiodic motion, and a diffuse set of dots represents chaotic or stochastic orbits. ($\ddot{x} + \sin x = A \sin wt$)

Frequency Spectra of Nonlinear Oscillators

In the case of (1.2-15) when $A = 0$, the first oscillator can exhibit periodic oscillations as in a linear system, but the frequency depends on the initial conditions. Thus, for example, the Duffing oscillator

$$\ddot{x} + \omega_0^2 x + \beta x^3 = 0 \qquad (1\text{-}2.17)$$

(discussed above) has a continuous frequency spectrum shown in Figure 1-14a. The spectrum shown in Figure 1-14b is for a particle bouncing between two stationary walls (sometimes called a *Fermi oscillator*), where the frequency depends on the initial velocity.

If we now couple this oscillator to a second oscillator, we see that it is very easy to get a resonance by choosing the right initial conditions.

In a classical undamped linear oscillator with natural frequency ω_0 and driving frequency Ω, resonance occurs only when $\Omega = \omega_0$. (In many mechanical and civil engineering applications, resonance means that a small oscillator can easily drive a large structure into unwanted large-amplitude oscillations.) However, in a nonlinear oscillator, such

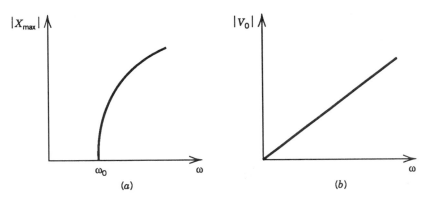

Figure 1-14 (*a*) Frequency spectrum for the free vibrations of the Duffing oscillator [Eq. (1-2.17)] without damping. X_{max} represents the initial amplitude. (*b*) Frequency spectrum of a mass oscillating between two walls. V_0 represents the initial velocity.

as the Duffing example (1-2.17), one can often achieve resonance at either multiples or integer fractions (i.e., harmonics or subharmonics) of the driving frequency Ω by simply choosing the right amplitude A that produces a frequency of oscillation $\omega(A)$ such that

$$\omega(A) = p\Omega/q, \quad \text{or} \quad q\omega = p\Omega \qquad (1-2.18)$$

Thus theoretically, the consequence of a continuous frequency spectrum for the free oscillations is an *infinite number of possible resonances* in the driven oscillator problem (1-2.15). The same can be said for two coupled oscillators (1-2.14).

For two coupled oscillators (e.g., two pendulums in Figure 1-15), we identify a measure of the amplitude of each oscillator J_1, J_2 and the phase or relative time in the cycle of oscillation θ_1, θ_2, such that $\omega_1 = \dot{\theta}_1$, $\omega_2 = \dot{\theta}_2$ are the frequencies. Then nonlinear resonance can occur when

$$n\dot{\theta}_1 = m\dot{\theta}_2 \qquad (1-2.19)$$

which can be satisfied by choosing the proper initial conditions J_{10}, J_{20}. Thus, the resonance condition can be rewritten as

$$n\omega_1(J_{10}) - m\omega_2(J_{20}) = 0 \qquad (1-2.20)$$

As in the linear case, resonance means that energy can be easily exchanged between two systems, which can lead to interesting and perhaps even chaotic dynamics. A classic experiment in this phenome-

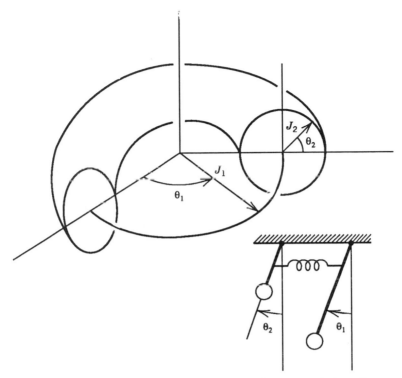

Figure 1-15 Schematic picture of the dynamics of two coupled oscillators.

non is a compound pendulum with a 2 : 1 frequency ratio. Experimental models are described by Rott (1970) and in Appendix C.

In general, when there are three or more coupled nonlinear oscillators, each of which has some identifiable phase or angle variable θ_i such that $\dot{\theta}_1 = \omega_i$ is the frequency, then nonlinear resonance can occur when the following relation holds:

$$n\omega_1 + m\omega_2 + \cdots + p\omega_N = 0 \tag{1-2.21}$$

where n, m, p, and so on, are positive or negative integers.

The consequences of nonlinear resonance in coupled oscillators are most profound and form the basis for understanding chaos in conservative or nondissipative dynamic systems.

Torus Map

The motion of two coupled oscillators can sometimes be visualized as a particle moving on a toroidal surface as shown in Figure 1-12a. If the particle motion on the torus consists of a position vector $\mathbf{r}(t)$

$$\mathbf{r} = \mathbf{R} + \boldsymbol{\rho}$$

Here the motion around the major axis given by $\mathbf{R}(t)$ occurs at frequency ω_1, whereas the motion around the minor axis described by $\boldsymbol{\rho}(t)$ has a frequency ω_2.

The motion $x(t)$ in Eq. (1-2.10) can then be thought of as two projections of this particle motion; that is, $x(t) = X(t) + \xi(t)$. $X(t)$ is the scalar projection of $\mathbf{R}$ onto the fixed plane Σ shown in Figure 1-12, and $\xi(t)$ is the projection $\boldsymbol{\rho}$ onto the horizontal plane.

Another useful description of the dynamics is to only look at the points of penetration of the toroidal orbit onto the fixed plane Σ as it slices through one side of the torus. This is known as a *synchronous point mapping* (e.g., see Minorsky, 1962) or, in modern terms, a *Poincaré map* as shown in Figure 1-12. To obtain an analytical expression for this map, we define another state variable $V(t)$ as simply the time derivative of the multifrequency motion $x(t) = A_1 \cos \omega_1 t + A_2 \cos \omega_2 t$. Then one chooses a phase angle of the first oscillator (e.g., $\omega_1 t_n = 2\pi n$). The phase plane dynamics are then described by two discrete time expressions ($v_n = \dot{x}(t_n)$)

$$\begin{aligned} x_{n+1} &= f(x_n, v_n) \\ v_{n+1} &= g(x_n, v_n) \end{aligned} \tag{1-2.22}$$

Finally, by rescaling the vertical axis and using polar coordinates, one obtains a first-order difference equation—that is, define

$$\tan \varphi_n = \frac{v_n/\omega_1}{x_n} \tag{1-2.23}$$

and

$$\varphi_{n+1} = \varphi_n + 2\pi \frac{\omega_2}{\omega_1} \tag{1-2.24}$$

More generally, one often finds a map of the form

$$\varphi_{n+1} = \varphi_n + F(\varphi_n) \tag{1-2.25}$$

with

$$F(\varphi_n + 2\pi) = F(\varphi_n) \qquad (1\text{-}2.26)$$

This map is known as a *circle map* or is sometimes known as a *twist map*. One can see that in the simplest case when $F = 2\pi\omega_2/\omega_1$ and when ω_2/ω_1 is irrational, the succession of points on the orbit will trace out a circle in the $(x_n, v_n/\omega_1)$ plane. On the other hand, if $\omega_1/\omega_2 = p/q$ (p/q are integers), then for $\Delta\varphi = \varphi_{n+q} - \varphi_n$, the orbit will visit precisely "p" points around the circle.

This discussion may seem like a complicated way to view the motion of two oscillators, but this model of toroidal motion and the resulting circle map has become an important conceptual as well as practical analytical tool to analyze complex dynamics of coupled systems, as will be shown in a later chapter.

Local Geometric Theory of Dynamics

Modern ideas about nonlinear dynamics are often presented in geometric terms or pictures. For example, the motion of an undamped oscillator, $\ddot{x} + \omega_0 x = 0$, can be represented in the phase plane $(x, \dot{x})$ by an ellipse. In this picture, time is implicit and the time history runs clockwise around the ellipse. The size of the ellipse depends on the given initial conditions for $(x, \dot{x})$.

More generally for nonlinear problems, one first finds the equilibrium points of the system and examines the motion around each equilibrium point. The local motion is characterized by the nature of the eigenvalues of the linearized system. Thus, if the dynamical model can be represented by a set of first-order differential equations

$$\dot{\mathbf{x}} = \mathbf{f}(\mathbf{x}) \qquad (1\text{-}2.27)$$

where $\mathbf{x}$ represents a vector whose components are the state variables, then the *equilibrium points* are defined by $\dot{\mathbf{x}} = 0$, or

$$\mathbf{f}(\mathbf{x}_e) = 0 \qquad (1\text{-}2.28)$$

For example, in the case of the harmonic oscillator, there is just one equilibrium pint at the origin $\mathbf{x} = (x, v \equiv \dot{x})$, $x_e = 0$, $v_e = 0$. To determine the nature of the dynamics about $\mathbf{x} = \mathbf{x}_e$, one expands the function $\mathbf{f}(\mathbf{x})$ in a Taylor series about each equilibrium point $\mathbf{x}_e$ and examines the dynamics of the linearized problem.

To illustrate the method, consider the set of two first-order equations:

$$\dot{x} = f(x, y)$$

$$\dot{y} = g(x, y)$$

(1-2.29)

When time does not appear explicitly in the functions $f(\;)$ and $g(\;)$, the problem is called *autonomous*. The equilibrium points must satisfy two equations: $f(x_e, y_e) = 0$ and $g(x_e, y_e) = 0$. Introducing small variables about each equilibrium point, that is,

$$x = x_e + \eta \quad \text{and} \quad y = y_e + \xi$$

the linearized system can be written in the form

$$\frac{d}{dt}\begin{Bmatrix} \eta \\ \xi \end{Bmatrix} = \begin{bmatrix} \dfrac{\partial f}{\partial x} & \dfrac{\partial f}{\partial y} \\ \dfrac{\partial g}{\partial x} & \dfrac{\partial g}{\partial y} \end{bmatrix} \begin{Bmatrix} \eta \\ \xi \end{Bmatrix}$$

(1-2.30)

where the derivatives are evaluated at the point (x_e, y_e).

Some authors use the notation $\nabla \mathbf{F}$ or $D\mathbf{F}$, where $\mathbf{F} = (f, g)$, to represent the matrix of partial derivatives in Eq. (1-2.30). The nature of the motion about each equilibrium point is determined by looking for eigensolutions

$$\begin{Bmatrix} \eta \\ \xi \end{Bmatrix} = \begin{Bmatrix} \alpha \\ \beta \end{Bmatrix} e^{st}$$

(1-2.31)

where α and β are constants. The motion is classified according to the nature of the two eigenvalues of $D\mathbf{F}$ [i.e., whether s is real or complex and whether $\text{Real}(s) > 0$ or < 0.]

Sketches of trajectories in the phase plane for different eigenvalues are shown in Figure 1-16. For example, the *saddle point* is obtained when both eigenvalues s are real, but $s_1 < 0$ and $s_2 > 0$. A *spiral* occurs when s_1 and s_2 are complex conjugates.

The *stability* of the linearized system (1-2.30) depends on the sign of $\text{Real}(s)$. When one of the real parts of s_1 and s_2 is positive, the motion about the equilibrium point is *unstable*. If the roots are not pure imaginary numbers, then theorems exist to show that the local motion of the linearized system is qualitatively similar to the original nonlinear system (1-2.29). Pure oscillatory motion in the linearized system ($s = \pm i\omega$) requires further analysis to establish the stability of

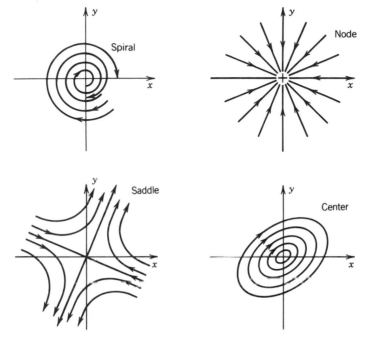

Figure 1-16 Classical phase plane portraits near four different types of equilibrium points for a system of two time-independent differential equations.

the nonlinear system. These ideas for a second-order system can be generalized to higher-dimensional phase spaces (e.g., see Arnold, 1978 or Guckenheimer and Holmes, 1983).

Bifurcations

As parameters are changed in a dynamical system, the stability of the equilibrium points can change as well as the number of equilibrium points. The study of these changes in nonlinear problems as system parameters are varied is the subject of *bifurcation theory*. Values of these parameters at which the qualitative or topological nature of motion changes are known as *critical* or *bifurcation* values.

As an example, consider the solutions to the undamped Duffing oscillator

$$\ddot{x} + \alpha x + \beta x^3 = 0 \qquad (1\text{-}2.32)$$

One can first plot the equilibrium points as a function of α. As α changes from positive to negative, one equilibrium point splits into three points. Dynamically, one center is transformed into a saddle point at the origin and two centers (Figure 1-17). This kind of bifurcation is known as a *pitchfork*. Physically, the force $-(\alpha x + \beta x^3)$ can be derived from a potential energy function. When α becomes negative, a one-well potential changes into a double-well potential problem. This represents a qualitative change in the dynamics, and thus $\alpha = 0$ is a critical bifurcation value.

Another example of a bifurcation is the emergence of limit cycles in physical systems. In this case, as some control parameter is varied, a pair of complex conjugate eigenvalues $s_1, s_2 = \pm i\omega + \gamma$ cross from the left-hand plane ($\gamma < 0$, a stable spiral) into the right-hand plane ($\gamma > 0$, an unstable spiral) and a periodic motion emerges known as

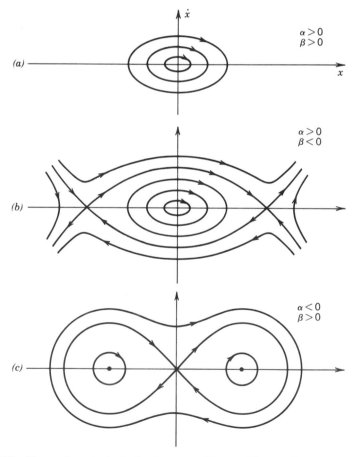

Figure 1-17   Phase plane trajectories for an oscillator with a nonlinear restoring force [Duffing's equation, Eq. (1-2.32)]: (*a*) Hard spring problem; $\alpha, \beta > 0$. (*b*) Soft spring problem; $\alpha > 0$, $\beta < 0$. (*c*) Two-well potential; $\alpha < 0$, $\beta > 0$.

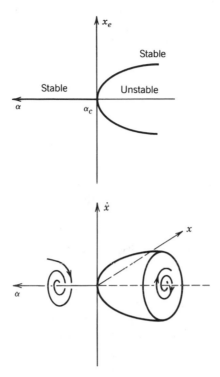

Figure 1-18   Bifurcation diagrams: (*a*) Pitchfork bifurcation for Duffing's equation [Eq. (1-2.32)]—transition from one to two stable equilibrium positions. (*b*) Hopf bifurcation—transition from stable spiral to limit cycle oscillation.

a *limit cycle*. This type of qualitative change in the dynamics of a system is known as a *Hopf bifurcation* and is illustrated in Figure 1-18.

The theory we have just described is called a *local* analysis because it only tells what happens dynamically in the vicinity of each equilibrium point. The pièce de resistance in classical dynamical analysis is to piece together all the local pictures and describe a *global* picture of how trajectories move between and among equilibrium points.

Such analysis is tractable when bundles of different trajectories corresponding to different initial conditions move more or less together as a laminar fluid flow. Such is the case when the phase space has only *two* dimensions. However, when there are three or more first-order equations, the bundles of trajectories can split apart and get tangled up into what we now call *chaotic motions*.

Strange Attractors

From this brief review, one can see that there are three classic types of dynamical motion:

1. Equilibrium
2. Periodic motion or a limit cycle
3. Quasiperiodic motion

These states are called *attractors*, because if some form of damping is present the transients decay and the system is "attracted" to one of the above three states. The purpose of this book is to describe another class of motions in nonlinear vibrations that is not one of the above classic attractors. This new class of motions is chaotic in the sense of not being predictable when there is a small uncertainty in the initial condition, and is often associated with a state of motion called a *strange attractor*.

The classic attractors are all associated with classic geometric objects in phase space, the equilibrium state is associated with a point, the periodic motion or limit cycle is associated with a closed curve, and the quasiperiodic motion is associated with a surface in a three-dimensional phase space. The "strange attractor," as we shall see in later chapters, is associated with a new geometric object (new relative to what is now taught in classical geometry) called a *fractal set*. In a three-dimensional phase space, the fractal set of a strange attractor looks like a collection of an infinite set of sheets or parallel surfaces, some of which are separated by distances which approach the infinitesimal. This new attractor in nonlinear dynamics requires new mathematical concepts and a language to describe it as well as new experimental tools to record it and give it some quantitative measure. The relationship between bifurcations and chaos is discussed in a recent book by Thompson and Stewart (1986).

1.3 MAPS AND FLOWS

Mathematical models in dynamics generally take one of three forms: differential equations (or *flows*), difference equations (called *maps*), and *symbol dynamic equations*.

The term *flow* refers to a bundle of trajectories in phase space originating from many contiguous initial conditions. The continuous time history of a particle is the most familiar example of a flow to those in engineering vibrations. However, certain qualitative and quantitative information may be obtained about a system by studying the evolution of state variables at discrete times. In particular, in this book we shall discuss how to obtain difference evolution equations from continuous time systems through the use of the Poincaré section.

These Poincaré maps can sometimes be used to distinguish between various qualitative states of motion such as periodic, quasiperiodic, or chaotic. In some problems not only time is restricted to discrete values, but knowledge of the state variables may be limited to a finite set of values or categories such as red or blue or zero or one. For example, in the double-well potential of Figure 1-2*b*, one may be interested only in whether the particle is in the left or right well. Thus, an orbit in time may consist of a sequence of symbols LRRLRLLLR ... . A periodic orbit might be LRLR ... or LLRLLR ... . In the new era of nonlinear dynamics, all three types of models are used to describe the evolution of physical systems. [See Crutchfield and Packard (1982) or Wolfram (1986) for a discussion of symbol dynamics.]

In a periodically forced vibratory system, a Poincaré map may be obtained by stroboscopically measuring the dynamic variables at some particular phase of the forcing motion. In an n-state variable problem, one can obtain a Poincaré section by measuring the $n - 1$ variables when the nth variable reaches some particular value or when the phase space trajectory crosses some arbitrary plane in phase space as shown in Figure 1-19 (see also Chapters 2 and 5). If one has knowledge of the time history between two penetrations of this plane, one can relate the position at t_{n+1} to that at t_n through given functions. For example, for the case shown in Figure 1-19,

$$\xi_{n+1} = f(\xi_n, \eta_n) \quad \text{and} \quad \eta_{n+1} = g(\xi_n, \eta_n) \tag{1-3.1}$$

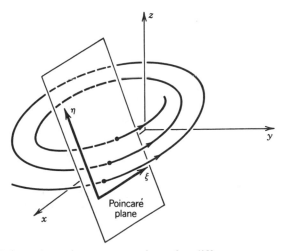

Figure 1-19 Poincaré section: construction of a difference equation model (map) from a continuous time dynamical model.

The mathematical study of such maps is similar to that for differential equations. One can find equilibrium or fixed points of the map, and one can classify these fixed points by the study of linearized maps about the fixed point. If $\mathbf{x}_{n+1} = \mathbf{f}(\mathbf{x}_n)$ is a general map of say n variables represented by the vector $\mathbf{x}$, then a fixed point satisfies

$$\mathbf{x}_e = \mathbf{f}(\mathbf{x}_e) \tag{1-3.2}$$

The iteration of a map is often written $\mathbf{f}(\mathbf{f}(\mathbf{x})) = \mathbf{f}^{(2)}(\mathbf{x})$. Using this notation, an "m-cycle" or m-periodic orbit is a fixed point that repeats after m iterations of the map, that is,

$$\mathbf{x}_0 = \mathbf{f}^{(m)}(\mathbf{x}_0) \tag{1-3.3}$$

Implied in these ideas is the notion that periodic motions in continuous time history show up as fixed points in the difference equations obtained from the Poincaré sections. Thus, the most generally accepted paradigms for the study of the transition from periodic to chaotic motions is the study of simple one-dimensional and two-dimensional maps. (See Chapter 3 for a discussion of maps.)

Three Paradigms for Chaos

Perhaps the simplest example of a dynamic model that exhibits chaotic dynamics is the logistic equation or population growth model (e.g., see May, 1976):

$$x_{n+1} = ax_n - bx_n^2 \tag{1-3.4}$$

The first term on the right-hand side represents a growth or birth effect, whereas the nonlinear term accounts for the limits to growth such as availability of energy or food. If the nonlinear term is neglected ($b = 0$), the linear equation has an explicit solution:

$$x_{n+1} = ax_n; \qquad x_n = x_0 a^n \tag{1-3.5}$$

This solution is stable for $|a| < 1$ and unstable for $|a| > 1$. In the latter case, the linear model predicts unbounded growth, which is unrealistic.

The nonlinear model (1-3.4) is usually cast in a nondimensional form:

$$x_{n+1} = \lambda x_n(1 - x_n) \tag{1-3.6}$$

This equation has at least one equilibrium point, $x = 0$. For $\lambda > 1$, two equilibrium points exist [i.e., solutions of the equation $x = \lambda x(1 - x)$]. To determine the stability of a map $x_{n+1} = f(x_n)$, one looks at the value of the slope $|f'(x)|$ evaluated at the fixed point. The fixed point is unstable if $|f'| > 1$. In the case of the logistic equation [Eq. (1-3.6)] when $1 < \lambda < 3$, there are two fixed points, namely, $x = 0$ and $x = (\lambda - 1)/\lambda$; the origin is unstable and the other point is stable.

For $\lambda = 3$, however, the slope at $x = (\lambda - 1)/\lambda$ becomes greater than 1 ($f' = 2 - \lambda$) and both equilibrium points become unstable. For parameter values of λ between 3 and 4, this simple difference equation exhibits many multiple-period and chaotic motions. At $\lambda = 3$, the steady solution becomes unstable, but a two-cycle or double-period orbit becomes stable. This orbit is shown in Figure 1-20. The value of x_n repeats every two iterations.

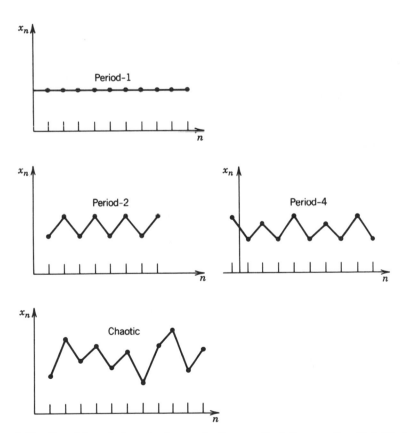

Figure 1-20   Possible solutions to the quadratic map [logistic equation (1-3.6)]. *Top*: Steady period-1 motion. *Middle*: Period-2 and period-4 motions. *Bottom*: Chaotic motions.

For further increases of λ, the period-2 orbit becomes unstable and a period-4 cycle emerges, only to bifurcate to a period-8 cycle for a higher value of λ. This period-doubling process continues until λ approaches the value $\lambda_\infty = 3.56994 \ldots$. Near this value, the sequence of period-doubling parameter values scales according to a precise law:

$$\frac{\lambda_{n+1} - \lambda_n}{\lambda_n - \lambda_{n-1}} \to \frac{1}{\delta}, \, \delta = 4.66920 \ldots \tag{1-3.7}$$

The limit ratio is called the *Feigenbaum number*, named after the physicist who discovered the properties of this map in 1978. (See Gleick, (1987), for the story of this discovery)

Beyond λ_∞, chaotic iterations can occur; that is, the long-term behavior does not settle down to any simple periodic motion. There are also certain narrow windows $\Delta\lambda$ for $\lambda_\infty < \lambda < 4$ for which periodic orbits exist. Periodic and chaotic orbits of the logistic map are shown in Figure 1-21 by plotting x_{n+1} versus x_n.

This map is not only useful as a paradigm for chaos, but it has been shown that other maps $x_{n+1} = f(x_n)$, in which $f(x)$ is double or multiple valued, behave in a similar manner with the same scaling law (1-3.7). Thus, the phenomenon of period doubling or bifurcation parameter scaling has been called a *universal* property for certain classes of one-dimensional difference equation models of dynamical processes.

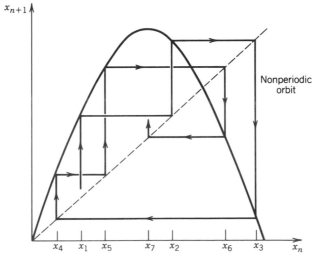

Figure 1-21 Graphical solution to a first-order difference equation. The example shown is the quadratic map (1-3.6).

Period doubling and Feigenbaum scaling (1-3.7) have been observed in many physical experiments (see Chapter 4). This suggests that for many continuous time history processes, the reduction to a difference equation through the use of the Poincaré section has the properties of the quadratic map (1-3.4)—hence the importance of maps to the study of differential equations. [See Chapter 3 for a further discussion of the logistic equation (1-3.4).]

Henon and Horseshoe Maps

Of course, most physical systems require more than one state variable, and it is necessary to study higher-dimensional maps. One extension of the Feigenbaum problem (1-3.6) is a two-dimensional map proposed by Henon (1976), a French astronomer:

$$x_{n+1} = 1 - ax_n^2 + y_n$$

$$y_{n+1} = \beta x_n$$

(1-3.8)

Note that if $\beta = 0$, we recover the quadratic map. When $|\beta| < 1$, the map contracts areas in the xy plane. It also stretches and bends areas in the phase plane as illustrated in Figure 1-22. This stretching, contraction, and bending or folding of areas in phase space is analogous to the making of a horseshoe. Multiple iterations such as *horseshoe*

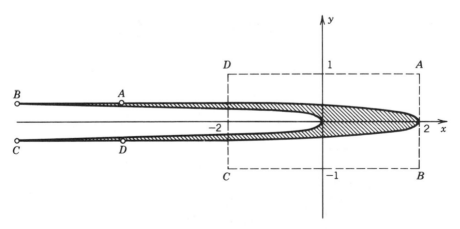

Figure 1-22   Transformation of a rectangular collection of initial conditions under an iteration of the second-order set of difference equations called a *Henon map* (1-3.8) showing stretching, contraction, and folding which leads to chaotic behavior ($\alpha = 1.4$, $\beta = 0.3$).

maps lead to complex orbits in phase space and loss of information about initial conditions and chaotic behavior.

An illustration of the ability of a simple map to produce complex motions is provided in Figure 1-23. In one iteration of the map, a rectangular area is stretched in the vertical direction, contracted in the horizontal direction, and folded or bent into a horseshoe and placed over the original area. Thus, points originally in the area get mapped back onto the area, except for some points near the bend in the horseshoe. If one follows a group of nearby points after many iterations of this map, the original neighboring cluster of points gets dispersed to all sectors of the rectangular area. This is tantamount to a loss of information as to where a point originally started from. Also, the original area gets mapped into a finer and finer set of points, as shown in Figure 1-23. This structure has a fractal property that is a characteristic of a chaotic attractor which has been labeled "strange." This fractal property of a strange attractor is illustrated in the Henon map, Figure 1-24. Blowups of small regions of the Henon attractor reveal finer and finer structure. This self-similar structure of chaotic

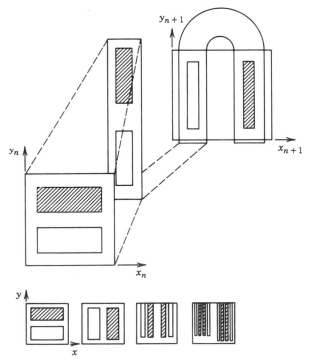

Figure 1-23 The horseshoe map showing how stretching, contraction, and folding leads to fractal-like properties after many iterations of the map.

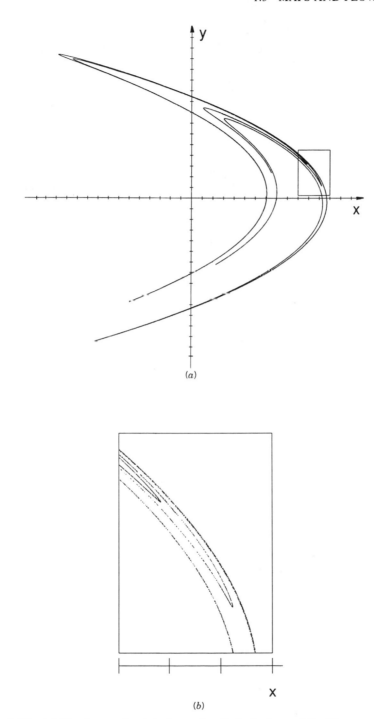

Figure 1-24 (a) The locus of points for a chaotic trajectory of the Henon map (α = 1.4, β = 0.3). (b) Enlargement of strange attractor showing finer fractal-like structure.

attractors can often be revealed by taking Poincaré maps of experimental chaotic oscillators (see Chapters 2 and 5). The fractal property of self-similarity can be measured using a concept of fractal dimension, which is discussed in Chapter 7.

It is believed by some mathematicians that horseshoe maps are fundamental to most chaotic differential and difference equation models of dynamic systems (e.g., see Guckenheimer and Holmes, 1983). This idea is the centerpiece of a method developed to find a criterion for when chaotic vibrations are possible in a dynamical system and when predictability of future time history becomes sensitive to initial conditions. This Melnikov method has been used successfully to develop criteria for chaos for certain problems in one-degree-of-freedom nonlinear oscillation (e.g., see Chapter 6).

The Lorenz Attractor and Fluid Chaos

For many readers, the preceding discussion on maps and chaos may not be convincing as regards unpredictability in real physical systems. And were it not for the following example from fluid mechanics, the connection between maps, chaos, and differential equations of physical systems might still be buried in mathematics journals. In 1963, an atmospheric scientist named E. N. Lorenz of M.I.T. proposed a simple model for thermally induced fluid convection in the atmosphere.[4] Fluid heated from below becomes lighter and rises, whereas heavier fluid falls under gravity. Such motions often produce convection rolls similar to the motion of fluid in a circular torus as shown in Figure 1-25. In Lorenz's mathematical model of convection, three state variables are used (x, y, z). The variable x is proportional to the amplitude of the fluid velocity circulation in the fluid ring, while y and z measure the distribution of temperature around the ring. The so-called Lorenz equations may be derived formally from the Navier–Stokes partial differential equations of fluid mechanics (1-1.3) (e.g., see Chapter 4). The nondimensional forms of Lorenz's equations are

$$\dot{x} = \sigma(y - x)$$

$$\dot{y} = \rho x - y - xz \qquad (1\text{-}3.9)$$

$$\dot{z} = xy - \beta z$$

---

[4] Lorenz credits Saltzman (1962) with actually discovering nonperiodic solutions to the convection problem in which he used a system of the five first-order equations. Mathematicians, however, chose instead to study Lorenz's simpler third-order set of equations (1-3.9). Thus flows the course of scientific destinies.

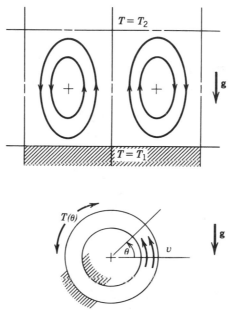

Figure 1-25 *Top*: Sketch of fluid streamlines in a convection cell for steady motions. *Bottom*: One-dimensional convection in a circular tube under gravity and thermal gradients.

The parameters σ and ρ are related to the Prandtl number and Rayleigh number, respectively, and the third parameter β is a geometric factor. Note that the only nonlinear terms are xz and xy in the second and third equations.

For $\sigma = 10$ and $\beta = 8/3$ (a favorite set of parameters for experts in the field), there are three equilibria for $\rho > 1$ for which the origin is an unstable saddle (Figure 1-26). When $\rho > 25$, the other two equilibria become unstable spirals and a complex chaotic trajectory moves between all three equilibria as shown in Figure 1-27. It was Lorenz's insistence in the years following 1963 that such motions were not artifacts of computer simulation but were inherent in the equations themselves that led mathematicians to study these equations further (e.g., see Sparrow, 1982). Since 1963, hundreds of papers have been written about these equations, and this example has become a classic model for chaotic dynamics. These equations are also similar to those that model the chaotic behavior of laser devices (e.g., see Haken, 1985).

Systems of other third-order equations have since been found to exhibit chaotic behavior. For example, the forced motion of a nonlinear oscillator can be written in a form similar to that of (1-3.9); New-

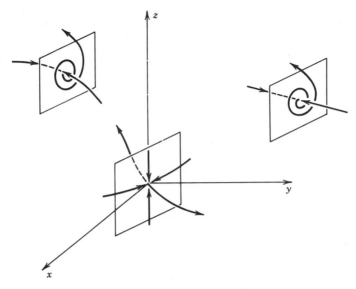

Figure 1-26 Sketch of local motion near the three equilibria for the Lorenz equations [Eqs. (1-3.9)].

ton's law for a particle under a force $F(x, t)$ is written

$$m\ddot{x} = F(x, t) \tag{1-3.10}$$

To put (1-3.10) into a form for phase space study, we write $y = \dot{x}$. Furthermore, if the mass is periodically forced, one can reduce the second-order nonautonomous system (1-3.10) to an autonomous system of third-order equations. Thus, we assume

$$F(x, t) = m(f(x, y) + g(t))$$
$$g(t + \tau) = g(t)$$

By defining $z = \omega t$ and $\omega = 2\pi/\tau$, the resulting equations become

$$\dot{x} = y$$
$$\dot{y} = f(x, y) + g(z) \tag{1-3.11}$$
$$\dot{z} = \omega$$

A specific case that has strong chaotic behavior is the Duffing oscillator $F = -(ax + bx^3 + cy)$ (see Chapters 2 and 4).

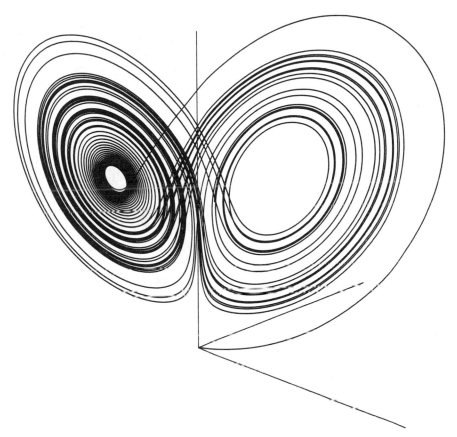

Figure 1-27 Trajectory of a chaotic solution to the Lorenz equations for thermofluid convection [Eqs. (1-3.9)] (numerical integration).

It is worth noting that for a two-dimensional phase space, solutions to autonomous systems cannot exhibit chaos because the solution curves of the "flow" cannot cross one another. However, in the forced oscillator or the three-dimensional phase space, these curves can become "tangled" and chaotic motions are possible.

Quantum Chaos

The focus of this book is unpredictability in classical Newtonian physics. But, what about the possibility of quantum chaos? We have all learned in elementary physics that as one approaches the microscopic scale, the motion of a particle must be described by a wave packet whose amplitude gives the probability of locating the particle. Quantum mechanics tells one about the motion of these wave or probability packets, but not about the precise motion of the particle. There is a transition region where both classical and quantum descriptions should give approximately the same answer. So, when a classical system exhibits chaotic dynamics near the quantum limit, the question naturally arises as to what would be the quantum description of this classical chaos.

This should be a fundamental question in physics today. There are those who think it is (e.g., see Ford, 1988). But for others there is a belief that we still do not know if we have posed the question properly; that is, perhaps quantum chaos is a concept full of redundancy or is an oxymoron. After all, what does unpredictable unpredictability mean? Still some experiments both physical and computational have been performed (e.g., see Pool, 1989 and Koch, 1990) in an attempt to settle the question of quantum chaos. At the time of this writing (1991), however, the subject is still in debate.

The existence or nonexistence of quantum chaos will not be resolved in this book. The reader is referred to the publications of some of the notable participants in the debate (e.g., Jensen, 1989 and Ford, 1988). One gets the feeling, however, that in the spirit of T. Kuhn's theory of scientific revolutions, that physicists are still looking for the right "paradigm" for quantum chaos—that is, a kind of Lorenz quantum model.

Closing Comments

Dynamics is the oldest branch of physics. Yet 300 years after publication of Newton's *Principia*, new discoveries are still emerging. The ideas of Euler, Lagrange, Hamilton, and Poincaré that followed, once conceived in the context of planetary mechanics, have now transcended all areas of physics. As the new science of dynamics gave birth to the calculus in the 17th century, so today modern nonlinear dynamics has ushered in new ideas of geometry and topology, such as fractals, which the 21st-century scientist must master to grasp the subject fully.

The ideas of chaos go back in Western thought to the Greeks. But these ideas centered on the order in the world that emerged from a formless chaotic, fluid world in prehistory. G. Mayer-Kress (1985) of Los Alamos National Laboratory has pointed out that the idea of chaos in Eastern thought, such as Taoism, was associated with patterns within patterns, eddies within eddies as occur in the flow of fluids (e.g., see the Japanese kimono design in Figure 1-28).

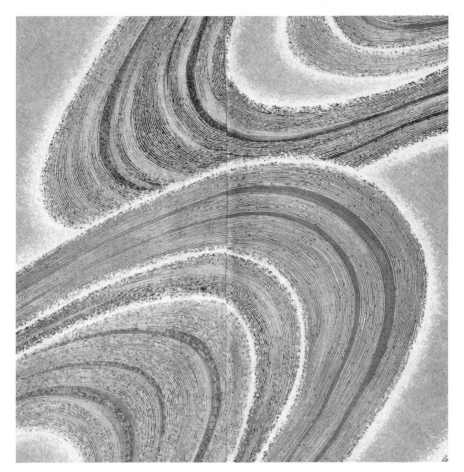

Figure 1-28 Fractal-like pattern in a Japanese kimono design. (Courtesy of Mitsubishi Motor Corp.)

The view that order emerged from an underlying formless chaos and that this order is recognized only by predictable periodic patterns was the predominant view of 20th-century dynamics until the last two decades. What is replacing this view is the concept of chaotic events resulting from orderly laws, not a formless chaos, but one in which there are underlying patterns, fractal structures, governed by a new mathematical view of our "orderly" world.

Since the first edition of this book was published, new discoveries continue to appear, such as multifractals, spatial complexity, and hyperchaos. The range of applications continues to grow. And, while much has been learned since Feigenbaum's period-doubling paper of 1978, there is still an air of excitement in this field with regard to both the mathematical ideas and the new applications.

PROBLEMS

1-1 Consider an unforced bilinear oscillator, without damping, whose linear stiffness changes from k_1 for $x < a_0$ to k_2 for $x \geq a_0$.

    (a) Sketch the flow lines in the phase plane $(x, y \equiv \dot{x})$ for different initial conditions.

    (b) Sketch the frequency spectrum versus amplitude for k_1, $k_2 > 0$, $k_1 < k_2$ and k_1, $k_2 > 0$, $k_1 > k_2$. What happens when $k_1 < 0$ and $k_2 > 0$?

1-2 Sometimes the roll dynamics of a ship can be modeled by a particle of mass m moving in a potential field with $V(x) = ax^2 - bx^3$; $a, b > 0$.

    (a) Find the fixed points of the motion and establish the local stability.

    (b) Sketch the flow lines for different initial conditions when there is no damping.

    (c) Sketch the flow lines when there is small linear damping. [See, for example, the work of Thompson et al. (1990).]

1-3 Write the equation of motion for a ball bouncing on a horizontal surface with coefficient of restitution $\varepsilon < 1$.

    (a) Sketch the phase plane motion.

    (b) Find a difference equation (map) that relates the velocity after impact at the $(n + 1)$st bounce to that at the nth bounce. Is this a linear or nonlinear problem? What if the table vibrates with $A \cos \Omega t$?

1-4 Consider the two-dimensional, planar motion of a ball bouncing on a concave surface which is described by the function $y = F(x)$. Assuming no loss of energy on impact, find the equations (a map) that relate the velocity and horizontal distance after impact at the $(n + 1)$st bounce to that of the nth bounce.

1-5 Using a calculator or personal computer, experimentally find the first four critical values for period doubling in the logistic equation, $x_{n+1} = \lambda x_n(1 - x_n)$. Find the first two approximations to the Feigenbaum number.

1-6   The following equation can be derived from the dynamics of an electron or proton in a circular accelerator (e.g., see Helleman, 1980a):

$$y_{n+1} + by_{n-1} = 2cy_n + 2y_n^2, \qquad |b| \leq 1$$

(a) Rewrite this equation as a set of first-order difference equations.

(b) For $b = 1$, show that this equation is equivalent to a special Henon map [Eq. (1-3.8)].

(c) For $b = 0$, show that this equation reduces to the logistic map [Eq. (1-3.6)].

1-7   Write down a set of three first-order differential equations which has a saddle-spiral fixed point at the origin and has bounded motion. Sketch a few flow lines in a two-dimensional projection. (A saddle-spiral has three eigen values of the form $s_1 = -\gamma$ $s_{2,3} = \alpha \pm i\beta$.)

1-8   The harmonic oscillator with a cubic nonlinearity $\ddot{x} + \omega_0^2 x + \alpha x^3 = 0$ is known to have periodic solutions whose period decreases with increase in amplitude. One classical technique known as *harmonic balance* has been used to find the relationship between the period T and amplitude A by assuming that for A not too large the nonlinear oscillation remains approximately sinusoidal, i.e., $x \approx A \cos \Omega t$ ($T = 2\pi/\Omega$). Assume this form and put it into the above nonlinear equation and require only that the $\cos \Omega t$ terms vanish. Show that the frequency spectrum is given by (see Figure 1-14a)

$$\Omega^2 = \omega_0^2 + \tfrac{3}{4}\alpha A^2$$

1-9   A dynamical system of even greater simplicity that exhibits period doubling and chaos is the set of three first-order differential equations (Rössler, 1976a,b):

$$\dot{x} = -y - z, \qquad \dot{y} = x + ay, \qquad \dot{z} = b + z(x - c)$$

(a) Show that for $b = 2$, $c = 4$, and $0 < a < 2$, this system has two fixed points.

(b) For z small, show that the motion near the origin is an unstable spiral.

(c) Suppose you know that these outward spiraling trajectories return to the origin. Can you make a sketch of the three-dimensional motion? [See also Chapter 12 of Thompson and Stewart (1986).]

1-10 Use a small computer with numerical integration software (e.g., using a Runge–Kutta algorithm), choose $b = 2$, $c = 4$ for the Rössler attractor in problem 1-9, and explore the parameter regime $0.3 \leq a \leq 0.4$ and look for period doubling. For the case $a = 0.398$, plot a return map x_{n+1} vs. x_n. Compare with the logistic equation (1-3.6).

2

HOW TO IDENTIFY CHAOTIC VIBRATIONS

"What will prove altogether remarkable is that some very simple schemes to produce erratic numbers behave identically to some of the erratic aspects of natural phenomena."

Mitchell Feigenbaum, 1980

Theorists and experimentalists approach a dynamical problem from different sides: The former is given the equations and looks for solutions, whereas the latter is given the solution and is looking for the equations or mathematical model.

In this chapter we present a set of diagnostic tests that can help identify chaotic oscillations and the models that describe them in physical systems. Although this chapter is written primarily for those not trained in the mathematical theory of dynamics, theoreticians may find it of interest to see how theoretical ideas about chaos are realized in the laboratory. In a later chapter (Chapter 6), we present some predictive criteria as well as more sophisticated diagnostic tests for chaos. This, however, requires some mathematical background, such as the theory of fractal sets (Chapter 7) and Lyapunov exponents (Chapter 6).

Engineers often have to diagnose the source of unwanted oscillations in physical systems. The ability to classify the nature of oscillations can provide a clue as to how to control them. For example, if the system is thought to be *linear*, large periodic oscillations may be traced to a *resonance* effect. However, if the system is *nonlinear*, a *limit cycle* may be the source of periodic vibration, which in turn may be traced to some *dynamic instability* in the system.

In order to identify nonperiodic or chaotic motions, the following checklist is provided:

(a) Identify nonlinear elements in the system.
(b) Check for sources of random input.
(c) Observe time history of measured signal.
(d) Look at phase plane history.
(e) Examine Fourier spectrum of signal.
(f) Take Poincaré map or return map of signal.
(g) Vary system parameters (look for bifurcations and routes to chaos).

In later chapters we discuss more advanced techniques. These include measuring two properties of the motion: fractal dimension and Lyapunov exponents. Also, probability density functions can be measured.

In the following, we go through the above-cited checklist and describe the characteristics of chaotic vibrations. To focus the discussion, the vibration of the buckled beam (double-well potential problem) is used as an example to illustrate the characteristics of chaotic dynamics.

A diagnosis of chaotic vibrations implies that one has a clear definition of such motions. However, as research uncovers more complexities in nonlinear dynamics, a rigorous definition seems to be limited to certain classes of mathematical problems. For the experimentalist, this presents a difficulty because his or her goal is to discover what mathematical model best fits the data. Thus at this stage of the subject, we will use a collection of diagnostic criteria as well as a variety of classes of chaotic motions (see Table 2-1). The experimentalist is encouraged to use two or more tests to obtain a consistent picture of the chaos.

To help sort out the growing definitions and classes of chaotic motions, we list the most common attributes without mathematical formulas, but with the most successful diagnostic tools in parentheses.

Characteristics of Chaotic Vibrations

Sensitivity to changes in initial conditions [often measured by Lyapunov exponent (Chapter 6) and fractal basin boundaries (Chapter 7)]

TABLE 2-1   Classes of Motion in Nonlinear Deterministic Systems

Regular Motion—Predictable: Periodic oscillations, quasiperiodic motion; not sensitive to changes in parameters or initial conditions

Regular Motion—Unpredictable: Multiple regular attractors (e.g., more than one periodic motion possible); long-time motion sensitive to initial conditions

Transient Chaos: Motions that look chaotic and appear to have characteristics of a strange attractor (as evidenced by Poincaré maps) but that eventually settle into a regular motion

Intermittent Chaos: Periods of regular motion with transient bursts of chaotic motion; duration of regular motion interval unpredictable

Limited or Narrow-Band Chaos. Chaotic motions whose phase space orbits remain close to some periodic or regular motion orbit; spectra often show narrow or limited broadening of certain frequency spikes

Large-Scale or Broad-Band Chaos—Weak: Dynamics can be described by orbits in a low-dimensional phase space $3 \leq n < 7$ (1–3 modes in mechanical systems), and usually one can measure fractal dimensions < 7; chaotic orbits traverse a broad region of phase space; spectra show broad range of frequencies especially below the driving frequency (if one is present)

Large-Scale Chaos—Strong: Dynamics must be described in a high-dimensional phase space; large number of essential degrees of freedom present, spatial as well as temporal complexity; difficult to measure reliable fractal dimension; dynamical theories currently unavailable

Broad spectrum of Fourier transform when motion is generated by a single frequency [measured by fast Fourier transform (FFT) using modern electronic spectrum analyzers]

Fractal properties of the motion in phase space which denote a strange attractor [measured by Poincaré maps, fractal dimensions (Chapter 7)]

Increasing complexity of regular motions as some experimental parameter is changed—for example, period doubling [often the Feigenbaum number can be measured (Chapters 1, 3, and 6)]

Transient or intermittent chaotic motions; nonperiodic bursts of irregular motion (intermittency) or initially randomlike motion that eventually settles down into a regular motion [measurement techniques are few but include the average lifetime of the chaotic burst or transient as some parameter is varied; the scaling behavior might suggest the correct mathematical model (see Chapter 6)]

Nonlinear System Elements

A chaotic system must have nonlinear elements or properties. A *linear system cannot exhibit chaotic vibrations*. In a linear system, periodic

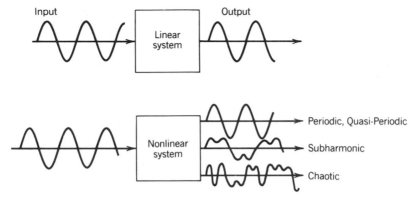

Figure 2-1 Sketch of the input–output possibilities for linear and nonlinear systems.

inputs produce periodic outputs of the same period after the transients have decayed (Figure 2-1). (Parametric linear systems are an exception.) In mechanical systems, nonlinear effects include the following:

1. Nonlinear elastic or spring elements
2. Nonlinear damping, such as stick–slip friction
3. Backlash, play, or bilinear springs
4. Most systems with fluids
5. Nonlinear boundary conditions

Nonlinear elastic effects can reside in either material properties or geometric effects. For example, the relation between stress and strain in rubber is nonlinear. However, while the stress–strain law for steel is usually linear below the yield stress, large displacement bending of a beam, plate, or shell may exhibit nonlinear relations between the applied forces or moments and displacements. Such effects in mechanics due to large displacements or rotations are usually called *geometric nonlinearities*.

In electromagnetic systems, nonlinear properties arise from the following:

1. Nonlinear resistive, inductive, or capacitive elements
2. Hysteretic properties of ferromagnetic materials
3. Nonlinear active elements such as vacuum tubes, transistors, and lasers
4. Moving media problems: for example, $\mathbf{v} \times \mathbf{B}$ voltages, where $\mathbf{v}$ is a velocity and $\mathbf{B}$ is the magnetic field

5. Electromagnetic forces: for example, $\mathbf{F} = \mathbf{J} \times \mathbf{B}$, where $\mathbf{J}$ is current and $\mathbf{F} = \mathbf{M} \cdot \nabla\mathbf{B}$, where $\mathbf{M}$ is the magnetic dipole strength

Common electric circuit elements such as diodes and transistors are examples of nonlinear devices. Magnetic materials such as iron, nickel, or ferrites exhibit nonlinear constitutive relations between the magnetizing field and the magnetic flux density. Some investigators have created negative resistors with bilinear current–voltage relations by using operational amplifiers and diodes (see Chapter 5).

The task of identifying nonlinearities in the system may not be easy: first, because we are often trained to think in terms of linear systems; and second, the major components of the system could be linear but the nonlinearity arises in a subtle way. For example, the individual elements of a truss structure could be linearly elastic, but the way they are fastened together could have play and nonlinear friction present; that is, the nonlinearities could reside in the boundary conditions.

In the example of the buckled beam, identification of the nonlinear element is easy (Figure 2-2). Any mechanical device that has more than one static equilibrium position either has play, backlash, or nonlinear stiffness. In the case of the beam buckled by end loads (Figure 2-2a), the geometric nonlinear stiffness is the culprit. If the beam is buckled by magnetic forces (Figure 2-2b), the nonlinear magnetic forces are the sources of chaos in the system.

Random Inputs

In classical linear random vibration theory, one usually treats a model of a system with random variations in the applied forces or model parameters of the form

$$[m_0 + m_1(t)]\ddot{x} + [c_0 + c_1(t)]\dot{x} + [k_0 + k_1(t)]x = f_0(t) + f_1(t)$$

where $m_1(t)$, $c_1(t)$, $k_1(t)$, and $f_1(t)$ are assumed to be random time functions with given statistical measures such as the mean or standard deviation. One then attempts to calculate the statistical properties of $x(t)$ in terms of the given statistical measures of the random inputs. In chaotic vibrations there are no *assumed* random inputs; that is, the applied forces or excitation are assumed to be deterministic.

By definition, chaotic vibrations arise from deterministic physical systems or nonrandom differential or difference equations. Although noise is always present in experiments, even in numerical simulations,

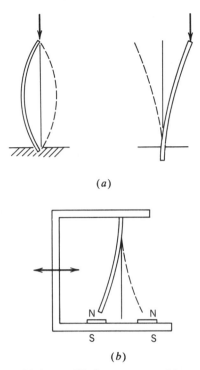

Figure 2-2   Nonlinear, multiple equilibrium state problems: (*a*) buckling of a thin elastic beam column due to axial end loads and (*b*) buckling of an elastic beam due to nonlinear magnetic body forces.

it is presumed that large nonperiodic signals do not arise from very small input noise. Thus, a large output-signal-to-input-noise ratio is required if one is to attribute nonperiodic response to a deterministic system behavior.

Observation of Time History

Usually, the first clue that the experiment has chaotic vibrations is the observation of the signal amplitude with time on a chart recorder or oscilloscope (Figure 2-3). The motion is observed to exhibit no visible pattern or periodicity. This test is *not foolproof*, however, because a motion could have a long-period behavior that is not easily detected. Also, some nonlinear systems exhibit quasiperiodic vibrations where two or more incommensurate periodic signals are present.

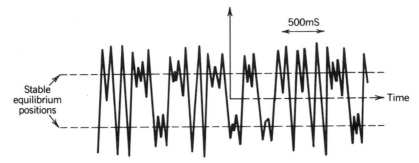

Figure 2-3   Time history of chaotic motions of a buckled elastic beam showing jumps between the two stable equilibrium states.

Phase Plane

Consider a one-degree-of-freedom mass with displacement $x(t)$ and velocity $v(t)$. Its equation of motion, from Newton's law, can be written in the form

$$\dot{x} = v$$

$$\dot{v} = \frac{1}{m} f(x, v, t) \tag{2-1}$$

where m is the mass and f is the applied force. The phase plane is defined as the set of points (x, v). (Some authors use the momentum mv instead of v.) When the motion is periodic (Figure 2-4a), the phase plane orbit traces out a closed curve which is best observed on an analog or digital oscilloscope. For example, the forced oscillations of a linear spring–mass–dashpot system exhibit an elliptically shaped orbit. However, a forced nonlinear system with a cubic spring element may show an orbit which crosses itself but is still closed. This can represent a subharmonic oscillation as shown in Figure 2-4a.

Systems for which the force does not depend explicitly on time—for example, $f = f(x, v)$ in Eq. (2-1)—are called *autonomous*. For autonomous nonlinear systems (no harmonic inputs), periodic motions are referred to as *limit cycles* and also show up as closed orbits in the phase plane (see Chapter 1).

Chaotic motions, on the other hand, have orbits which never close or repeat. Thus, the trajectory of the orbits in the phase plane will tend to fill up a section of the phase space as in Figure 2-4b. Although

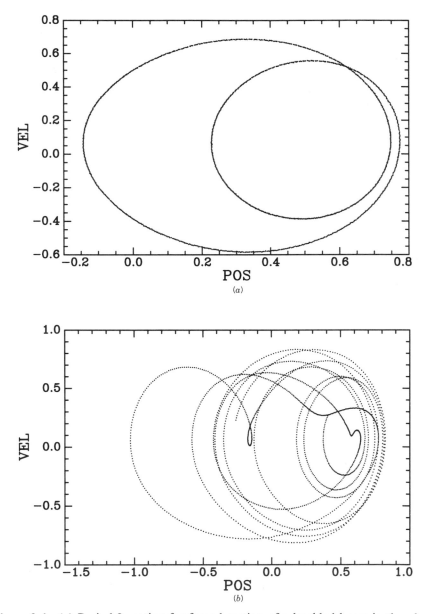

Figure 2-4 (*a*) Period-2 motion for forced motion of a buckled beam in the phase plane (bending strain versus strain rate). (*b*) Chaotic trajectory for forced motion of a buckled beam.

this wandering of orbits is a clue to chaos, continuous phase plane plots provide very little information and one must use a modified phase plane technique called *Poincaré maps* (see below).

Often, one has only a single measured variable $v(t)$. If $v(t)$ is a velocity variable, then one can integrate $v(t)$ to get $x(t)$ so that the phase plane consists of points $[\int_0^t v \, dt, v(t)]$.

Pseudo-Phase Space Method. Another technique that has been used when only one variable is measured is the time-delayed pseudo-phase-plane method (also called the *embedding space method*). For a one degree-of-freedom system with measurement $x(t)$, one plots the signal versus itself but delayed or advanced by a fixed time constant: $[x(t), x(t + T)]$. The idea here is that the signal $x(t + T)$ is related to $\dot{x}(t)$ and should have properties similar to those in the classic phase plane $[x(t), \dot{x}(t)]$. In Figure 2-5 we show a pseudo-phase-plane orbit for a harmonic oscillator for different time delays. If the motion is chaotic, the trajectories do not close (Figure 2-6). The choice of T is not crucial, except to avoid a natural period of the system. When the state variables are greater than two (e.g., position, velocity, time, or forcing phase), the higher-dimensional pseudo-phase-space trajectories can be constructed using multiple delays. For example, a three-dimensional space can be constructed using a vector with components $(x(t), x(t + T), x(t + 2T))$. More will be said about this technique in Chapter 5.

Fourier Spectrum and Autocorrelation

One of the clues to detecting chaotic vibration is the appearance of a broad spectrum of frequencies in the output when the input is a single-frequency harmonic motion or is dc (Figure 2-7). This characteristic of chaos becomes more important if the system is of low dimension (e.g., one to three degrees of freedom). Often, if there is an initial dominant frequency component ω_0, a precursor to chaos is the appearance of subharmonics ω_0/n in the frequency spectrum (see below). In addition to ω_0/n, harmonics of this frequency will also be present of the form $m\omega_0/n$ ($m, n = 1, 2, 3, \ldots$). An illustration of this test is shown in Figure 2-7. Figure 2-7a shows a single spike in both the driving force and the response of a buckled beam. Figure 2-7b shows a broad spectrum, indicating possible chaotic motions.

One must be cautioned against concluding that multiharmonic outputs imply chaotic vibrations, because the system in question might have many hidden degrees of freedom of which the observer is un-

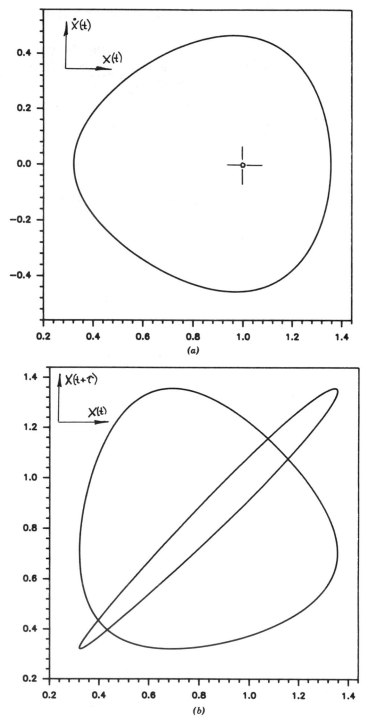

Figure 2-5 (*a*) Phase-plane trajectory of Duffing oscillator (1-2.4); $\alpha = -1$, $\beta = 1$. (*b*) Pseudo-phase-plane trajectory for the periodic oscillator in (*a*) for two delay times.

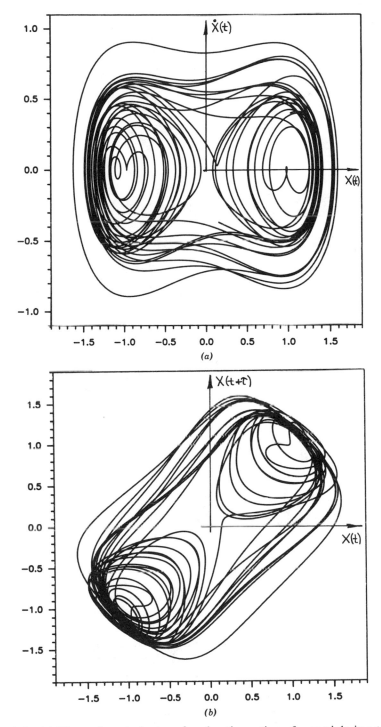

Figure 2-6 (*a*) Phase-plane trajectory for chaotic motion of a particle in a two-well potential (buckled beam) under periodic forcing (1-2.4); $\alpha = 1$, $\beta = 1$. (*b*) Pseudo-phase-plane trajectory of chaotic motion in (*a*).

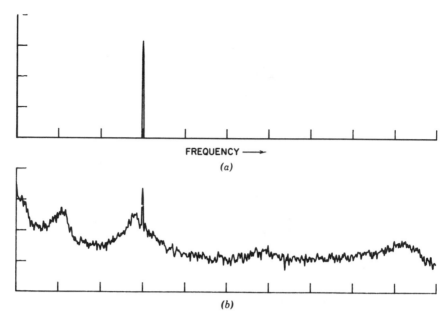

Figure 2-7 (*a*) Frequency spectrum of buckled elastic beam for low-amplitude excitation—linear periodic response. (*b*) Frequency spectrum of buckled elastic beam for larger excitation—broad-band response of beam due to chaotic vibration.

aware. In large-degree-of-freedom systems, the use of the Fourier spectrum may not be of much help in detecting chaotic vibrations unless one can observe changes in the spectrum as one varies some parameter such as driving amplitude or frequency.

Another useful measure of the predictability of the motion is an autocorrelation function

$$A(\tau) = \frac{1}{T}\int_{T}^{-T} f(t)f(t + \tau)\, dt$$

where T is very large compared to the dominant periods in the motion. An autocorrelation of a periodic signal produces a periodic function $A(\tau)$ as shown in Figure 2-8*a*. But $A(\tau)$ for a chaotic or random signal shows $A(\tau) \rightarrow 0$, for $\tau > \tau_c$ where τ_c is some characteristic time (Figure 2-8*b*). Modern signal processing electronics can calculate $A(\tau)$ as well as the Fourier transform in real time as the data is gathered. Hence, both tools are useful for experimental bifurcation studies because one can look for qualitative changes in $A(\tau)$ as some parameter is varied. The characteristic time τ_c is a measure of the time the motion can be

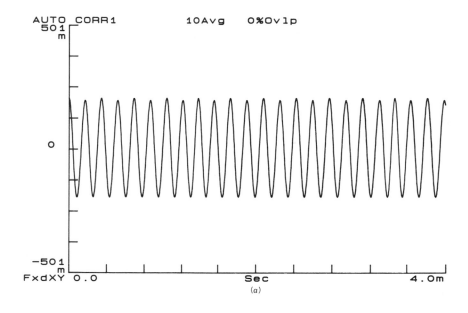

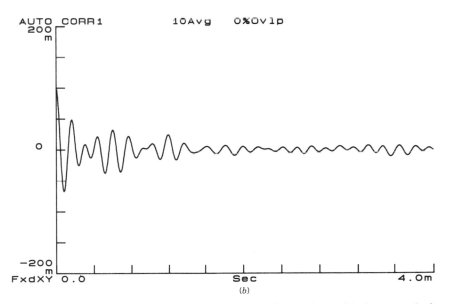

Figure 2-8 (*a*) Autocorrelation function for periodic motions. (*b*) Autocorrelation function for a chaotic signal.

predicted in the future, and is believed by some researchers to be related to the Lyapunov exponent (see Chapter 6).

Poincaré Maps and Return Maps

In the mathematical study of dynamical systems, a map refers to a time-sampled sequence of data $\{x(t_1), x(t_2), \ldots, x(t_n), \ldots, x(t_N)\}$ with the notation $x_n = x(t_n)$. A simple deterministic map is one in which the value of x_{n+1} can be determined from the values of x_n. This is often written in the form (see Chapter 3)

$$x_{n+1} = f(x_n) \tag{2-2}$$

This can be recognized as a *difference equation*. The idea of a map can be generalized to more than one variable. Thus, x_n could represent a vector with M components $x_n = (Y1_n, Y2_n, \ldots, YM_n)$ and Eq. (2-2) could represent a system of M equations.

For example, suppose we consider the motion of a particle as

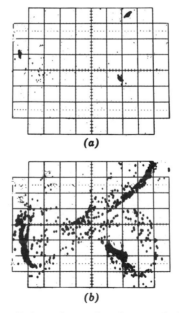

(a)

(b)

Figure 2-9   (*a*) Phase-plane Poincaré map showing a period-3 subharmonic motion of a periodically forced buckled beam. (*b*) Chaotic motion near a period-3 subharmonic.

displayed in the phase plane $(x(t), \dot{x}(t))$. However, if instead of looking at the motion continuously, we look only at the dynamics at discrete times, then the motion will appear as a sequence of dots in the phase plane (Figures 1-19 and 2-9). If $x_n \equiv x(t_n)$ and $y_n \equiv \dot{x}(t_n)$, this sequence of points in phase plane represents a *two-dimensional* map:

$$x_{n+1} = f(x_n, y_n)$$

$$y_{n+1} = g(x_n, y_n)$$

(2-3)

When the sampling times t_n are chosen according to certain rules, to be discussed below, this map is called a *Poincaré map*.

Poincaré Maps for Forced Vibration Systems. When there is a driving motion of period T, a natural sampling rule for a Poincaré map is to choose $t_n = nT + \tau_0$. This allows one to distinguish between periodic motions and nonperiodic motions. For example, if the sampled harmonic motion shown in Figure 2-4a is synchronized with its period, its "map" in the phase plane will be two points. If the output, however, were a subharmonic of period 3, the Poincaré map would consist of a set of three points as shown in Figure 2-9a.

Another nonchaotic Poincaré map is shown in Figure 2-10, where the motion consists of two *incommensurate* frequencies

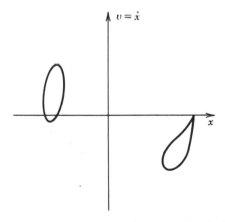

Figure 2-10   Phase-plane Poincaré map showing a quasiperiodic motion of a periodically forced two-degree-of-freedom beam in a two-well magnetic potential.

$$x(t) = C_1\sin(\omega_1 t + d_1) + C_2\sin(\omega_2 t + d_2) \qquad (2\text{-}4)$$

where ω_1/ω_2 is an irrational number. If one samples at a period corresponding to either frequency, the map in the phase plane will become a continuous closed figure or orbit. This motion is sometimes called *almost-periodic* or *quasiperiodic* motion or "motion on a torus" and is not considered to be chaotic (see also Figure 1-12).

Finally if the Poincaré map does not consist of either a finite set of points (Figure 2-9a) or a closed orbit (Figure 2-10), the motion may be

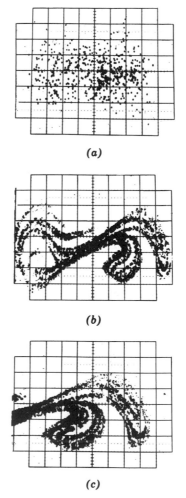

(a)

(b)

(c)

Figure 2-11   (a) Poincaré map of chaotic motion of a buckled beam with low damping. (b, c) Poincaré map of chaotic motion of a buckled beam for higher damping showing fractal-like structure of a strange attractor. [From Moon (1980a) with permission of ASME, copyright 1980.]

chaotic (Figure 2-11). Here we must distinguish between damped and undamped systems. In undamped or lightly damped systems the Poincaré map of chaotic motions often appear as a cloud of unorganized points in the phase plane (Figure 2-11a). Such motions are sometimes called *stochastic* (e.g., see Lichtenberg and Lieberman, 1983). In damped systems the Poincaré map will sometimes appear as an infinite set of highly organized points arranged in what appear to be parallel lines as shown in Figure 2-11b,c. In numerical simulations, one can enlarge a portion of the Poincaré map (see Figure 2-12) and observe further structure. If this structured set of points continues to exist after several enlargements then the map is said to be *fractal-like*, and one says the motion behaves as a *strange attractor*. This embedding of structure within structure is often referred to as a *Cantor set* (see Chapter 7).

The appearance of fractal-like or Cantor-set-like patterns in the Poincaré map of a vibration history is a strong indicator of chaotic motions. The classes of patterns of Poincaré maps are listed in Table 2-2.

Poincaré Maps in Autonomous Systems. Steady-state vibrations can also be generated without periodic or random inputs if the motion originates from a dynamic instability such as wind-induced flutter in an elastic structure (Figure 2-13) or a temperature-gradient-induced convective motion in a fluid or gas (e.g., Benard convection, Figure 1-25). One is then led to ask how to choose the sampling times in a Poincaré map. Here the discussion gets a little abstract.

Consider the lowest-order chaotic system governed by three first-order differential equations (e.g., the Lorenz equations of Chapter 1). In an electromechanical system the variables $x(t)$, $y(t)$, and $z(t)$ could represent displacement, velocity, and control force as in a feedback-

---

TABLE 2-2   Classification of Poincaré Maps

Finite number of points: periodic or subharmonic oscillation

Closed curve: quasiperiodic, two incommensurate frequencies present

Open curve: suggest modeling as a one-dimensional map; try plotting $x(t)$ versus $x(t + T)$

Fractal collection of points: strange attractor in three phase-space dimensions

Fuzzy collection of points: (i) dynamical systems with too much random or noisy input; (ii) strange attractor but system has very small dissipation—use Lyapunov exponent test; (iii) strange attractor in phase space with more than three dimensions—try multiple Poincaré map; (iv) quasiperiodic motion with three or more dominant incommensurate frequencies

controlled system. We then imagine the motion as a trajectory in a three-dimensional phase space (Figure 2-14). A Poincaré map can be defined by constructing a two-dimensional oriented surface in this space and looking at the points (x_n, y_n, z_n) where the trajectory pierces this surface. For example, we can choose a plane $n_1 x + n_2 y + n_3 z = c$ with normal vector $\mathbf{n} \equiv (n_1, n_2, n_3)$. As a special case, choose points

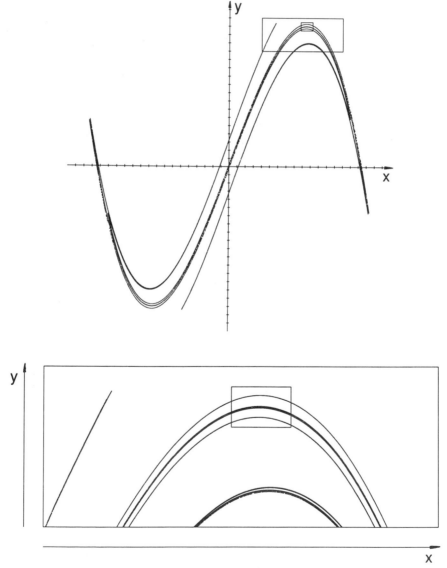

Figure 2-12   Poincaré map of chaotic vibration of a forced nonlinear oscillation showing self-similar structure at finer and finer scales.

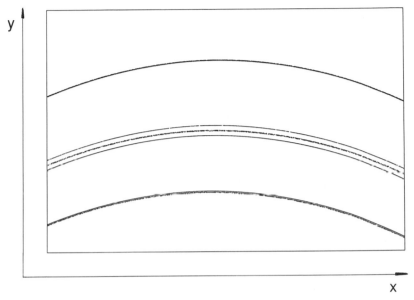

Figure 2-12   (*Continued*)

where $x = 0$. Then the Poincaré map consists of points which pierce this plane with the same sense; that is, if $s(t)$ represents a unit vector along the trajectory, $s(t_n) \cdot n$ must always have the same sign.

This definition of the Poincaré map actually includes the case when the system is periodically forced. Consider, for example, a forced nonlinear oscillator with equations of motion

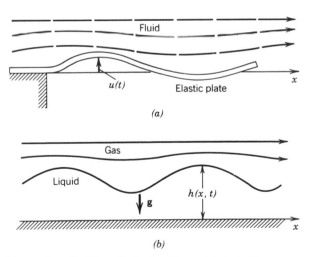

Figure 2-13   Examples of self-excited vibrations: (*a*) fluid flow over an elastic plate and (*b*) gas flow over a liquid interface.

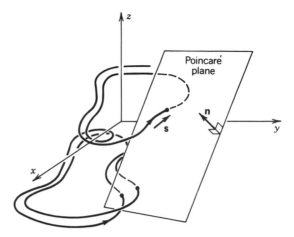

Figure 2-14   Sketch of time evolution trajectories of a third-order system of equations and a typical Poincaré plane.

$$\dot{x} = y \tag{2-5}$$

$$\dot{y} = F(x, y) + f_0\cos(\omega t + \phi_0) \tag{2-6}$$

Then this system can be made to look like an autonomous one by defining

$$z = \omega t + \phi_0 \tag{2-7}$$

and

$$\dot{x} = y \tag{2-8}$$

$$\dot{y} = F(x, y) + f_0\cos z \tag{2-9}$$

$$\dot{z} = \omega \tag{2-10}$$

Thus, a natural sampling time is chosen when $z = $ constant. This system can be thought of as a cylindrical phase space where the values of z are restricted: $0 \leq z \leq 2\pi$. A picture of several Poincaré maps is then given as in Figure 2-15 for different values of z (see also Chapter 5, Figure 5-7).

Reduction of Dynamics to One-Dimensional Maps. In Chapter 1 we saw that simple one-dimensional maps or difference equations of the form $x_{n+1} = f(x_n)$ can exhibit period-doubling bifurcations and chaos

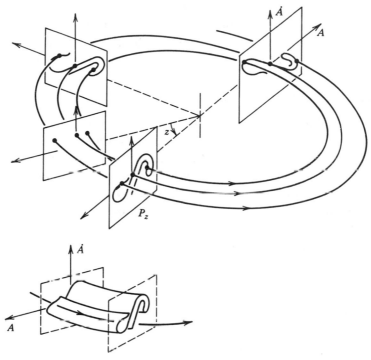

Figure 2-15 Sketch of a strange attractor for a forced nonlinear oscillator: product space of the Poincaré plane and the phase of the forcing excitation.

when the function $f(x)$ has at least one maximum (or minimum), as in Figure 1-21. Period-doubling phenomena have been observed in so many different complex physical systems (fluids, lasers, p–n electronic junctions) that in many cases the dynamics may sometimes be modeled as a one-dimensional map. This is especially possible in systems with significant dissipation. To check this possibility, one samples some dynamic variable using a Poincaré section as discussed above; that is, $x_n = x(t = t_n)$. Then one plots each x_n against its successor value x_{n+1}. This is sometimes called a *return map*. Two criteria must be met to declare the system chaotic. First, the points x_{n+1} versus x_n must appear to be clustered in some apparent functional relation; and second, this function $f(x)$ must be multivalued—that is, it must have a maximum or a minimum. If this be the case, one then attempts to fit a polynomial function to the data and uses this mapping to do numerical experiments or analysis along the lines of the quadratic map (Chapters 1 and 3). Examples of this technique may be found in Shaw (1984) in the problem of a dripping faucet and in Rollins and Hunt (1982) in an experiment with a varactor diode in an electrical circuit (see also Chapter 3 for a discussion of these problems). This technique is discussed further in Chapter 5. An example using experimental data from the vibration of a levitated magnet over a high-temperature superconductor is shown in Figure 2-16 (see Moon, 1988).

Bifurcations: Routes to Chaos

Periodic to Chaotic Motions Through Parameter Changes. In conducting any of these tests for chaotic vibrations, one should try to vary one or more of the control parameters in the system. For example, in the case of the buckled structure (Figure 2-2), one can vary either the forcing amplitude or forcing frequency, or in the case of the nonlinear circuit, one can vary the resistance. The reason for this procedure is to see if the system has steady or periodic behavior for some range of the parameter space. In this way, one can have confidence that the system is in fact deterministic and that there are no hidden inputs or sources of truely random noise.

In changing a parameter, one looks for a pattern of periodic responses. One characteristic precursor to chaotic motion is the appearance of subharmonic periodic vibrations. There may in fact be many patterns of prechaos behavior. Several models of prechaotic behavior have been observed in both numerical and physical experiments (see Gollub and Benson, 1980 or Kadanoff, 1983).

Period-Doubling Route to Chaos. Period doubling in physical systems have been observed experimentally in all branches of classical physics, chemistry, and biology as well as in many technical devices. Although this route to chaos in ubiquitous in science, it is by no means the only path to unpredictable dynamics. In the period-doubling phenomenon, one starts with a system with a fundamental periodic motion. Then as some experimental parameter is varied, say λ, the motion undergoes a bifurcation or change to a periodic motion with twice the period of the original oscillation. As λ is changed further, the system bifurcates to periodic motions with twice the period of the previous oscillation. One outstanding feature of this scenario is that the critical values of λ at which successive period doublings occur obey the following scaling rule (see also Chapters 1 and 3):

$$\frac{\lambda_n - \lambda_{n-1}}{\lambda_{n+1} - \lambda_n} \rightarrow \delta = 4.6692016 \qquad (2\text{-}11)$$

as $n \rightarrow \delta$. (This is called the *Feigenbaum number*, named after the

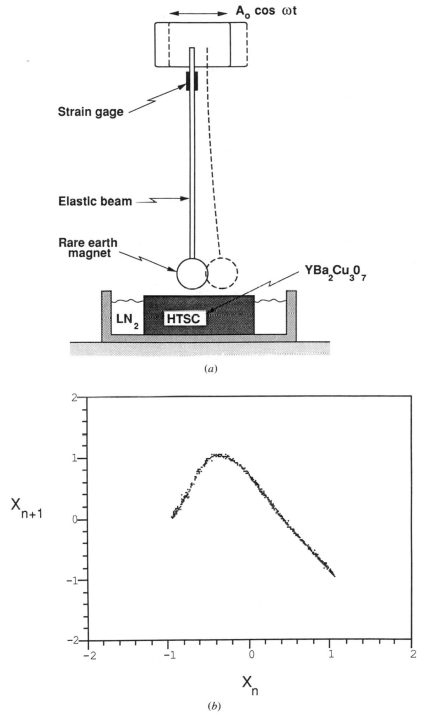

Figure 2-16 (*a*) Sketch of vibrating magnet near a high-temperature superconducting material. (*b*) Return map based on the amplitude signal from the strain gage. [From Moon (1988) with permission of North-Holland Publishing Company.]

physicist who discovered this scaling behavior.) In practice, this limit approaches δ by the third or fourth bifurcation.

This process will accumulate at a critical value of the parameter, after which the motion becomes chaotic.

This phenomenon has been observed in a number of physical systems as well as numerical simulations. The most elementary mathematical equation that illustrates this behavior is a first-order difference equation (see Chapters 1 and 3):

$$x_{n+1} = \lambda x_n(1 - x_n) \tag{2-12}$$

As the system parameter is changed beyond the critical value, chaotic motions exist in a band of parameter values. However, these bands may be of finite width; that is, as the parameter is varied, periodic windows may develop. Periodic motions in this regime may again undergo period-doubling bifurcations, again leading to chaotic motions (see Section 6.3).

The period-doubling model for the route to chaos is an elegant, aesthetic model and has been described in many popular articles. However, while many physical systems exhibit properties similar to those of (2-12), many other systems do not. Nevertheless, when chaotic vibrations are suspected in a system, it is worthwhile checking to see if period doubling is present.

Bifurcation Diagrams. A widely used technique for examining the prechaotic or postchaotic changes in a dynamical system under parameter variations is the bifurcation diagram (an example is shown in Figure 2-17). Here some measure of the motion (e.g., maximum amplitude) is plotted as a function of a system parameter such as forcing amplitude or damping constant. If the data are sampled using a Poincaré map, it is very easy to observe period doubling and subharmonic bifurcations as shown in the experimental data for a nonlinear circuit from a paper by Bryant and Jeffries (1984*a,b*) at the University of California, Berkeley. However, when the bifurcation diagram loses continuity, it may mean either quasiperiodic motion or chaotic motion and further tests are required to classify the dynamics.

Quasiperiodic Route to Chaos. Although period doubling is the most celebrated scenario for chaotic vibration, there are several other schemes that have been studied and observed. In one proposed by Newhouse et al. (1978), they imagine a system which undergoes successive dynamic instabilities before chaos. For example, suppose a

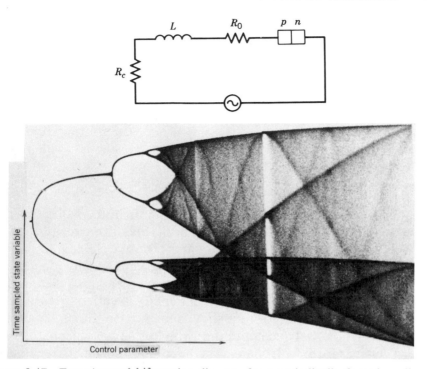

Figure 2-17 Experimental bifurcation diagram for a periodically forced nonlinear circuit with a p–n junction; periodically sampled current versus drive ampltiude voltage. [From Van Buskirk and Jeffries (1985) with permission of The American Physical Society, copyright 1985.]

system is initially in a steady state and becomes dynamically unstable after changing some parameter (e.g., flutter). As the motion grows, nonlinearities come into effect and the motion becomes a limit cycle. Such transitions are called *Hopf bifurcations* in mathematics (e.g., see Abraham and Shaw, 1983). If after further parameter changes the system undergoes two more Hopf bifurcations so that three simultaneous coupled limit cycles are present, chaotic motions become possible.

Thus, the precursor to such chaotic motion is the presence of two simultaneous periodic oscillations. When the frequencies of these oscillations, ω_1 and ω_2, are not commensurate, the observed motion itself is not periodic but is said to be *quasiperiodic* [see Eq. (2-4)]. As discussed above, the Poincaré map of a quasiperiodic motion is a closed curve in the phase plane (Figure 2-10). Such motions are imagined to take place on the surface of a torus where the Poincaré map represents a plane which cuts the torus (see Figure 2-18). If ω_1 and ω_2 are incommensurate, the trajectories fill the surface of the torus. If

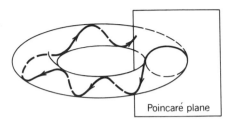

Figure 2-18   Sketch illustrating the coupled motion of two oscillators and the Poincaré plane used to detect a quasiperiodic route to chaos.

ω_1/ω_2 is a rational number, the trajectory on the torus will eventually close, although it might perform many orbits in both angular directions of the torus before closing. In the latter case the Poincaré map will become a set of points generally arranged in a circle. Chaotic motions are often characterized in such systems by the breakup of the quasiperiodic torus structure as the system parameter is varied (Figure 2-19).

Evidence for the three-frequency transition to chaos have been observed in flow between rotating cylinders (Taylor–Couette flow) where vortices form with changes in the rotation speed. Three Fourier spectra from one such experiment are shown in Figure 2-20. In the top figure, one periodic motion appears to be present. In the middle figure, two major motions are evident. In the bottom figure, we have the sign of an increase in broad-band noise which is characteristic of chaotic behavior.

Intermittency.   In a third route to chaos, one observes long periods of periodic motion with bursts of chaos. This scenario is called *intermittency*. As one varies a parameter, the chaotic bursts become more frequent and longer (e.g., see Manneville and Pomeau, 1980). Evidence for this model for prechaos has been claimed in experiments on convection in a cell (closed box) with a temperature gradient (called *Rayleigh–Benard convection*) (see Figure 2-21). Some models for intermittency predict that the average time of the regular or laminar phase of the motion $\langle\tau\rangle$ will scale in a precise way as some system parameter is varied; for example,

$$\langle\tau\rangle \approx \frac{1}{(\lambda - \lambda_c)^{1/2}} \qquad (2\text{-}13)$$

where λ_c is the value at which the periodic motion becomes chaotic.

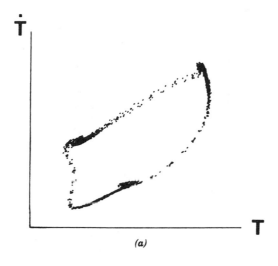

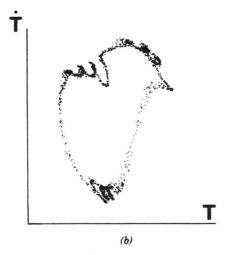

Figure 2-19 (*a*) Poincaré section of a quasiperiodic motion in Rayleigh–Benard thermal convection with a frequency ratio close to $\omega_1/\omega_2 = 2.99$. (*b*) Breakup of the torus surface prior to the onset of chaos. [From Bergé (1982).]

It should be noted that for some physical systems, one may observe all three patterns of prechaotic oscillations and many more depending on the parameters of the problem. The benefit in identifying a particular prechaos pattern of motion with one of these now "classic" models is that a body of mathematical work on each exists which may offer better understanding of the chaotic physical phenomenon under study.

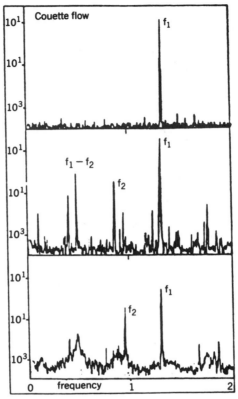

Figure 2-20   Evidence for the three-frequency transition to chaos in the flow between rotating cylinders (Taylor–Couette flow); the rotation difference increases from top to bottom. [From Swinney and Gullub (1978).]

Quasiperiodicity and Mode-Locking

One phenomenon that sometimes appears in searching for patterns of dynamic behavior in periodically driven systems is *mode-locking*. This behavior typically occurs in physical systems with natural limit cycle generators, such as negative resistance circuits, unstable control systems, aeroelastic oscillators, biochemical and chemical oscillators,

Figure 2-21   Sketch of the time history for intermittent-type chaos.

and thermofluid convections, such as a Rayleigh–Benard cell. Suppose, for example, such a system exhibits a natural periodic limit cycle of frequency ω_1 and is then perturbed by an external periodic force of frequency ω_2. Then one can ask, at what frequency will the combined system oscillate?

A very nice review of the phenomena of mode-locking and quasiperiodicity is given in Glazier and Libchaber (1988).

The early observation of mode-locking or frequency-locking goes back to the Dutch physicist Christiaan Huygens (1629–1695), who observed how two pendulum clocks attached to a common structure become synchronized. Glazier and Libchaber (1988) give many modern references of experimental observations of mode-locking in physical and biological systems including chemistry, solid-state physics, fluid mechanics, and biology.

In certain problems, mode locking can be observed fairly easily. One example is shown in Figure 2-22, which shows experimental data of a biological oscillator excited by a periodic electrical stimulus

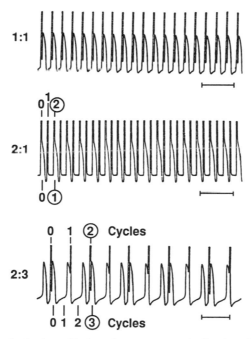

Figure 2-22 Phase-locked oscillations between a periodic electrical stimulus (upper spikes on traces), and self-oscillations of a group of chick heart cells. (*Top*) $N:M = 1:1$. (*Middle*) $N:M = 2:1$. (*Bottom*) $N:M = 2:3$. [From Guevera et al. (1990) with permission of Harcourt Brace Jovanovich, Inc.]

(Guevara et al., 1990). Without the stimulus, the oscillator, which is a collection of cells from a chick heart, will produce a periodic train of electrical pulses called *action potentials*. When the periodic external stimulus is applied in very short time durations, one can observe the two types of signals in the output as shown in Figure 2-22. In a mode-locked condition, for every N stimuli pulses there are M action potentials. Shown in Figure 2-22 are ratios $N:M$ of $1:1, 2:1, 2:3$.

If the frequency of the stimulus is changed in a small way, the ratio $N:M$ is preserved—this is the mode-locking. If the change in the control frequency is sufficiently large, then the motion may become quasiperiodic or may become locked in another $N:M$ ratio. Thus, there is a *finite width* of the *control frequency* $\omega_{N:M} = \Delta f_2$ in which the $N:M$ ratio is fixed. This width, $\omega_{N:M}$, then depends on the strength or amplitude of the control stimulus. Plotted in the plane of control amplitude and frequency, each fixed $N:M$ mode-locking regime looks like a wedge shaped region as shown in Figure 2-23.

These mode-locking regimes are called *Arnold tongues* in honor of the Soviet dynamicist who provided the mathematical theory. The data in Figure 2-23 are from an experiment involving thermofluid convection in mercury in a small box with external excitation provided by a magnetic body force. One can see that the width of the tongues grows with the strength of the periodic stimulus. At a certain amplitude these tongues overlap, creating hysteresis and the possibility of cha-

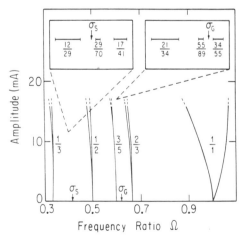

Figure 2-23 Experimental Arnold torques in the driving amplitude–frequency plane for thermal convection in mercury (Rayleigh–Bénard motion) driven periodically by passing electric current through the mercury in the presence of a small magnetic field. The insets show the relative widths near the golden mean σ_G and the silver mean σ_S. [From Glazier and Libchaber (1988) with permission of IEEE.]

otic dynamics. Within each tongue one can also have period doubling in which the mode-locked ratio goes to $2N:2M$, $4N:4M$, and so on.

The phenomena of mode-locking and quasiperiodicity can often be modeled by a one-dimensional map called the *circle map* (1-2.25) (see also Chapter 3):

$$\theta_{n+1} = \theta_n + \Omega + \kappa \sin \theta_n \quad (\text{mod } 2\pi)$$

where the angular variable θ is periodic in 2π radians. This model comes from a geometric view of the driven pendulum as the motion of a particle on a torus as discussed in Chapter 1 and in Figure 2-18.

Transient Chaos

Sometimes chaotic vibrations appear for some parameter changes but eventually settle into a periodic or quasiperiodic motion after a short time. According to Grebogi et al. (1983b), such transient chaos is a consequence of a crisis or the sudden disappearance of sustained chaotic dynamics. Thus, experiments and numerical simulation should be allowed to run for a time after one thinks the system is in chaos even if the Poincaré map seems to be mapping out a fractal structure characteristic of strange attractors.

Transient problems are important in technical dynamics. New methods of characterizing transient chaotic behavior and unpredictability have recently received attention. See e.g. the work of Tel (1990).

Conservative Chaos

Although much of the new excitement about nonlinear dynamics has focused on chaotic dynamics in dissipative systems, chaotic behavior in nondissipative or so-called conservative systems had been known for some time. In fact, the search for solutions to the equations of celestial mechanics in the late 19th century led mathematicians like Poincaré to speculate that many dynamic problems were sensitive to initial conditions and hence were unpredictable in the details of the motions of orbiting bodies.

The study of chaotic dynamics in energy-conserving systems, while not the principal focus of this book, has received much attention in the literature and sometimes is found under the heading of "Hamiltonian Dynamics," which refers to the methods of Hamilton (and also Jacobi) that are used to solve nonlinear problems in multi-degree-of-freedom nondissipative systems [e.g., see Chapter 1; also see the excellent monographs by Lichtenberg and Lieberman (1983) and by Rasband (1990)].

Examples of conservative systems in the physical world include orbital problems in celestial mechanics and the behavior of particles in electromagnetic fields. Hence, much of the work in this field has been done by those interested in plasma physics, astronomy, and astrophysics (e.g., see Sagdeev et al., 1988 and Zaslavsky et al., 1991).

Although most earth-bound dynamics problems have some energy loss, some, like structural systems or microwave cavities, have very little damping and over a finite period of time can behave like a conservative or Hamiltonian system. An example might be the vibration of an orbiting space structure. Also, conservative system dynamics provides a limiting case for small damping dynamic analysis. Thus, while we do not attempt to present a rigorous or lengthy summary of Hamiltonian dynamics, it is useful to discuss the general features of these problems.

Typically, energy-conserving systems can exhibit the same types of bounded vibratory motion as lossy systems including periodic, subharmonic, quasiperiodic, and chaotic motions. One of the main differences, however, between vibrations in lossy and lossless problems is that chaotic orbits in lossy systems exhibit a fractal structure in the phase space whereas chaotic orbits in lossless systems do not.

Chaotic orbits in conservative systems tend to visit all parts of a subspace of the phase space uniformly; that is, they exhibit a uniform probability density over restricted regions in the phase space. Thus, lossless systems exhibit Poincaré maps different from those of lossy problems. However, the use of Lyapunov exponents as a measure of nearby orbit divergence is still valid. An example of a system with no dissipation is the ball bouncing on an elastic table where the table is moving and the impact is assumed to be lossless or elastic. Details of this problem are discussed in Chapter 3.

Lyapunov Exponents and Fractal Dimensions

The tests for chaotic vibrations described in this chapter are mainly qualitative and involve some judgment and experience on the part of the investigator. Quantitative tests for chaos are available and have been used with some success. Two of the most widely used criteria are the Lyapunov exponent (see Chapter 6) and the fractal dimension (see Chapter 7). In summary, these two indicators are currently interpreted as follows:

1. Positive Lyapunov exponents imply *chaotic dynamics*.
2. Fractal dimension of the orbit in phase space implies the existence of a *strange attractor*.

The Lyapunov exponent test can be used for both dissipative or nondissipative (conservative) systems, whereas the fractal dimension test only makes sense for dissipative systems.

The Lyapunov exponent test measures the sensitivity of the system to changes in initial conditions. Conceptually, one imagines a small ball of initial conditions in phase space and looks at its deformation into an ellipsoid under the dynamics of the system. If d is the maximum length of the ellipsoid and d_0 is the initial size of the initial condition sphere, the Lyapunov exponent λ is interpreted by the equation

$$d = d_0 2^{\lambda(t - t_0)}$$

One measurement, however, is not sufficient, and the calculation must be averaged over different regions of phase space. This average can be represented by

$$\lambda = \lim_{N \to \infty} \frac{1}{N} \sum_{1}^{N} \frac{1}{(t_i - t_{0i})} \ln_2 \frac{d_i}{d_{0i}}$$

A more detailed discussion is given in Chapter 6 along with references.

The fractal dimension is related to the discussion of the horseshoe map in Chapter 1. There we saw that in a chaotic dynamic system, regions of phase space are stretched, contracted, folded, and remapped onto the original space. This remapping for dissipative systems leaves gaps in the phase space. This means that orbits tend to fill up less than an integer subspace in phase space. The fractal dimension is a measure of the extent to which orbits fill a certain subspace, and a noninteger dimension is a hallmark of a *strange attractor*. There are many definitions of fractal dimension, but the most basic one is derived from the notion of counting the number of spheres N of size ε needed to cover the orbit in phase space. Basically, $N(\varepsilon)$ depends on the subspace of the orbit. If it is a periodic or limit cycle orbit, then $N(\varepsilon) \approx \varepsilon^{-1}$. When the motion lies on a strange attractor, $N(\varepsilon) \approx \varepsilon^{-d}$ or

$$d = \lim_{\substack{N \to \infty \\ \varepsilon \to 0}} \frac{\ln N}{\ln(1/\varepsilon)}$$

Further discussion is given in Chapter 7.

Although both quantitative tests can be automated using computer control, experience and judgment are still required to provide a conclusive assessment as to whether the motion is *chaotic* or a *strange attractor*. Finally, almost all physical examples of *strange attractors* have been found to be chaotic; that is, noninteger *d* implies λ > 0. However, a few mathematical models and physical problems have been studied where one does not imply the other.

Strange–Nonchaotic Motions

As more research is done in the field of modern nonlinear dynamics, discoveries are made of new categories of motions. We have seen the list grow from periodic, subharmonic, quasiperiodic, intermittent, chaotic, and hyperchaotic to spatially and temporally chaotic dynamics. However, nearly a decade ago J. Yorke and coworkers at the University of Maryland suggested that a new type of motion was possible, namely, strange nonchaotic motions—that is, motions which were geometrically fractal in the phase space, but whose Lyapunov exponents were *not* positive.

At the time, almost all the dynamic models with physical relevance and dissipation exhibited at least one positive Lyapunov when the attractor looked fractal. And to some extent the terms *chaotic* and *strange attractor* were often used interchangeably. The original example of Grebogi, Ott et al., in 1984, of a strange–nonchaotic attractor, however, looked to many to be a singular case, not relevant to physical problems. However, through a series of papers this group has amassed convincing evidence for this type of dynamic, which is especially relevant to the systems which exhibit multiple sources of oscillation (either forced or autonomous) and exhibit quasiperiodic motion [e.g., see Grebogi et al. (1984), Ding et al. (1989a,b), and Ditto et al. (1990)].

We will not discuss the theory of such motions in great depth, but we will summarize one of the experimental examples for which strange–nonchaotic motions are thought to occur.

Strange–nonchaotic attractors are difficult to diagnose experimentally because reliable methods for calculating Lyapunov exponents are not readily available. However, another tool that has been used is the scaling properties of the Fourier spectrum of the time series (Ditto et al., 1990). Define $|S(w)|$ as Fourier transform of the signal sampled at one of the forcing frequencies. Then the spectral distribution function $N(s)$ is defined as the number of peaks in the $|S(w)|$ with amplitude greater than s. For two-frequency quasiperiodic attractors, $N \sim \ln s$.

For three-frequency quasiperiodic attractors, $N \sim [\ln s]^2$, and for strange–nonchaotic motions, $N \sim s^{-a}$, $1 < a < 2$.

The experiment described in Ditto et al. (1990) consists of a thin elastica clamped at the base and initially buckled under gravity. The material used was an amorphous magnetostrictive ribbon called MET-GLAS which exhibits large changes in the effective Young's modulus in a magnetic field. The ribbon was placed in a vertical magnetic field which had two frequency components

$$\mathbf{B} = B_1\cos \omega_1 t + B_2\cos \omega_2 t$$

where B_1, $B_2 \sim 0.5$–0.9 Oe. The data recorded the change in curvature of the ribbon near the clamped end. Evidence for strangeness was obtained by taking a Poincaré section triggered on one of the driving frequencies. The two-frequency quasiperiodic motion exhibited a closed circular figure. However, the surface of section of the so-called strange–nonchaotic motion showed a fractal-like pattern. The spectral distribution function (Figure 2-24) showed the theoretical scaling $N \sim s^{-a}$ for strange–nonchaotic motions with $a = 1.25$.

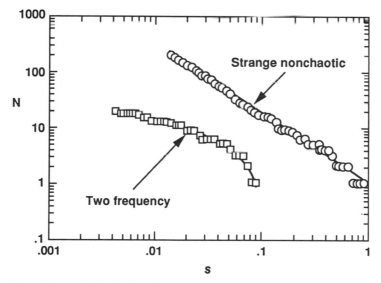

Figure 2-24 Spectral distribution function $N(s)$ for a buckled magnetoelastic cantilever beam driven by a vertical magnetic field $H = H_1\cos \omega_1 t + H_2\cos \omega_2 t$ with $\omega_2 = \gamma\omega_1$, $\gamma = (\sqrt{5} - 1)/2$. The upper curve ω is for strange–nonchaotic motion ($H_1 = 0.71$, $H_2 = 0.80$), and the lower curve is for quasiperiodic motion ($H = 0.71$, $H_2 = 0.53$). [From Ditto et al. (1990).]

It is clear that further research will be needed to better define these peculiar motions. But there is now evidence that the new world of nonlinear dynamics has a growing list of new species of complex motion.

For a numerical study of strange–nonchaotic motion in a Van der Pol oscillator, the reader should see the paper by Kapitaniak et al. (1990).

PROBLEMS

2-1 Use the definition of a linear operator to show that the equation for the motion of a damped pendulum $\ddot{q} + c\dot{q} + b \sin q = 0$ is *not* linear.

2-2 Consider a ball bouncing between two walls (neglect gravity) for which one wall has a small periodic motion. Show that the dynamics is *not* governed by a linear operator.

2-3 Sketch the power spectrum density of
  (a) $(\cos \omega t)^3$
  (b) $\left(\cos \dfrac{\omega t}{2}\right)^4$

2-4 Consider the output of a nonlinear oscillator to have a primary frequency component plus a subharmonic:

$$x(t) = A \cos \frac{\omega}{2} t + B \cos(\omega t + \phi_0)$$

If one defines a second state variable $\dot{x} = y$, then when will the orbit as plotted in the phase plane (x, y) show a double loop? Will this be any different if the signal had primary and harmonic components (i.e., ω, 2ω)?

2-5 Suppose that the output of a dynamical system has two frequency components, that is,

$$x(t) = A \cos \omega_1 t + B \cos(\omega_2 t + \phi)$$

  (a) If one takes a Poincaré map on the phase $\omega_1 t$, show that the map $(x, y = \dot{x})$ is an ellipse when ω_1/ω_2 is incommensurate.

(b) Describe the map dynamics if $\omega_1/\omega_2 = p/q$, where p and q are integers.

(c) What does the map look like in the phase plane (x, y) if there is a third frequency ω_3?

2-6   A return map, x_{n+1} versus x_n, shows a bilinear relationship

$$F(x) = \begin{cases} a + bx, & x < 1 \\ c - dx, & x > 1 \end{cases}$$

If $b > 0$, $d > 0$, show that the Lyapunov exponent is positive when $b, d > 1$. Sketch a few iterations of the map in the (x_{n+1}, x_n) plane.

2-7   A damped harmonic oscillator (series circuit of inductor, capacitor and resistor) can be represented mathematically by the following differential equation and solution:

$$\ddot{x} + 2\gamma\dot{x} + x = 0$$
$$x(t) = C_0 e^{-2\gamma t} \cos(\sqrt{1 - \gamma^2}\, t + \varphi_0)$$

Sketch the solution in the phase plane $(x, y = \dot{x})$. Define a Poincaré section when $x > 0$, $y = 0$. Assume that y is small $(\gamma^2 \ll 1)$ and show that the resulting map takes the form $x_{n+1} = \lambda x_n$, and $\lambda < 1$. Can you estimate λ?

2-8   *Pseudo-Phase Plane.* The equation for a linear harmonic oscillator (spring and mass or inductor and capacitor circuit) is given by $\ddot{x} + x = 0$. The solutions for this equation can be represented in the phase plane by an ellipse written in parametric form:

$$x = A \sin t, \qquad y = \dot{x} = A \cos t$$

Derive an expression for the solution $x = A \sin t$ in terms of pseudo-phase-plane variables (x, x'), where $x' = x(t + T)$. Plot this expression for different values of A.
[*Answer:* $(x' - x \cos T)^2 = (A^2 - x^2)\sin^2 T$.]

3

MODELS FOR CHAOS; MAPS AND FLOWS

All the richness in the natural world is not a consequence of complex physical law, but arises from the repeated application of simple laws.
L. P. Kadanoff[1]

3.1 INTRODUCTION

To those of us educated in the physical sciences, the infinitesimal differential calculus was the first abstract mathematical tool that one struggled with on the road to understanding mathematical physics. Later, we learned to model the world with the calculus of differential equations. In this view, the physical world was reduced to sets of differential equations: the Navier–Stokes equation of fluid mechanics; the equations for elasticity for solids and structures such as beams, plates, and shells; Maxwell's equations for electromagnetics; and the heat equation for thermal problems. The solutions of these differential equations are drawn in phase space as continuous orbits. A bundle of such trajectories, corresponding to different initial conditions, generates orbits that look like the flow of a fluid—hence the modern term, *flows*, to describe the dynamics of continuous time systems. Thus it comes with some surprise to many in the physical sciences that some

---

[1] L. P. Kadanoff is a physicist at the University of Chicago. Quote taken from *Physics Today*, March 1991, page 9.

85

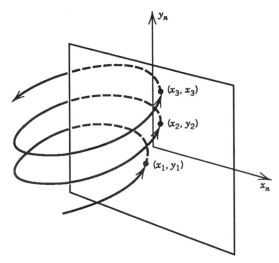

Figure 3-1   Sketch showing the relation between a continuous time orbit in a 3-D phase space and a 2-D discrete time point mapping.

dynamical phenomena can be exactly represented by finite time *difference* equations, or *maps* as they are now called.

However, for students of the biological and social sciences, the dynamic laws are more often posed as relationships between events at discrete time intervals, and the use of difference equations is more natural (e.g., see May, 1976, 1987). After all, the events of birth and death or the publication of unemployment statistics occur at specific times.

Today, however, the study of modern dynamics in the physical sciences requires a working knowledge of iterated maps, especially to understand the basic nature of chaotic behavior. As illustrated in Chapter 2, the Poincaré section of the dynamics of a physical system described by three first-order differential equations naturally leads to a discrete time map on the plane as illustrated in Figure 3-1. Thus, we begin with the study of two coupled nonlinear difference equations or second-order maps.

The Geometry of Mappings: Maps on the Plane

The dynamics of maps can be described both geometrically and algebraically. We will first take a look at the qualitative properties of maps and then look at more specific examples for more quantitative information.

By a mapping we mean a transformation of a contiguous set of points from one set of positions to another. The dynamic evolution of a particular initial point under repeated application of this transformation is called an *iterated map* or simply a *map*. We begin with a discussion of maps on the plane or of two-dimensional (2-D) maps which are written as a set of two coupled difference equations.

For most of the maps discussed in this book we assume that they can be written in explicit form,[2] that is,

$$x_{n+1} = F(x_n, y_n)$$
$$y_{n+1} = G(x_n, y_n)$$

(3-1.1)

We also assume that each point (x_n, y_n) has a unique iterate and that each iterate (x_{n+1}, y_{n+1}) has a unique antecedent. This implies that an inverse mapping can be found:

$$x_n = F^{-1}(x_{n+1}, y_{n+1})$$
$$y_n = G^{-1}(x_{n+1}, y_{n+1})$$

(3-1.2)

For example, in the case of the most general form of a quadratic, polynomial, area-preserving mapping of the plane (see Henon, 1969), one can find F^{-1}, G^{-1} quite easily; that is, suppose

$$x_{n+1} = F(x_n, y_n) = x_n \cos \alpha - (y_n - x_n^2)\sin \alpha$$
$$y_{n+1} = G(x_n, y_n) = x_n \sin \alpha - (y_n - x_n^2)\cos \alpha$$

(3-1.3)

Then it can be shown that the inverse is given by

$$x_n = x_{n+1}\cos \alpha + y_{n+1}\sin \alpha$$
$$y_n = -x_{n+1}\sin \alpha + y_{n+1}\cos \alpha + (x_{n+1}\cos \alpha + y_{n+1}\sin \alpha)^2$$

Area-preserving mappings arise in the dynamics of conservative or so-called Hamiltonian dynamics.

When the functions F, G are continuous and differentiable, the change in the transformation of a small differential area under one

---

[2] Although this form is convenient for mathematical analysis, there are many problems where the relation between x_{n+1} and x_n is implicit. One example is a mass bouncing between two periodically oscillating walls.

iteration of the mapping is measured by the Jacobian, defined by the determinate

$$J = \begin{vmatrix} \dfrac{\partial F}{\partial x} & \dfrac{\partial F}{\partial y} \\[2mm] \dfrac{\partial G}{\partial x} & \dfrac{\partial G}{\partial y} \end{vmatrix} \qquad (3\text{-}1.4)$$

Sometimes the following notation is used:

$$J = \frac{\partial(F, G)}{\partial(x, y)}$$

Area-preserving maps have $J = 1$ as one can check by the above example of Henon (1969). When dissipation is present in physically relevant maps, $J < 1$, or areas contract under the mapping.

Impact Oscillator Maps

There are many examples in physics and engineering where the dynamics of a particle are linear within finite time intervals and where energy is put in or taken out in very short time periods between these linear dynamic intervals. Such examples arise in particle accelerators, or in the motion of two gears with play between the teeth.

As an example of how a 2-D map can arise in dynamics, we consider the motion of an oscillator under periodic impacts (Figure 3-2).

An analysis of a general impact oscillator has been given by Helleman (1980a) based on the motion of a proton in intersecting storage rings. A similar problem involving a particle in a plasma may be found in Sagdeev et al. (1988). The basic equation in these problems is a linear oscillator which is periodically impacted at time intervals of $2\pi n/\Omega$ in which the momentum change depends on the motion:

$$\ddot{x} + \omega_0^2 x = f(x) \sum \delta(t - 2\pi n/\Omega) \qquad (3\text{-}1.5)$$

where $\delta(t - \tau_0)$ is the classical delta function whose integral is unity. Between impulses the solution is proportional to $A \cos(\omega_0 t + \varphi_0)$. Piecing together these linear solutions using the above equation, one

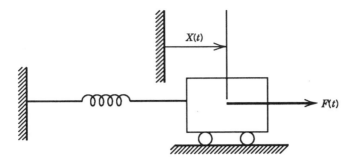

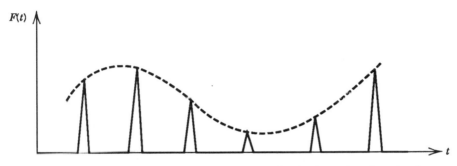

Figure 3-2 A linear oscillator with periodic, amplitude-dependent impulse forces.

can show that the displacement after impact x_n and the velocity after each impact $v_n \equiv \dot{x}(t_n^+)$ satisfy the following difference equations:

$$x_{n+1} = x_n \cos \beta + v_n(\sin \beta)/\omega_0$$
$$v_{n+1} = -x_n \omega_0 \sin \beta + v_n \cos \beta + f(x_{n+1})$$

(3-1.6)

where $\beta = 2\pi\omega_0/\Omega$. It is easy to show that this is an area-preserving map, that is, $J = 1$.

Classification of Map Dynamics

We describe here a few typical motions of discrete time dynamical systems. In particular we focus on 2-D maps in the plane. As discussed above, they arise quite naturally from a Poincaré map of three-state variable continuous time dynamics.

Fixed Points. As the term implies, iteration of the map at a fixed point or *equilibrium* point $\mathbf{x}_e$ brings the system to the same point in one time cycle, that is,

$$\mathbf{x}_e = F(\mathbf{x}_e) \tag{3-1.7}$$

For example, in the case of the *cubic map*

$$F = y, \qquad G = ax - bx^3 + cy \tag{3-1.8}$$

given by Holmes (1979) in the study of chaos in a buckled beam, the fixed points are found from the equations

$$x_e = y_e, \qquad y_e = ax_e - bx_e^3 + cy_e \tag{3-1.9}$$

Cycle Points. *Cycle points* are similar to fixed points except that the dynamics undergo several iterations before returning to the fixed point, that is,

$$\mathbf{x}_{n+m} = \mathbf{x}_n \quad \text{or} \quad \mathbf{x}_n = F^{(m)}(\mathbf{x}_n) \tag{3-1.10}$$

Note that, as illustrated in Figure 3-3, each of the m points in the cycle is a cycle point.

The so-called Standard map appears in many applications in physics, including accelerator particle dynamics (Lichtenberg and Lieber-

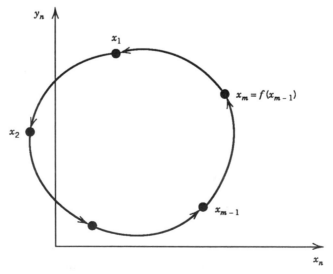

Figure 3-3   An m-cycle orbit of a 2-D map.

man, 1983) as well as the dynamics of a bouncing particle on a vibrating surface (Holmes, 1982). In this map the state variables (φ_n, v_n) represent the time or phase of impact (mod 2π) and the velocity after impact. Thus, the map operates on a cylindrical phase plane:

$$\varphi_{n+1} = \varphi_n + v_n \quad (\text{mod } 2\pi)$$

$$v_{n+1} = \alpha v_n - \gamma\cos(\varphi_n + v_n)$$

(3-1.11)

where $0 < \alpha \leq 1$. When $\alpha = 1$ there is no energy dissipation and the map is area-preserving.

If this transformation is defined by T, then the fixed points of the period-2 motion are the fixed points of T^2 where the superscript indicates that the mapping is applied twice. The equations which determine these points for the standard map are given by

$$\varphi_1 = \varphi_0 + v_0$$

$$v_1 = \alpha v_0 - \gamma\cos(\varphi_0 + v_0)$$

$$\varphi_2 = \varphi_0 = \varphi_1 + v_1$$

$$v_2 = v_0 - \alpha v_1 - \gamma\cos(\varphi_1 + v_1)$$

Thus we have four equations in four unknowns. The details of the solution can be found in Holmes (1982).

Two examples of cycle points are shown graphically in Figure 3-4a,b. One is for the standard map and the other is for the quadratic map discussed by Henon (1969). In the case of the quadratic map there are two sets of cycle-3 points. These are the fixed points of the transformation T^3. The set with the ellipses around each point are called *centers* and are stable, whereas the set with what looks like two crossed curves going through them are *saddle points* and are unstable.

Quasiperiodic Motions

As discussed in Chapter 1, when two oscillators have incommensurate frequencies, the combined motion in a map describes an elliptic-type orbit in the map as shown in Figure 1-13. In the case of multiple-period fixed points or cycles, each of the fixed points in the cycle is surrounded by ellipse-shaped curves. In a quasiperiodic motion close

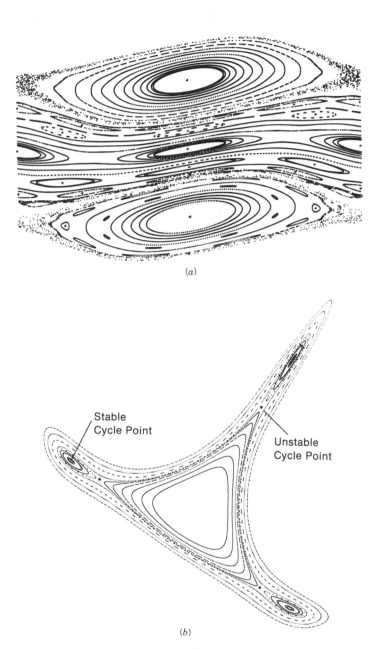

(a)

(b)

Figure 3-4 (a) Standard map (no dissipation) showing cycle points. (b) Quadratic map of Henon (1969) showing three-cycle points.

to these fixed points, the orbit will visit each of the ellipses in turn, slowly mapping out a closed curve after many iterations of the map.

Stochastic Orbits

In the case of conservative or area-preserving maps (Hamiltonian system), a chaotic orbit often occurs near a saddle point and the orbit is characterized by a uniformly dense collection of points with no apparent order. An example is shown in Figure 3-5. These chaotic orbits can have positive Lyapunov exponents, indicating a sensitive dependence on initial conditions, but the orbit does not exhibit fractal structure as in a dissipative map.

Fractal Orbits

These motions are typical for dissipative maps; that is,

$$\left| \frac{\partial(x_{n+1}, y_{n+1})}{\partial(x_n, y_n)} \right| < .1$$

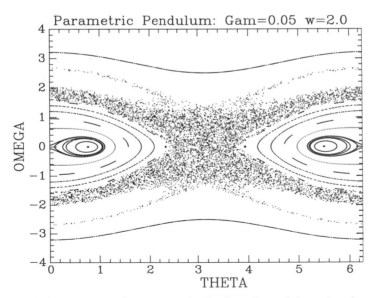

Figure 3-5 Poincaré map of a parametrically forced pendulum showing periodic orbits (*points*), quasiperiodic orbits (*closed curves*), and stochastic orbits (*diffuse set of points*). Relative change in length, $\gamma = 0.05$; (forcing frequency) $\omega = 2.0$.

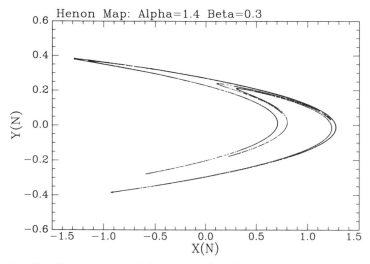

Figure 3-6   The Henon map (1-3.8) showing fractal structure. $\alpha = 1.4$; $\beta = 0.3$.

Examples include the Henon map [Eq. (1-3.8)] shown in Figures 1-24, 3-6, and the standard map [Eq. (3-1.11)] shown in Figure 3-11. Although the motion is unpredictable, the orbit traces out an infinite set of curves which may be viewed by looking at finer and finer scales as in Figure 1-24.

3.2 LOCAL STABILITY OF 2-D MAPS

As in the analysis of dynamics governed by ordinary differential equations, the dynamic behavior of systems described by sets of difference equations can be analyzed by looking at linearized maps near the fixed points. Let us assume that the origin is a fixed point; then for a 2-D map [Eq. (3.1-1)] the linear map takes the form

$$x_{n+1} = ax_n + by_n$$
$$y_{n+1} = cx_n + dy_n \tag{3-2.1}$$

where

$$\begin{bmatrix} a & b \\ c & d \end{bmatrix} = \begin{bmatrix} \dfrac{\partial F}{\partial x} & \dfrac{\partial F}{\partial y} \\ \dfrac{\partial G}{\partial x} & \dfrac{\partial G}{\partial y} \end{bmatrix}$$

and the derivatives of the nonlinear map functions are evaluated at
(0, 0).

The analysis of linear maps is straightforward and can be found in
several texts (e.g., see Bender and Orzag, 1978). One begins by guess-
ing at a power-law type of solution which for linear difference equations takes the form

$$\begin{Bmatrix} x_n \\ y_n \end{Bmatrix} = \lambda^n \begin{Bmatrix} e_1 \\ e_2 \end{Bmatrix} \tag{3-2.2}$$

Substitution of this solution into the linear equations (3-2.1) leads
to the following eigenvalue problem:

$$\begin{bmatrix} (a - \lambda) & b \\ c & (d - \lambda) \end{bmatrix} \begin{Bmatrix} e_1 \\ e_2 \end{Bmatrix} = 0$$

or

$$\lambda^2 - (a + b)\lambda + ad - cb = 0 \tag{3-2.3}$$

These equations establish the stability criteria for a linear map.

$|\lambda| < 1$:   Solution is stable
          Nonlinear system is stable
$|\lambda| > 1$:   Solution is unstable
          Nonlinear system is unstable
$|\lambda| = 1$:   Solution is neutrally stable
          Stability of nonlinear system depends on nonlin-
          ear terms in map.

Similar to the analysis for differential equations, one must find
the corresponding eigenvectors along with determination of the two
eigenvalues, λ. When these vectors exist, they establish the directions
along which simple multiplicative dynamics $x_{n+1} = \lambda x_n$ takes place.
For an arbitrary initial condition, however, the total solution takes the
form

$$x_n = c_1 \lambda_1^n e_{11} + c_2 \lambda_2^n e_{21} \tag{3-2.4}$$

when λ_1, λ_2 are distinct. Here (e_{11}, e_{21}) is the eigenvector correspond-

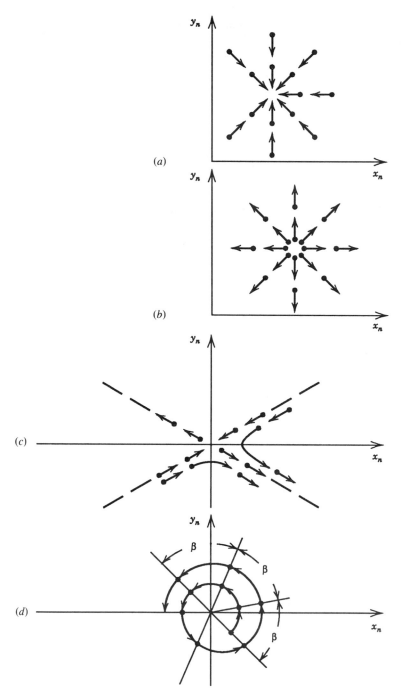

Figure 3-7   Fixed points of 2-D maps: (*a*) stable node, (*b*) unstable node, (*c*) saddle point, (*d*) spiral points.

ing to the eigenvalue λ_1. Rules for special cases for identical eigenvalues can be found in Bender and Orzag (1978).

A few classic examples of local fixed point dynamics of 2-D maps are worth mentioning and are illustrated in Figure 3-7.

I. Stable Node, λ_1, $\lambda_2 < 1$ (Both Real). If $0 < \lambda_1 < 1$, then for initial conditions along the corresponding eigenvector direction the motion moves toward the origin without changing sign. However, if λ_1 or $\lambda_2 < 0$, then the motion can flip between positive and negative eigen-directions while still moving closer to the fixed point.

II. Unstable Node, λ_1, $\lambda_2 > 1$ (Both Real). The same comments in case I hold except the motion moves away from the fixed point and the motion is called *unstable*.

III. Saddle Point, $\lambda_1 < 1$, $\lambda_2 > 1$ (Both Real). Initial conditions along the λ_1 eigenvector result in inward motion; however, any small component along the λ_2 eigenvector takes the total solution away from the origin and the solution is *unstable*.

IV. λ_1, $\lambda_2 = \alpha \pm i\beta = \rho e^{\pm i\varphi}$. These fixed points are known as stable or unstable *spirals* depending on whether $\rho < 1$ or $\rho > 1$. Sometimes the term stable or unstable *focus* is used. Note that although these solutions are analogous to oscillatory motion in differential equations, one can have oscillating map solutions in cases I, II, and III as well as in spiral solutions.

When $|\lambda| > 1$ or $|\lambda| < 1$, these points are known as *hyperbolic* points and the stability of the linearized map gives a good picture of the stability of the nonlinear map in the neighborhood of the fixed point.

When $|\lambda| = 1$, however, these points are sometimes called *elliptic* points, and the stability of the nonlinear map depends on the terms of higher order than the linear ones.

When one has multiple-period fixed points or an N-cycle point, the stability can be determined by looking at the linearized map of the T^N map at one of the cycle points.

Finally, to get a complete picture of the dynamics of nonlinear maps, one must look at how sets of points transform under the map. This is the subject of the next section.

3.3 GLOBAL DYNAMICS OF 2-D MAPS

In the previous section we looked at the dynamics of specific trajectories in the neighborhood of a fixed point. Here we examine how the map transforms a set of contiguous points under one or more iterations. In particular, it is a special type of 2-D map called a *horseshoe* or *baker's* map that produces stretching and folding and is thought to be the fundamental mechanism for the creation of chaotic dynamics.

In the following we will look at simple maps that produce translation, dilatation, shearing, stretching, and folding. The first four effects can be accomplished with a linear map.

Linear Transformation

A general linear transformation takes the form

$$\mathbf{x}_{n+1} = \mathbf{b} + \mathbf{A} \cdot \mathbf{x}_n \tag{3-3.1}$$

where the constant vector $\mathbf{b}$ and the elements of the matrix $\mathbf{A}$ are assumed to be real. The constant $\mathbf{b}$ represents a uniform translation. To examine the effect of $\mathbf{A}$, consider the case $\mathbf{b} = 0$, and consider the following special cases (see Figure 3-8).

(i) $$\mathbf{A} = \begin{bmatrix} r & 0 \\ 0 & r \end{bmatrix} \quad \text{(dilatation)} \tag{3-3.2}$$

In this case, an area contracts or expands uniformly such that circles remain circles. This is similar to the uniform thermal expansion of a material due to heating. Areas change by a factor $\det \mathbf{A} = r^2$ under each iteration of the map.

(ii) $$\mathbf{A} = \begin{bmatrix} s & 0 \\ 0 & c \end{bmatrix} \quad \text{(stretching } s > 1, \text{ contraction } c < 1) \tag{3-3.3}$$

Under this transformation, a small square is deformed into a rectangular shape, with the x-axis direction being stretched by a factor s > 1 while the y-axis direction is contracted by a factor c < 1. The change of area, $\det \mathbf{A} = sc$, depends on the relative amounts of stretching and contracting. For conservative dynamics, $sc = 1$.

(iii) $$\mathbf{A} = \begin{bmatrix} \cos \alpha & -\sin \alpha \\ \sin \alpha & \cos \alpha \end{bmatrix} \quad \text{(rotation)} \tag{3-3.4}$$

This operation rotates areas around the origin by an angle α. Areas are preserved—that is, $\det \mathbf{A} = 1$.

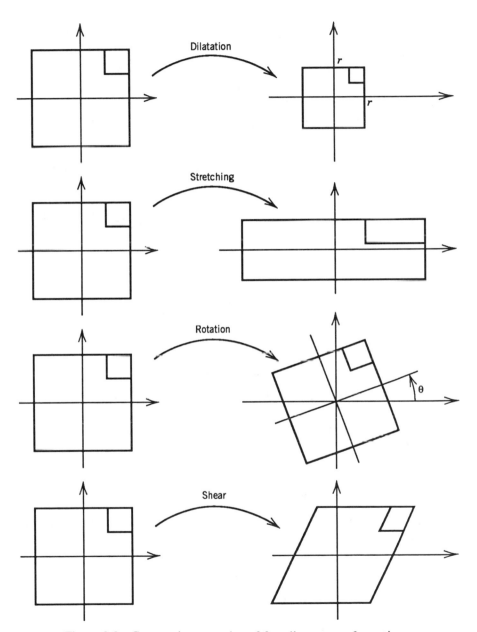

Figure 3-8 Geometric properties of four linear transformations.

(iv) $A = \begin{bmatrix} 1 & c \\ 0 & 1 \end{bmatrix}$ (pure shear transformation)   (3-3.5)

This is a horizontal shearing operation which moves the points in the upper half-plane to the right and which moves those in the lower half-plane to the left. A square area becomes a rhombus, and areas remain preserved under this transformation—that is, det $A = 1$. However, this deformation is identical to case ii—that is, a combined stretching and contraction along 45° directions with $sc = 1$.

(v) $A = \begin{bmatrix} 1 & 0 \\ 0 & -1 \end{bmatrix}$ (reflection about y-axis)   (3-3.6)

Our final example is a simple reflection about the y-axis. This can be generalized by using a composition of two linear mappings. First define T_1 as a reflection about the y-axis, then define T_2 as a rotation α to obtain a mapping

$$x_{n+1} = T_2 \circ T_1(x_n)$$

or

$$A = \begin{bmatrix} \cos\alpha & -\sin\alpha \\ \sin\alpha & \cos\alpha \end{bmatrix} \begin{bmatrix} 1 & 0 \\ 0 & -1 \end{bmatrix} = \begin{bmatrix} \cos\alpha & \sin\alpha \\ \sin\alpha & -\cos\alpha \end{bmatrix}$$

Generalizations of the other mappings (i–iv) can be done in a similar way and are left as an exercise.

Folding in 2-D Maps—Horseshoes

A simple quadratic map that produces folding as in Figure 1-22 is a map similar to but slightly different from Eq. (3-1.3):

$$x_{n+1} = x_n$$
$$y_{n+1} = y_n - \beta x_n^2$$
(3-3.7)

This map is area-preserving—that is, $J = 1$. However, a thin rectangular area centered on the x-axis is deformed into the shape of a horseshoe under one iteration.

Composition of Maps—Henon Map

When two transformations T_1, T_2 are performed on a set of points, we denote the operation by $T_2 \circ T_1$ and call this a composition.

The now classic paradigm of a strange attractor generated by a simple 2-D quadratic difference equation is the Henon map (Henon, 1976) introduced in Chapter 1:

$$x_{n+1} = 1 + y_n - \alpha x_n^2$$
$$y_{n+1} = \beta x_n \tag{3-3.8}$$

This map is area-contracting provided that $0 < \beta < 1$.

This map can be seen as a composition of five transformations $T_5 \circ T_4 \circ T_3 \circ T_2 \circ T_1$ defined as follows (see Figure 3-9):

$$T_1: \quad \mathbf{b} = (0, 1) \quad \text{(translation)}$$

$$T_2: \quad x_{n+1} = x_n, \quad y_{n+1} = y - \alpha x_n^2 \quad \text{(folding)}$$

$$T_3: \quad \mathbf{A} = \begin{bmatrix} 0 & 1 \\ -1 & 0 \end{bmatrix} \quad \text{(rotation, } \alpha = -\pi/2)$$

$$T_4: \quad \mathbf{A} = \begin{bmatrix} 1 & 0 \\ 0 & -1 \end{bmatrix} \quad \text{(reflection)}$$

$$T_5: \quad \mathbf{A} = \begin{bmatrix} 1 & 0 \\ 0 & \beta \end{bmatrix} \quad \text{(contraction)}$$

As discussed in Chapter 1, repeated applications of this map leads to a multiple stretching and folding and the fractal properties of a strange attractor.

The Horseshoe Map

The previous example of a horseshoe mapping uses a continuous quadratic polynomial function. However, a conceptually simpler mapping was proposed by Smale (1962) as a model for complex dynamics, as previously discussed in Chapter 1 (Figure 1-23). A similar transformation is the baker's map (see Chapter 6 (6-4.26)) which is analogous to rolling, cutting and folding of pastry or bread dough (Figure 6-33).

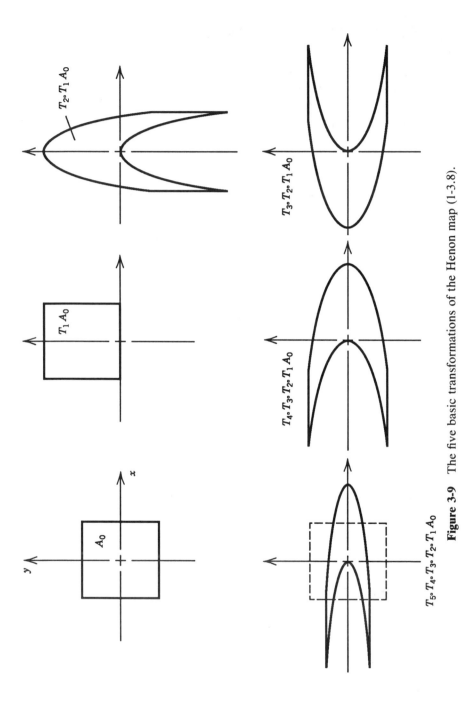

Figure 3-9 The five basic transformations of the Henon map (1-3.8).

3.4 SADDLE MANIFOLDS, TANGLES, AND CHAOS

As a prelude to Chapter 6 ("Criteria for Chaotic Vibrations"), it can be said that one of the keys to the kingdom of chaos is the search for the existence of horseshoe-like maps. In many systems, strange attractors are organized around saddle-type fixed points—that is, points at which orbits come into a small region of phase space along a *stable manifold* and go out along paths close to an *unstable manifold*. This saddle-type fixed point creates the contraction and stretching mechanisms for a horseshoe map. The nonlinear aspect of the map is required to produce the folding or bending mechanism of the horseshoe map. Repeated iteration of the stretching, contracting, and folding of regions of phase space lead to unpredictability and, in the case of dissipative systems, lead to the fractal nature of the dynamics in the map.

Thus the calculation of fixed points and determining the nature of their stability is no mere academic exercise, but can be used along with computational tools to give a clue to the nature of the chaotic attractor. This is illustrated for the case of the standard map [Eqs. (3.1-11)] which has been used to study the dynamics of a ball bouncing on a vibrating surface (e.g., see Guckenheimer and Holmes, 1983). In this model, x_n represents the phase in the driving period at which the ball hits the table (mod 2π), and y represents the velocity of the ball after impact. The Poincaré map is taken at the point of contact resulting in the following difference equations:

$$x_{n+1} = x_n + y_n \quad (\text{mod } 2\pi)$$
$$y_{n+1} = \alpha y_n - \gamma \cos(x_n + y_n)$$

$$(3\text{-}4.1)$$

This map is a variation of impact oscillator maps [Eqs. (3.1-6)].

The Jacobian of this map is α, which represents the loss of energy on impact. Here $\alpha < 1$ and areas are contracted with each iteration of the map.

One can show that fixed points of this map are given by

$$y_e = 2\pi n$$

$$x_e = \cos^{-1}\left[\frac{(\alpha - 1)2\pi n}{\gamma}\right]$$

$$(3\text{-}4.2)$$

[See Holmes (1982) for a more complete description.]

The stability of each fixed point can be ascertained by looking at the eigenvalues of the linearized Jacobian matrix of the map

$$[\nabla F] \left\{ \begin{matrix} \bar{x} \\ \bar{y} \end{matrix} \right\} = \lambda \left\{ \begin{matrix} \bar{x} \\ \bar{y} \end{matrix} \right\} \qquad (3\text{-}4.3)$$

In this problem there are two eigenvalues at each fixed point λ_1, λ_2 and there are two eigenvectors whose direction is given by

$$\bar{y} = (\lambda - 1)\bar{x}$$

To take a specific example, consider the fixed point $(x_e, y_e) = (\pi/2, 0)$ of the map (3-4.1) for which

$$\lambda^2 - (1 + \alpha + \gamma)\lambda + \alpha = 0$$

One can show that both eigenvalues are real and that $\lambda_1 > 1$, $\lambda_2 < 1$, which implies a saddle-type fixed point. Figure 3-10 shows the direction of the stable and unstable manifolds at the point $(\pi/2, 0)$. However, to determine the shape of these manifolds far from the fixed point, one must use numerical methods.

To numerically calculate the unstable manifold, we note that a map does not generate a continuous orbit. To generate the manifold, one must generate a large number of orbits, each originating near the fixed point with coordinates satisfying $\bar{y}_0 = (\lambda_1 - 1)\bar{x}_0$.

To numerically determine the stable manifold, we solve the inverse map, choose initial conditions along the stable direction of the saddle, and iterate backwards in time. The inverse map is given by

$$y_{n-1} = \frac{1}{\alpha} y_n + \frac{\gamma}{\alpha} \cos x_n$$
$$x_{n-1} = x_n - y_{n-1}(x_n, y_n) \qquad (3\text{-}4.4)$$

for $n = 0, -1, -2, \ldots$. A large number of initial conditions (say 100–200), each chosen close to the fixed point and orbits iterated (say 10–20 times), will generate a collection of points that lie approximately on each of the two manifolds of the saddle point.

The results of such a calculation for the $(\pi/2, 0)$ fixed point of the standard map are shown in Figure 3-10 for a dissipation parameter of $\alpha = \frac{1}{2}$ and forcing amplitude values $\gamma = 2, 3.3, 6$. The first case shows these two manifolds to be nonintersecting. However, for $\gamma = 3.3$, we

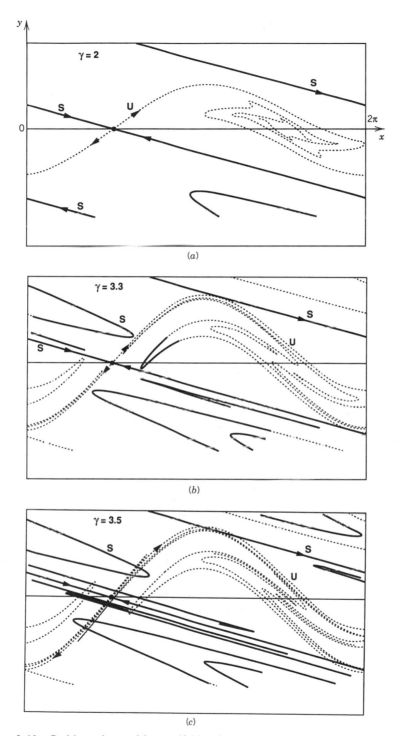

Figure 3-10 Stable and unstable manifolds of the standard map (3-4.1) at $(\pi/2, 0)$ for (a) $\gamma = 2$, (b) $\gamma = 3.3$, and (c) $\gamma = 6$.

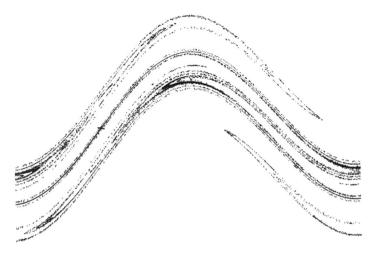

Figure 3-11 Collection of points from multiple iterations of the standard map (3-4.1) from a single initial condition ($\gamma = 6$).

see that they are almost touching. This is a critical case because there is a theorem attributed to Poincaré that says if they intersect once, they will intersect an infinite number of times (see Chapter 6 for more discussion). Note that for flows, such intersection is not allowed, except at fixed points, but for maps it's O.K.

It is this tangling of manifolds that is believed to create the horseshoe maps which are necessary for certain kinds of chaos. For $\gamma = 6$ it is clear that these two manifolds have indeed intersected many times.

In fact, in the region $\gamma \sim 6$, the standard map yields a strange attractor which can be verified by iterating the map (3-4.1) many times from one initial condition (Figure 3-11). What is remarkable about Figure 3-11 is that the shape of the strange attractor for $\gamma = 6$ is very close to the shape of the unstable manifold for the same value. This is believed by theorists to be more than coincidental—namely, that for many chaotic attractors the chaos is organized by the unstable manifold of a saddle-type fixed point.

3.5 FROM 2-D TO 1-D MAPS

When time is a continuous variable, the description of chaotic dynamics requires three state variables, and a Poincaré section naturally leads to a two-dimensional iterated map. The question then arises: Under what conditions can this two-dimensional description be further reduced to a single, first-order difference equation or 1-D map? To illustrate this question and give a clue to its answer, we offer two examples, namely, the kicked rotor oscillator and the Henon map.

The Kicked Rotor

As we have seen in the examples of the horseshoe map (Figure 1-23) or the logistic equation [Eq. (1-3.6)] in Chapter 1, the nature of the chaotic dynamics is best uncovered by taking a Poincaré section of a continuous time flow in phase space. However, for most differential equation models of physical systems, it is impossible to obtain analytical results. Again, we look at a variation of the impact map. In the example considered here, we imagine a rotor with rotary intertia J and damping c which is subject to both a steady torque $c\omega_0$ and a periodic series of *pulsed torques* as shown in Figure 3-12 (see also Schuster, 1984). The equation of motion representing the change in angular momentum of the rotor is given by

$$J\dot{\omega} + c\omega = c\omega_0 + T(\theta) \sum_{n=-\infty}^{+\infty} \delta(t - n\tau) \qquad (3\text{-}5.1)$$

The term $\delta(t - n\tau)$ represents the classical delta function which is zero everywhere except at the $t = n\tau$ and whose area is unity. Thus, for times $n\tau - \varepsilon < t < n\tau + \varepsilon$, where $\varepsilon \ll 1$, the angular momentum change is given by

$$J(\omega^+ - \omega^-) = T(\theta(n\tau)) \qquad (3\text{-}5.2)$$

For example, if the torque is created by a vertical force as shown in Figure 3-12, then the pulsed torque is proportional for $T(\theta) = F_0\sin\theta$.

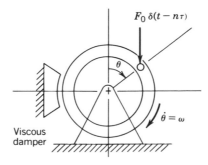

Figure 3-12 Rotor with viscous damping and periodically excited torque studied by Zaslavsky (1978).

When $T(\theta) = 0$, Eq. (3-5.1) has a steady solution $\omega = \omega_0$, $\theta = \omega_0 t$.

To obtain a Poincaré map, we take a section right before each pulsed torque. Thus, we define $\theta_n = \theta(t = n\tau - \varepsilon)$, $\varepsilon \to 0^+$. One can relate (θ_n, ω_n) to $(\theta_{n+1}, \omega_{n+1})$ by solving the linear differential equation between pulses and using the jump in angular momentum condition (3-5.2) across the pulse. Between pulses, the rotation rate has the following behavior:

$$\omega = \omega_0 + ae^{-ct/J} \tag{3.5-3}$$

Carrying out this procedure, one can derive the following exact Poincaré map for the system (3-5.1):

$$\omega_{n+1} = \frac{c\tau}{J}\omega_0 + \omega_n - \frac{c}{J}(\theta_{n+1} - \theta_n) + \frac{1}{J}T(\theta_n)$$

$$\theta_{n+1} = \omega_0\tau + \theta_n + \frac{J}{c}(1 - e^{-c\tau/J})\left(\omega_n + \frac{1}{J}T(\theta_n) - \omega_0\right) \tag{3-5.4}$$

These equations were first derived by the Soviet physicist Zaslavsky (1978) to treat the nonlinear interaction between two oscillators in plasma physics. In this mechanical analog of this problem, ω_0 represents the frequency of one uncoupled oscillator [see also Ott (1981) for a derivation].

This two-dimensional map is often nondimensionalized using

$$x_n = \frac{\theta_n}{2\pi} \pmod 1$$

$$y_n = \frac{\omega_n - \omega_0}{\omega_0}$$

For $T(\theta) = F_0\sin\theta$ and $\varepsilon = F_0/J\omega_0$, Eqs. (3-5.4) then become

$$y_{n+1} = e^{-\Gamma}(y_n + \varepsilon\sin 2\pi x_n) \tag{3-5.5}$$

$$x_{n+1} = \left\{x_n + \frac{\Omega}{2\pi} + \frac{\Omega}{2\pi\Gamma}(1 - e^{-\Gamma})y_n + \frac{K}{\Gamma}(1 - e^{-\Gamma})\sin 2\pi x_n\right\}$$

where the braces { } indicate that only the fractional part is used (i.e., mod 1 or $0 \le \theta \le 2\pi$). Also, $K = \varepsilon\Omega/2\pi$, $\Gamma = c\tau/J$, and $\Omega = \omega_0\tau$. Here y_n measures the departure of the speed from the unperturbed equilibrium speed $\omega = \omega_0$. Note that this map contracts areas for $\Gamma > 0$ and preserves areas for $\Gamma = 0$.

This system of two difference equations has been found to exhibit chaotic solutions only if the following conditions are satisfied when ε is small:

$$1 < \frac{\Gamma}{1 - e^{-\Gamma}} < K \tag{3-5.6}$$

A typical case is shown in Figure 3-13 for the parameters $\Gamma - 5$, $\varepsilon = 0.3$, $\Omega = 100$, and $K = 9$.

The problem of a kicked or pulsed double rotor with two degrees of freedom has been investigated by Kostelich et al. (1985, 1987).

Circle Map

A simpler version of the Zaslavsky map for two coupled oscillators can be obtained by letting the damping become larger, $\Gamma \gg 1$. In this limit, one can ignore the changes in ω or y (note that Δy is small in Figure 3-13). This leads to a one-dimensional map known as a *circle map:*

$$x_{n+1} = \left\{ x_n + \frac{\Omega}{2\pi} - \frac{K}{\Gamma} \sin 2\pi x_n \right\} \tag{3-5.7}$$

This equation has received extensive study (e.g., see Rand et al., 1982) and is a model for the quasiperiodic oscillation between two oscillators with uncoupled ($K = 0$) frequencies in the ratio Ω. In this

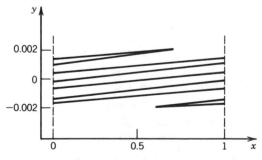

Figure 3-13 Strange attractor for the Zaslavsky map (3-5.4) for the kicked rotor in Figure 3-12: x represents angular rotation (mod 1), and y represents the angular velocity.

example, one can see the steps that lead from the physics to an exact 2-D map and then to an approximate 1-D map (3-5.7).

The Henon Map

Earlier we saw that the quadratic, dissipative map (3.3-8) contracted area proportional to the constant β. This results in a strange attractor with fractal structure and a fractal dimension, d, with $1 < d < 2$. However, as $\beta \rightarrow 0$, the coupling between the two variables x, y becomes weaker. Also, one can see from the chaotic attractor that the points seem to be distributed along a one-dimensional curve in the plane and $d \sim 1$ (see Figure 3-14).

In the limit of small β, the Henon map becomes asymptotic to a curve in the x–y plane given by

$$x = 1 - \frac{\alpha}{\beta^2} y^2$$

For small β, y becomes small and the Henon map approaches the one-dimensional quadratic map

$$x_{n+1} = 1 - \alpha x_n^2$$

which is similar to the logistic map (1-3.6).

Experimental Evidence for Reduction of 2-D to 1-D Maps

There is much experimental evidence to date to support the proposition that many chaotic physical phenomena can be approximately modeled by one-dimensional maps of the form (see Figure 2-16)

$$x_{n+1} = F(x_n)$$

This evidence comes in two forms: (i) calculation of $F(x_n)$ from measurement of a single state variable and (ii) reduction of 2-D Poincaré map data to a first-order map.

An example of the former is the experimental work of a group at the University of Texas (Roux et al., 1983) who have measured the dynamics of chemical reactors. One famous example is the Belousov–Zhabotinski reactor which involves over a dozen reactions. In spite of the complexity of this system, a simple experimental return map was measured as shown in Figure 3-15 and is similar in form to a quadratic map.

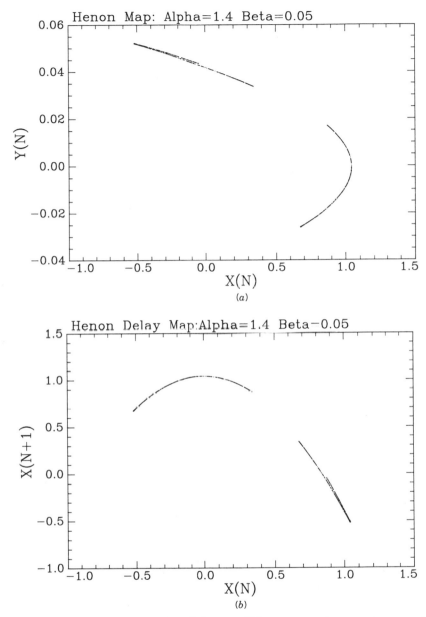

Figure 3-14 (*a,b*) Henon maps (1-3.8) for two different control parameters $\alpha = 1.4$, $\beta = 0.05$.

The second example is taken from control theory in which a mass is shuttled back and forth by a servomotor as shown in Figure 3-16*a* (Golnaraghi and Moon, 1991). The position is controlled by an error signal which is the difference between a desired periodic motion of the mass and the actual position.

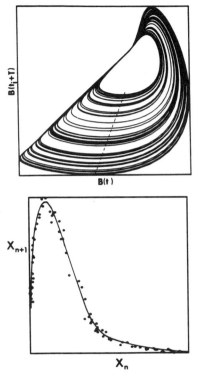

Figure 3-15   Representation of the chemical dynamics of the Belousov–Zhabotinski reaction using a 1-D map. [From Roux et al. (1983).]

Nonlinearity arises from the motion constraints near the end of its travel. When the control gain exceeds some limit, the mass becomes chaotic. However, the Poincaré map triggered off the periodic control signal appears to be one-dimensional, as shown in Figure 3-16b. A return map reveals a noninvertible 1-D piecewise linear map, as shown in Figure 3-16c. Another example for a friction oscillator is shown in Figure 5-17.

The reduction of dynamical behavior from complex systems to a 1-D map is more than just academic interest. The simple map, once determined, can serve as an information compression coding from which properties such as the probability density distribution or the Lyapunov exponent can be obtained by simply iterating the map.

Obtaining dynamical information from a 1-D map for a physical system is many orders of magnitude faster than trying to integrate the partial or ordinary differential equations that describe the underlying physics.

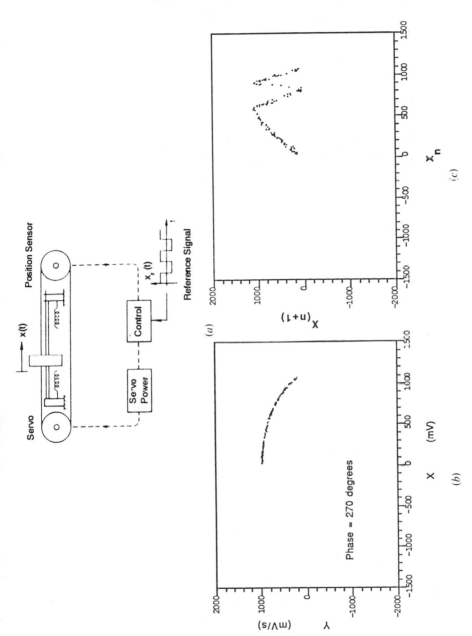

Figure 3-16 Experimental 1-D map for a chaotic control problem. (*a*) Sketch of mass with trilinear spring forces, periodic command force, and feedback control; (*b*) 2-D Poincaré map; (*c*) 1-D return map. [From Golnaraghi and Moon (1991).]

3.6 PERIOD-DOUBLING ROUTE TO CHAOS

If Chaos Theory is the study of the pathways from simple to complex dynamics, then period doubling must be considered one of the principal routes to chaotic behavior in physical systems with nonlinearities. Periodic doubling is a phenomenon in which the period of repetition of a cyclic dynamic process doubles with the change of some control parameter, and continues to double at successively closer parameter values until the period is so long it is practically aperiodic. This phenomenon has been reported in hundreds of published experimental papers and in dozens of different physical and even biological systems where strong nonlinearities are present.

Much of the theory of periodic doubling is based on the study of first-order difference equations

$$x_{n+1} = F(x_n) \tag{3-6.1}$$

and in particular the study of the quadratic map or so-called "logistic" equation of ecological modeling:

$$x_{n+1} = \lambda x_n (1 - x_n) \tag{3-6.2}$$

Here the growth or birth rate of a species in the next generation is proportional to λ, while the quadratic decay or "death" in the species is governed by the nonlinear term.

The core of the mathematical theory is associated with the work of Mitchell Feigenbaum, formerly of Los Alamos National Laboratory and now at Rockefeller University in New York City. Feigenbaum (1980), however, attributes the qualitative observation of periodic doubling in equations of the above form (3-6.2) to the earlier work of Metropolis, Stein, and Stein (1973). A later paper of Robert May (1976) of Princeton University described some of the qualitative results in equations of mathematical biology.

However, it was Feigenbaum (1978) who derived *quantitative* measures of the period doubling phenomenon. Furthermore, it was his assertion that these measures could be observed in physical systems with more complex mathematical description than (3.6-2) that led experimentalists on the quest for *universal* measures of the steps on the period-doubling route to chaos. It will take later historians of science to discover who first observed this phenomenon, but the avalanche of observations began around the time of observations reported in fluids by Libchaber and Maurer (1978) and in electronic circuits by Linsay (1981).

There are many excellent mathematical descriptions of this phenomenon, and we shall not attempt to describe all the details here. [The previously cited paper by Feigenbaum (1980) is an excellent readable source, however.] Because this book emphasizes the physical and experimental aspects of chaos, we shall focus mainly on those results of the theory that are relevant to the observation and understanding of period doubling in physical systems.

Qualitative Features of Period Doubling

One of the distinguishing features of this book is the attempt to describe the distinctive patterns of different dynamic phenomena in a visual way that dynamicists can recognize from computer or oscilloscope images of dynamic data. In the case of period doubling associated with continuous time dynamics, there are six common displays of this phenomena:

1. Time history
2. Phase plane
3. Poincaré map
4. Fourier transform
5. Autocorrelation
6. Probability density function

These different data processing and display techniques are common for many modern computer and dynamic signal analyzers. An example is shown in Figure 3-17 for a nonlinear electronic circuit [see Chapter 4 and Matsumoto et al. (1985)].

In the continuous time domain, the change from period-1 to period-2 motion clearly shows the doubling of the fundamental period. The phase plane shows a qualitative change from one orbit to two overlapping orbits. Also the Poincaré map changes from one to two periodic points.

The Fourier transform, usually performed experimentally with a discrete fast Fourier transform (FFT) chip, shows the halving of the fundamental frequency $\omega_0 = 2\pi/\tau_0$. In further period doublings, say in period $2n$, not only will $\omega_0/2n$ be present, but equally likely one will see other harmonics $m\omega_0/2n$. The autocorrelation function is related to the FFT and is qualitatively similar to the time domain representation.

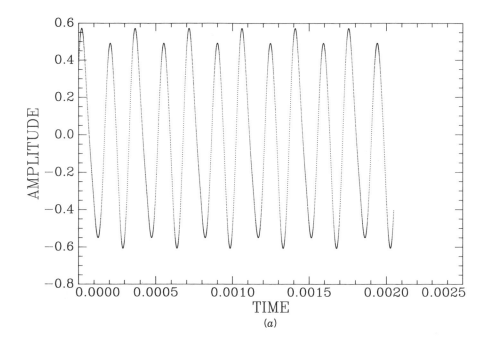

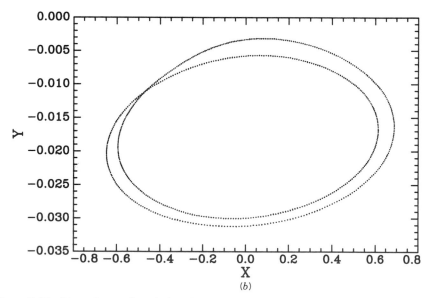

Figure 3-17 Experimental tools for observing period-doubling bifurcations. (*a*) Time history; (*b*) Phase plane; (*c*) Fourier transform; (*d*) Probability density function.

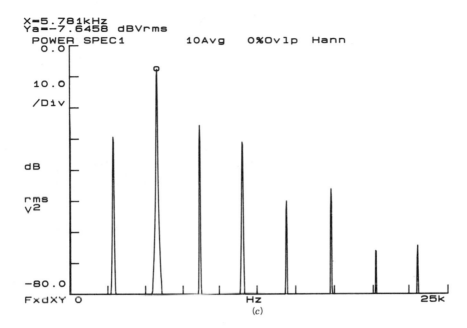

(c)

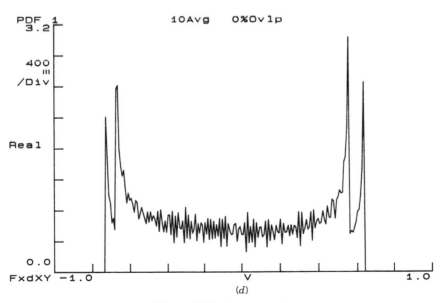

(d)

Figure 3-17 (*Continued*)

Finally, one can sometimes perform a probability density calculation with electronic spectrum analyzers or computers. In period doubling, the classic two-spike signature for a sine wave oscillation changes to a four-spike picture indicating four vertical tangents in the phase-space representation of the period-doubled waveform.

Poincaré Maps and Bifurcation Diagrams

A more informative representation of period doubling in a continuous time system can often be presented by plotting the discrete Poincaré map measures of the signal as one slowly changes one of the control parameters. The Poincaré map of the base period signal is represented by one point in the x-axis, whereas the Poincaré map for the period-doubled map shows up as two points on the x-axis.

In a bifurcation diagram, one plots the Poincaré map of the x-axis versus a control parameter λ. This can be easily done in computer simulation of iterated maps, such as for the logistic map shown in Figure 3-18. This diagram shows the values of the control parameter at which the motion changes from a period-n to a period-$2n$ oscillation as well as shows the regions where suspected chaos may be found.

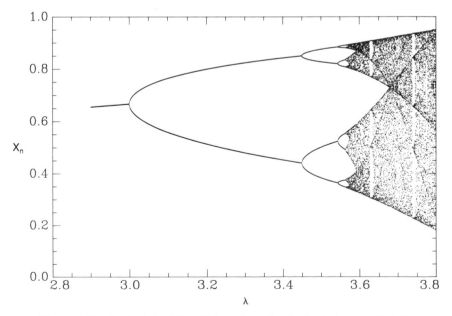

Figure 3-18   Period-doubling bifurcations for the logistic map (3-6.2).

Quantitative Measures of Period Doubling

In this section we illustrate the nature of the quantitative analysis of period-doubling bifurcations using the logistic map. The reader should consult more mathematical books on the subject for a more rigorous treatment (e.g., see Devaney, 1989).

A general quadratic map takes the form

$$x_{n+1} = a + bx_n + cx_n^2$$

When $b > 0$, $c < 0$, one can rescale this equation into the standard form (3-6.2)

$$x_{n+1} = \lambda x_n(1 - x_n) = F(x_n)$$

The basic analysis of Eq. (3-6.2) usually involves the following procedure.

1. Find fixed points of $F(x_n)$ and some of the two-cycle iterates $F^{2n} = F(F^{2n-2}(x_n))$.
2. Establish the stability of these fixed points.
3. Examine the relationship between the critical bifurcation values of the control parameter $(\lambda_1, \lambda_2, \lambda_3, \ldots, \lambda_n)$.
4. Determine the limit point λ_∞ at which the first chaotic dynamics results.
5. Look at scaling properties of the successive period-2n orbits.
6. Look at the scaling of the spectral properties of the period-2n orbits.

As described earlier, the fixed or equilibrium points of a discrete iterated map may be found by finding those values of the state variable where $x_{n+1} = x_n$ for a fixed control parameter, that is,

$$x_e = F(x_e)$$

In general, there is more than one equilibrium point. For the logistic map, these points are given by

$$x_e = 0, \qquad x_e = \frac{\lambda - 1}{\lambda}, (\lambda > 1)$$

Thus, for $\lambda < 1$ the only fixed point of the map is at the origin. When the control parameter λ is increased to a value greater than 1, a period-1 or period-n orbit may be possible. In an n-orbit, the map will produce a sequence of points

$$(x_0, x_1, x_2, \ldots, x_{n-1}, x_n = x_0)$$

In the case of a period-2 motion we have the orbit $(x_0, x_1, x_2 = x_0)$ where

$$x_1 = F(x_0) = F(F(x_1)) = F^2(x_1)$$
$$x_2 = F(x_1) = F(F(x_2)) = F^2(x_2)$$

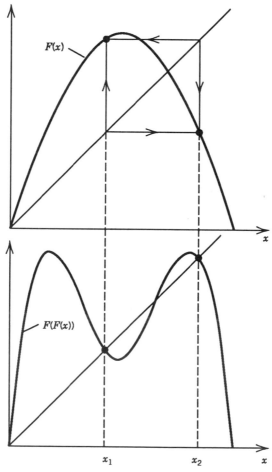

Figure 3-19 Comparison of the period-1 and period-2 maps for the logistic equation (3-6.2).

and

$$F^2(x) = \lambda^2 x(1 - x)[1 - \lambda x(1 - x)]$$

One of the fixed points of $F^2(x)$ is $x = 0$. However, two of the other three fixed points of $F^2(x)$ give us the orbit points of the period-2 map. This is clearly shown in Figure 3-19. From this figure, one can see that one of the fixed points of $F^2(x)$ is unstable $|dF^2/dx| > 1$ while the other two points give the stable orbit points of the period-2 cycle of $F(x)$.

This pattern continues as λ is increased through higher period-doubling bifurcations. At each critical value of λ, the stable fixed

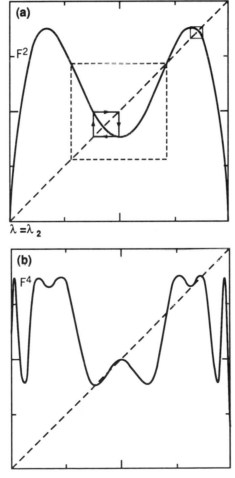

Figure 3-20 Comparison of period-2 and period-4 maps for the logistic equation (3-6.2). [From Feigenbaum (1980).]

points of the previous period-$2n$ map become unstable and the next higher-order period-$2(n + 1)$ map throws off two new fixed points which become the new orbit points of the next period-doubling orbit. An example for the transition from period 2 to period 4 is shown in Figure 3-20.

Scaling Properties of Period-Doubling Maps

Feigenbaum Numbers. Perhaps one of the notable discoveries of the theory of chaotic dynamics was a set of scaling relationships for the period-doubling route to chaos that could be tested in computer or physical experiments. Using both computational evidence and analysis, Feigenbaum proposed that the bifurcation values of the control parameter of the logistic map, λ_n, converge in an exponential way as one approaches the limiting value λ_∞ at which chaotic dynamics develop; that is

$$\lambda_\infty - \lambda_n = c\delta^{-n} \tag{3-6.3}$$

If one knows three consecutive values of the period-doubling bifurcation parameter, λ_{n-1}, λ_n, λ_{n+1}, then the number δ can be found:

$$\delta = \lim_{n \to \infty} \frac{\lambda_n - \lambda_{n-1}}{\lambda_{n+1} - \lambda_n}$$

$$\delta = 4.6692016 \ldots \tag{3-6.4}$$

What is remarkable is that Feigenbaum showed that this value is *universal* within a class of maps $x_{n+1} = F(x_n)$, where $F(x)$ has a smooth maximum. [Another technical requirement involves an operator on $F(x)$ called the *Schwartzean derivative*; see Feigenbaum (1980) or, e.g., Rasband (1990).] He also proposed that in physical systems that undergo such period-doubling phenomena, the above limit of δ should also hold. This is a remarkable assertion. However, the experimental evidence does not always provide one with a large number of measurable period-doubling bifurcations (five or six seem to be the experimental limit). In spite of this, there is good reason to believe that this scaling number is correct in physical problems.

Amplitude Scaling. Amplitude scaling has to do with a comparison of features of the $2(n + 1)$ period doubling as compared with the amplitude features of the $2n$ period doubling. There are several ways to describe this scaling. To choose one, let's examine the bifurcation diagram for the logistic map (3-6.2) shown in Figure 3-21 which plots the amplitudes of the cycle points as a function of the control parameter λ. One can see that within each period-doubling region there is a $2n$ cycle which has a cycle point at the midpoint $x = \frac{1}{2}$. These particular cycles have been called *supercycles*. Then we define an amplitude measure a_n as shown in Figure 3-21 where $a_1 = x_2 - x_1$, $a_2 = x_2 - x_4$, and so on. Feigenbaum discovered that the ratio a_n/a_{n+1} has a limiting value

$$\lim_{n \to \infty} \frac{a_n}{a_{n+1}} = 2.50 \dots \tag{3-6.5}$$

Another way to describe this scaling is to notice that each minor period-doubling pitchfork can be mapped onto the preceding pitchfork by simply blowing up the minor bifurcation region by the factor 2.50. This idea is related to the mathematical concept of renormalization [e.g., see Chapter 6 or Feigenbaum (1980)].

The third way to see this scaling is when period doubling occurs in a continuous time system. Here the period doubling produces ripple modulation on the fundamental periodic signal. The ratio of one of the

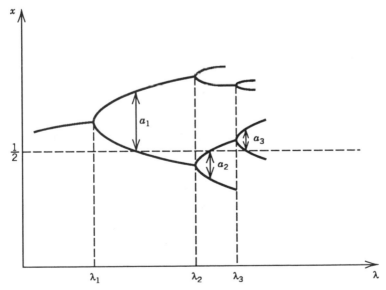

Figure 3-21 Bifurcation diagram of the logistic map (3-6.2) showing relative amplitudes of supercycles.

ripples in the period $2(n + 1)$ cycle to a ripple in the $2n$ cycle is given by the factor 2.5.

Again, the significance of the ratio is that if period doubling occurs in a dissipative system with an underlying hump map, then this property will result regardless of the particular shape of the 1-D mapping function $F(x)$.

Subharmonic Spectra Scaling

The scaling of the amplitudes in successive period-doubling regimes has as a consequence a related scaling property in the Fourier spectra of the dynamic signals. This is illustrated in Figure 3-22. Again Feigenbaum (1980) has shown that the successive subharmonic spectra peaks in the Fourier transform will be 8.2 db below the previous subharmonic peaks. This result has been confirmed in a number of experiments (e.g., see Libchaber et al., 1980) as shown in Figure 3-23.

Symbol Sequences in Period Doubling

In many experimental problems we have only qualitative information about the dynamics; for example, a rotor turns clockwise or counterclockwise or a particle moves to the left side or the right side in some mechanical system. In such problems one can use what is called

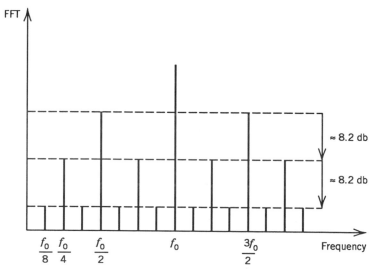

Figure 3-22   Universal scaling of the fourier transformation amplitudes for the logistic map (3-6.2).

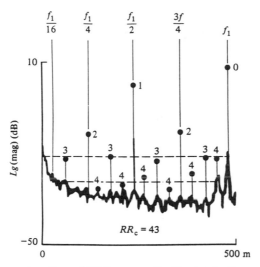

Figure 3-23 Universal scaling of FFT spectra for Rayleigh–Benard thermal convection. [From Libchaber et al. (1980).]

course-graining; that is, we divide up the state space into a course grid. In the case of the logistic map (3.6-2) the obvious course-graining are the two intervals I_1: $0 \leq x < \frac{1}{2}$ and I_2: $\frac{1}{2} < x \leq 1$. Then one can code the dynamics by a sequence of symbols L or R indicating whether the motion is in the I_1 or I_2 interval. Thus, a period-2 orbit might be LRLRLR ... , or a period-3 orbit might be LRRLRR....

In early work in the 1970s, Metropolis, Stein, and Stein (1973) showed that as the control parameter is varied for different hump maps, the sequence of symbol sequences is the same. A list of such symbol sequences is given in Table 3-1 for the logistic map and for the map derived from a tent map for a friction oscillator (Feeny and Moon, 1989). In the friction oscillator the course-graining or partitioning is associated with sticking (S) or slipping (N) motions.

This is then another indication of the existence of the underlying order in 1-D map models of deterministic dynamical systems which exhibit chaotic behavior.

Period Doubling in Conservative Systems

The scaling behavior of the control parameter values at which period doubling occurs and the resulting Feigenbaum number (3-6.4) were derived for a dynamical system with dissipation. When the system is nondissipative, however, one gets a different "Feigenbaum" number. This has been illustrated in a paper by Helleman (1980a) and elsewhere

TABLE 3-1   Symbol Sequences

Friction Oscillator[1]	$\ddot{x} + 2\zeta\dot{x} + x + \eta(x)f(\dot{x}) = \alpha \cos \Omega\tau$		Logistic Map[2]	$x_{n+1} = \lambda x_n(1 - x_n)$	
Period	Symbol Sequence	α	Period	Symbol Sequence	λ
1	N	1.36	2	R	3.2360680
4*	NSN	1.38	4*	RLR	3.4985617
8*	NSNNNSN	1.3925	8*	RLRRRLR	not listed
10	NSNNNSNSN	1.4064	10	RLRRRLRLR	not listed
6	NSNNN	1.415	6	RLRRR	3.6275575
7	NSNNNN	1.45415	7	RLRRRR	3.7017692
5	NSNN	1.4737	5	RLRR	3.7389149
7	NSNNSN	1.4909	7	RLRRLR	3.7742142
3	NS	1.535	3	RL	3.8318741
6*	NS•NS	1.551	6*	RLLRL	3.8445688
			7	RLLRLR	3.8860459
5	NSSN	1.5973	5	RLLR	3.9057065
			7	RLLRRR	3.9221934
			6	RLLRR	3.9375364
			7	RLLRRL	3.9510322
4	NSS	1.695	4	RLL	3.9602701
			7	RLLLRL	3.9689769
			6	RLLLR	3.9777664
			7	RLLLRR	3.9847476
5	NSSS	1.833	5	RLLL	3.9902670
			7	RLLLLR	3.9945378
			6	RLLLL	3.9975831
			7	RLLLLL	3.9993971

Legend: N = No-stick
S = Stick
R = Right half
L = Left half

Source: [1]Feeny and Moon (1992b); [2]Schuster (1988).

(e.g., see Benettin et al., 1980a) by an example for a two-dimensional conservative map:

$$x_{n+1} = y_n$$
$$y_{n+1} = -x_n + 2y_n(c + y_n) \tag{3-6.6}$$

Helleman derived this equation from the motion of a proton in a storage ring with periodic impulses. He thus found an exact map for a nonlinear, second-order forced system by integrating the equation of motion between impulses.

That the map is conservative can be proved by looking at the change of area of a small parallelepiped in one iteration of the map. The change in area is measured by the determinate of the Jacobian matrix, (3-1.4) which can be shown to be unity for this map.

It can be shown that iterations of this map yield a multiperiodic orbit for $c \geq -1$. In fact, the transitions from one to two periodic and from two to four periodic, and so on, are given by

$$c_1 = -1, \quad c_2 = 1 - \sqrt{5}, \quad c_3 = 1 - \sqrt{4 + \sqrt{5/4}}$$

as the reader can verify on a calculator or small personal computer.

For large k, this sequence of critical bifurcation parameters has the relationship

$$c_k \sim c_\infty + A/\delta_c^k \tag{3-6.7}$$

where $\delta_c = 8.7210$ from numerical experiments. This value differs from the Feigenbaum value of 4.6692 found for the logistic map. It is believed that δ_c is "universal" for all conservative maps.

3.7 THE MEASURE OF CHAOS; LYAPUNOV EXPONENTS

Chaos in deterministic systems implies a sensitive dependence on initial conditions. This means that if two trajectories start close to one another in phase space, they will move exponentially away from each other as the map is iterated. If d_0 is a measure of the initial distance between two starting points, at a later time the distance may be written in the form

$$d_n = d_0 2^{\Lambda n} \tag{3-7.1}$$

(The choice of base 2 is convenient, but arbitrary.) However, in most physical problems the motion is bounded, and d_n cannot go to infinity. Thus, the exponent Λ must be averaged over the trajectory, that is,

$$\Lambda = \lim_{N \to \infty} \frac{1}{N} \sum_{n=0}^{N} \log_2 \left| \frac{d_{n+1}}{d_{0n}} \right|$$

In the case of a one-dimensional map,

$$x_{n+1} = f(x_n)$$

an explicit rule can be derived. At the nth iteration choose

$$d_{0n} = \varepsilon$$

$$d_{n+1} = f(x_n + \varepsilon) - f(x_n) \simeq \left.\frac{df}{dx}\right|_n \varepsilon$$

Thus (3-7.1) becomes

$$\Lambda = \lim_{N\to\infty} \frac{1}{N} \sum_{n=0}^{N} \log_2 \left|\frac{df}{dx}(x_n)\right| \tag{3-7.2}$$

An illustrative example is the Bernoulli map

$$x_{n+1} = 2x_n \ (\text{mod } 1) \tag{3-7.3}$$

as shown in Figure 3-24. Here (mod 1) means

$$x(\text{mod } 1) = x - \text{Integer}(x)$$

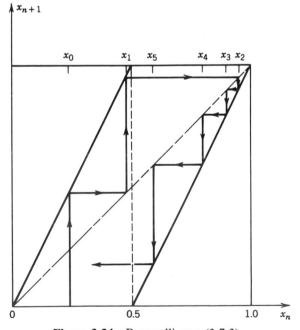

Figure 3-24   Bernoulli map (3-7.3).

This map is multivalued and is chaotic. Except for the switching value $x = \frac{1}{2}$, $|f'| = 2$. Applying the definition (3-7.3) we find $\Lambda = 1$. Thus, on the average, the distance between points on nearby trajectories grows as

$$d_n = d_0 2^n$$

The units of Λ are one bit per iteration. One interpretation of Λ is that information about the initial state is lost at the rate of one bit per iteration. To see this, write x_n in binary notation, that is,

$$x_n = \left(\frac{1}{2} + \frac{1}{4} + \frac{1}{16} + \frac{1}{128}\right) \equiv 0.1101001$$

and x (mod 1) means 1.101001 (mod 1) = 0.101001. Thus, the map $2x_n$ (mod 1) moves the decimal point to the right and drops the integer value. So if we start out with m significant decimal places of information, we lose one for each iteration; that is, we lose one bit of information. After m iterations we have lost knowledge of the initial state of the system.

Another example is the *tent map*:

$$\begin{array}{ll} x_{n+1} = 2rx_n, & x_n < \frac{1}{2} \\ x_{n+1} = 2r(1 - x_n), & x \geq \frac{1}{2} \end{array} \qquad (3\text{-}7.4)$$

As in the Bernoulli map (3-7.3), $|f'(x)| = 2r$ is a constant and the Lyapunov exponent is found to be (Lichtenberg and Lieberman, 1983, pp. 416–417)

$$\Lambda = \log 2r \qquad (3\text{-}7.5)$$

When $2r > 1$, $\lambda > 0$ and the motion is chaotic, but when $2r < 1$, $\lambda < 0$ and the orbits are regular; in fact, all points in $0 < x < 1$ are attracted to $x = 0$ (Schuster, 1984, p. 22).

Our final example is the logistic equation (3.6-2)

$$x_{n+1} = ax_n(1 - x_n)$$

This map may become chaotic when $a > 3.57$. This can be verified by numerical calculation of the Lyapunov exponent as a function of a as shown in Figure 3-25. Beyond $a = 3.57$, the Lyapunov exponent is

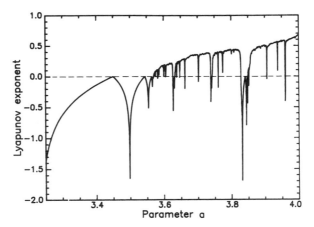

Figure 3-25 Lyapunov exponent for the logistic map $x_{n+1} = ax_n(1 - x_n)$ as a function of the control parameter a.

positive except within multiperiod windows $3.57 < a < 4$. When $a = 4$, it has been shown that $\Lambda = \ln 2$. It has been shown (see Schuster, 1984) that this value can be derived analytically by finding a transformation of the tent map for $r = 1$ into the logistic map.

Probability Density Function for Maps

Living and designing in the face of uncertainty is part of the everyday world. We often deal with natural unpredictability such as the weather with probability measures. Thus, it is natural to seek probabilistic descriptors when deterministic systems are in a state of dynamical chaos. Such statistical tools are widely used in the random excitation of linear systems. However, probabilistic mathematics for chaotic dynamics of nonlinear systems are not readily available. One exception is the case of systems governed by a first-order difference equation or map.

Our discussion of probability measures for one-dimensional maps can only be introductory. However, some mathematical language is necessary and useful if one wishes to read more advanced treatments of the subject. For a first-order map, this description involves a function $P(x)$ called the *probability density function* (PDF), where x is the state variable that governs the map

$$x_{n+1} = F(x_n)$$

Because x is a continuous variable, $P(x)\,dx$ is the probability that the dynamical orbit will occur between $(x, x + dx)$. The domain of the variable x over which $P(x) \neq 0$ is sometimes called the *support* of the probability measure. One complication in chaotic systems is that the support is sometimes fractal. In this case, $P(x)$ is not a continuous function. However, in practical systems there is always a small amount of noise which tends to smooth out the fractal nature of $P(x)$. Thus, when an integration of $P(x)$ makes sense, the probability of an orbit occurring between $x_1 \leq x_n \leq x_2$ is given by

$$P[x_1, x_2] = \int_{x_1}^{x_2} P(x)\,dx \qquad (3\text{-}7.6)$$

and

$$\int_{-\infty}^{\infty} P(x)\,dx = 1$$

In statistical theory, one can have time-varying probabilistic measures. However, in this book we assume that some chaotic attractor exists and that $P(x)$ is a so-called *invariant measure;* that is, it is a property of the attractor that does not change with time.

Several examples of invariant measures or PDFs for one-dimensional maps have been discovered analytically. Two such cases are the tent map and the logistic map:

Tent Map: $x_{n+1} = rx_n,$ $x_n < \tfrac{1}{2}$

       $x_{n+1} = 1 + r - rx_n,$ $x_n \geq \tfrac{1}{2}; r > 1$ $(3\text{-}7.7)$

       $P(x) = 1$

Logistic Map: $x_{n+1} = 4x_n(1 - x_n)$

$$P(x) = \frac{1}{\pi \sqrt{x(1 - x)}} \qquad (3\text{-}7.8)$$

Both maps are sometimes called *one-hump maps,* and the PDF can be derived from a functional equation. Consider a general one-hump map shown in Figure 3-26. Then orbits that arrive in the differential domain between $(x, x + dx)$ have two preimages x_1 and x_2 where

$$x = F(x_1) = F(x_2)$$

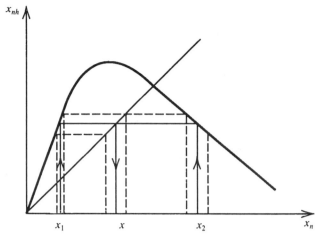

Figure 3-26   Hump map showing two preimage contributions to the probability density functions.

Assuming that $F(x)$ is continuous and differentiable at x_1, x_2, one can easily show that the PDF must satisfy the following functional equation:

$$P(x) = \frac{P(x_1)}{|F'(x_1)|} + \frac{P(x_2)}{|F'(x_2)|} \tag{3-7.9}$$

For the tent map with $r = 2$, $F'(x) = 2$ and

$$P(x) = \tfrac{1}{2}[P(x_1) + P(x_2)] \tag{3-7.10}$$

where $x_1 = 2x$, $x_2 = 2 - 2x$. It is easy to see that $P(x) = $ constant is a solution.

This idea can be extended to multihump maps using the idea that the probability of orbits arriving in the vicinity of x is the sum of the probabilities that they originated in the multiple preimages of the map $F(x)$.

Numerical Calculation of PDF

A fairly obvious way to calculate an approximation to $P(x)$ is to divide up or partition the domain of x into N cells of size Δx, and then run the map for several thousand iterations, counting the number of times N_i an orbit enters the ith cell. The set of numbers, sometimes called a *histogram* $\{P_i; i = 1, ..., N\}$, is given by

$$P_i = N_i/N$$

The histogram is then considered to be an approximation to $P(x)$. This technique is sometimes called *course-graining*. This method runs into problems when the *support* of $P(x)$ is fractal. However, the set of numbers $\{P_i\}$ may still be a good approximation to $P(x)$ when the map is subject to a small amount of noise, that is,

$$x_{n+1} = F(x_n) + \eta(x_n)$$

where $\eta(x_n)$ is a small random variable. Such noise is present not only in physical systems but also in computing machines, which are always limited to a finite precision.

The effect of small amounts of random noise on the dynamics of one-dimensional maps has been studied by several authors. Two graphs from a study by Crutchfield et al. (1982) shown in Figure 3-27 show the effect of noise on the PDF for the logistic map, when the control parameter is slightly larger than the critical parameter for chaos. Notice the smoothing of the peaks due to noise. These data were generated by partitioning the line $[0, 1]$ into 10^3 bins and using 10^6 iterations.

The extension of the analytical method such as (3-7.9) to two- or higher-dimensional maps and flows requires other mathematical tools such as the Fokker–Planck equations (e.g., see Gardiner, 1985). However, the course-graining numerical technique is often straightforward and can be implemented computationally as well as in experimental measurements (e.g., see Chapters 2 and 5).

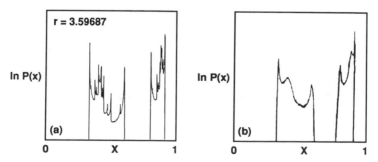

Figure 3-27   (a,b) The effect of a small amount of noise on the probability density function for the logistic map (3-6.2). [From Crutchfield et al. (1982).]

PDF and Lyapunov Exponents

As discussed above, the Lyapunov exponent is often calculated as a time average of the slope of the map function $F'(x) \equiv dF/dx$:

$$\Lambda = \frac{1}{N} \sum \ln|F'(x)| \tag{3-7.11}$$

where $\Lambda > 0$ defines chaos.

When one has a probability distribution function for the map, the Lyapunov exponent can sometimes be calculated using a space average, that is,

$$\Lambda = \int_0^1 P(x) \ln|F'(x)| \, dx \tag{3-7.12}$$

In the case of the tent map for $r = 2$, $P(x) = 1$, $|F'(x)| = 2$, and $\Lambda = \ln 2$. When the base of the logarithm is 2, then Λ represents the loss of 1 bit of information per cycle of the map.

3.8   3-D FLOWS; MODELS AND MAPS

Dynamical models using differential equations occur naturally in the physical sciences because the equations of classical physics have been formulated in terms of partial differential equations. With suitable spatial assumptions, the equations of mass balance, energy, momentum, and electromagnetics can be reduced to a set of coupled ordinary differential equations. The simplest set which can exhibit chaotic solutions is a set of three;

$$\dot{x} = F(x, y, z)$$
$$\dot{y} = G(x, y, z) \tag{3-8.1}$$
$$\dot{z} = H(x, y, z)$$

The dynamics can be visualized by constructing a "velocity" vector field at every point in the three-dimensional phase space (x, y, z); that is, $V = (\dot{x}, \dot{y}, \dot{z}) \equiv (F, G, H)$. The rate of change of volume in this phase space is given by the divergence of this vector field:

$$\nabla \cdot V = \frac{\partial F}{\partial x} + \frac{\partial G}{\partial y} + \frac{\partial H}{\partial z} \tag{3-8.2}$$

For dissipative problems $\nabla \cdot V < 0$, whereas for conservative problems $\nabla \cdot V = 0$.

In this section we examine a few examples of three-dimensional flows and show how the dynamics generates two- and one-dimensional maps. It is also a good exercise to interpret discrete time fixed points and map dynamics in terms of the original 3-D flow geometry. For example, a cycle-3 fixed point of a 2-D map may imply three loops of a period-3 subharmonic orbit in the 3-D flow. Or, if a closed orbit exists in a 2-D map, then the underlying flow may exhibit a quasiperiodic motion in 3-D that moves on a toroidal surface.

Lorenz Model for Fluid Convection

One of the first models shown to exhibit chaotic behavior in numerical simulation was a fluid convection model of E. Lorenz of M.I.T. (1963) as described briefly in Chapter 1. In this model, $x(t)$ represents a measure of the fluid velocity, and $y(t)$, $z(t)$ represent measures of the spatial temperature distribution in the fluid layer under gravity. The equations were derived in a more complex form by Saltzman (1961) (see also Chapter 4) and were simplified by Lorenz as follows:

$$\dot{x} = \sigma(y - x)$$
$$\dot{y} = rx - y - xz \tag{3-8.3}$$
$$\dot{z} = -bz + xy$$

These equations are derived from the energy and momentum balance relations for the fluid.

Here σ represents the Prandtl number, which is a ratio of kinematic viscosity to thermal conductivity; r is called a Rayleigh number and is proportional to the temperature difference between the upper and lower surfaces of the fluid; and b is a geometric factor. Note that the only nonlinear terms are two quadratic polynomials.

The general form of these equations also serves as a simple model for complex dynamics in certain laser devices (e.g., see Haken, 1985).

The dynamics of this system can be visualized in a 3-D phase space (x, y, z) (Figure 1-27) using a "velocity" field $(\dot{x}, \dot{y}, \dot{z})$ given by the right-hand side of (3-8.3). The divergence of this "velocity" field shows that differential volume elements of phase space are uniformly contracting, that is,

$$\nabla \cdot V = -(\sigma + b + 1) \tag{3-8.4}$$

A typical set of values for the study of this equation is $\sigma = 10$, $b = \frac{8}{3}, 1 < r \leq 28$, which are the ones studied by Lorenz. Unfortunately, these nondimensional groups do not relate to real geoconvection flow parameters. However, these parameters can be replicated in a laboratory convection experiment called a thermosyphon (see Chapter 4).

Many authors have reproduced and extended the original analyses of Lorenz. Below is a summary of some of these results using r as a control variable.

The fixed points of the flow are found by setting $\dot{x} = \dot{y} = \dot{z} = 0$ and are given by

$$(x, y, z) = (0, 0, 0)$$
$$(x_e, y_e, z_e) = (\pm\sqrt{b(r-1)}, \pm\sqrt{b(r-1)}, r-1) \tag{3-8.5}$$

The stability of the fixed point at the origin may be found by solving the following eigenvalue problem:

$$\det \begin{bmatrix} (\lambda + \sigma) & -\sigma & 0 \\ -r & (\lambda + 1) & 0 \\ 0 & 0 & (\lambda + b) \end{bmatrix} = 0$$

$$\lambda = -\frac{(\sigma + 1)}{2} \pm \frac{1}{2}[(\sigma + 1)^2 - 4\sigma(1 - r)]^{1/2} \tag{3-8.6}$$

For $r < 1$ the origin is the only fixed point, but when $r > 1$ two other fixed points are born as given above. The stability of these points can be studied by looking at the linearized equation near each of these fixed points.

Also, the global dynamics are bounded in a finite volume (sphere) of phase space (e.g., see Bergé et al., 1985).

As one increases the temperature difference between upper and lower surfaces (i.e., by increasing r), the following dynamical bifurcations occur:

I. $0 < r < 1$. There is only one stable fixed point at the origin.

II. $1 < r < 1.346$. Two new stable nodes are born and the origin becomes a saddle with a one-dimensional, unstable manifold (Figure 3-28).

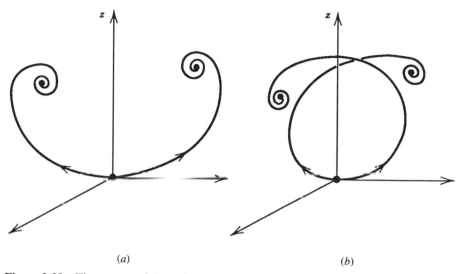

(a)   (b)

Figure 3-28   The nature of three fixed points of the Lorenz equations (1-3.9) for pre-chaos values of the control parameter r. (a) $1 < r < 1.346$, (b) $r > 1.346$.

III. $1.346 < r < 13.926$. At the lower value the stable nodes become stable spirals.

IV. $13.926 < r < 24.74$. Unstable limit cycles are born near each of the spiral nodes, and the basins of attraction of each of the two fixed points become intertwined. The steady-state motion is sensitive to initial conditions.

V. $24.74 < r$. All three fixed points become unstable. Chaotic motions result.

The dynamic orbit shown in Figure 1-27 is best viewed "live" on a computer graphics terminal. There one can see the interplay between the unstable spirals.

Lorenz Maps.   One of the remarkable features of the Lorenz study in 1963 was not only his use of computer simulation to demonstrate unpredictable dynamics, but also his use of a discrete time map to show the underlying nature of this chaos. Observing the two spiral motions alternating back and forth, Lorenz asked if the local z maximum of one circuit Z_n could be a predictor of the next relative maximum z value Z_{n+1}. The numerical data are reproduced in Figure 3-29, and it is clear that this is similar to a tent map (3-7.7) studied earlier.

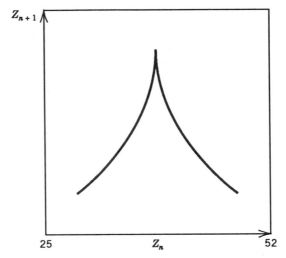

Figure 3-29 One-dimensional map of Lorenz based on the relative maximum values z_n of the time integration of the thermal convection model equation (3-8.3).

Note that this "mapping" is not a classic Poincaré map because the trajectories do not penetrate a surface of section.

In later work, however, researchers did find a Poincaré map (e.g., see Guckenheimer and Holmes, 1983 and Sparrow, 1982). This map is shown schematically in Figure 3-30. A plane $z = r - 1$ contains the

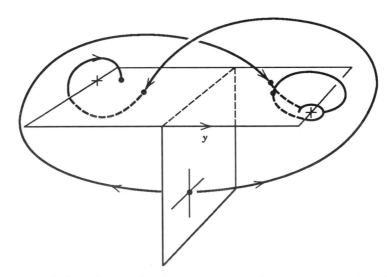

Figure 3-30 Poincaré plane for the construction of a return map using the flow generated by the Lorenz equation [Eq. (3-8.3)].

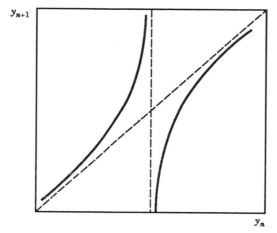

Figure 3-31 Poincaré map of the Lorenz equations based on the section shown in Figure 3-30.

two fixed points. When the trajectory penetrates this plane, the point is projected onto a line connecting the two fixed points from which one obtains a one-dimensional map similar to the one shown in Figure 3-31. This map is similar to the Bernoulli map (3-7.3). In both cases, it is remarkable that such complex dynamics can be reduced to one-dimensional maps.

Duffing's Equation and the "Japanese Attractor"

A classic differential equation that has been used to model the nonlinear dynamics of mechanical and electrical systems is the harmonic oscillator with a cubic nonlinearity:

$$\ddot{x} + \gamma\dot{x} + \alpha x + \beta x^3 = F \cos \Omega t \qquad (3\text{-}8.7)$$

This equation has been named after G. Duffing, a German electrical engineer/mathematician who studied it in the 1930s. With $\alpha = 0$, it is a model for a circuit with a nonlinear inductor (see Chapter 4); and with $\alpha < 0$, $\beta > 0$, it is a model for the postbuckling vibrations of an elastic beam column under compressive loads. We will focus on the case of $\alpha = 0$, which has been extensively studied by a group of engineers at Kyoto University for several decades first under the leadership of C. Hayashi (see Hayashi, 1985) and then under Professor Y. Ueda (e.g., see Ueda, 1979, 1991). This equation can be written as a set of three first-order nondimensional differential equations:

$$\dot{x} = y$$

$$\dot{y} = -ky - x^3 + B\cos 2\pi z \qquad (3\text{-}8.8)$$

$$\dot{z} = 1 \qquad (\text{mod } 1)$$

The (mod 1) indicates that the phase space has a cylindrical geometry.

If we construct a set of vectors $(\dot{x}, \dot{y}, \dot{z})$ on a three-dimensional grid, we can imagine the flow of a fluid. This is the modern geometric view of solutions of differential equations as a flow in a 3-D vector space in contrast to the algebraic methods many learned prior to the 1980s.

One of the ways to produce a map is to look at the penetration of these flow trajectories through the planes $z = 0, 1, 2, \ldots$ (in Cartesian representation Figure 3-32) or through $z = 0$ plane in the cylindrical space representation. Shown in Figure 3-32 are three trajectories: One is a fixed point, one lies on the stable manifold of the saddle point of the map, and the other goes through the unstable manifold of the map.

Ueda and his co-workers not only were one of the first to observe chaotic solutions of the Duffing equation (using analog computers), but were one of the first to relate the tangling of the stable and unstable manifolds of stable points of stroboscopic maps to the formation of a

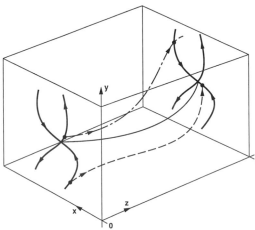

Figure 3-32 Three trajectories in the phase space of a periodically forced oscillator originating from stable and unstable manifolds and the saddle point of a Poincaré map.

strange attractor. An example of this is shown in Figure 3-33, which was obtained from analog and digital simulation of (3-8.7). Such maps, originating from differential equations, are not as easy to generate or analyze as those calculated directly from difference equations. Nevertheless, this example shows the importance of discrete time maps to the understanding of continuous time dynamics. It also shows the role of the unstable manifolds in Poincaré maps as organizing topologies for strange attractors.

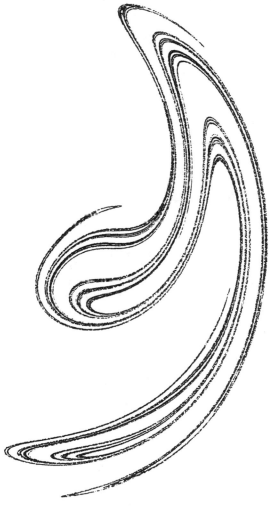

Figure 3-33   Strange attractors for periodically forced Duffing oscillator (3-8.8) for a circuit with a nonlinear inductor. [From Ueda (1979).]

A Map from a 4-D Conservative Flow

The following example illustrates two ideas relating to flows and maps. The first is the idea of chaos in lossless systems sometimes called *conservative* or *Hamiltonian* dynamics (see also Chapter 1). The second idea is how one can obtain a 2-D map from dynamics in a 4-D phase space. Another lesson from this example is that because there are no attractors in conservative systems, each initial condition results in a unique type of motion, namely, periodic, quasiperiodic, or stochastic (chaotic).

The example is illustrated in Figure 3-34, which shows a particle on a rotary base (Cusumano, 1990). The motion of the particle is confined to a vertical plane that is fixed to the rotating base. In Figure 3-34a the restoring force on the rotating particle is a linear spring, whereas in Figure 3-34b,c the restoring force is gravity. This problem introduces *inertial* nonlinearities, commonly known as *centripetal* and *Coriolis* accelerations which appear in Newton's laws of motion when written in polar, spherical, or path coordinates. Leaving the derivation of the equations of motion as an exercise, one can obtain two coupled second-order differential equations from a Lagrange's equation formulation of the problem with the spherical angles q_1, q_2 as generalized coordinates, nondimensionalized for unit mass and unit radius:

$$\ddot{q}_1 + \omega_1^2 q_1 - \tfrac{1}{2}(\dot{q}_2)^2 \sin 2q_1 = 0$$
$$[\varepsilon + \sin^2 q_1]\ddot{q}_2 + \varepsilon\omega_2^2 q_2 + \dot{q}_1\dot{q}_2 \sin 2q_1 = 0$$

(3-8.9)

where ε represents the ratio of base inertia to particle inertia $\varepsilon = J/ML^2$ and ω_1, ω_2 are the uncoupled natural frequencies of the oscillators when q_1, q_2 are small.

Ostensibly it would appear that the dynamics would naturally be described in a 4-D phase space. But, because energy is conserved, there is an added constraint which can be easily shown to be given by

$$[\varepsilon + \sin^2 q_1]\dot{q}_2^2 + \dot{q}_1^2 + \omega_1^2 q_1^2 + \varepsilon\omega_2^2 q_2^2 = 2E = \text{constant} \quad (3\text{-}8.10)$$

where E is the total kinetic and potential energies.

Using this expression, one can eliminate one of the four phase-space variables, say $\dot{q}_2$, to obtain a continuous time motion in a 3-D space $(q_1, \dot{q}_1, q_2)$. To obtain a 2-D map, one sets $q_2 = 0$, $\dot{q}_2 > 0$ to obtain a discrete time mapping in the plane (q_1, q_2). This procedure

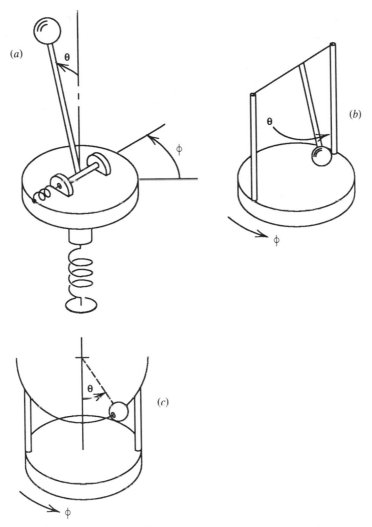

Figure 3-34   Three two-degree-of-freedom oscillators with inertial nonlinearities. [From Cusumano (1990).]

was carried out for the above equations by integrating them (in the form of first-order equations) and then saving (q_1, q_2) when $q_2 = 0$, $\dot{q}_2 > 0$. These calculations were part of a Ph.D. dissertation of J. Cusumano (1990) in his study of the out-of-plane chaotic dynamics of an elastic torsion beam (see Cusumano and Moon, 1990) in which this model was used as a simple two-mode approximation to the original partial differential equation.

The results of these calculations are shown in Figure 3-35 for increasing values of the energy E. Note that for a given energy there are many types of orbits, possibly depending on the initial conditions.

Of particular note are the motions near the origin. For small energy, there are quasiperiodic motions about the origin. However, for larger energy the origin becomes unstable and two stable out-of-plane quasi-periodic motions exist away from the origin. Finally, for higher energies the map shows a diffuse set of points which indicate stochastic or chaotic dynamics and which wander about the phase space in an

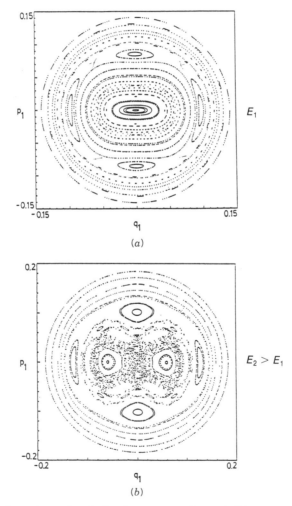

Figure 3-35 Poincaré map for the lossless system shown in Figure 3-34 for two values of the initial energy. [From Cusumano (1990).] (Note, $p_1 = \dot{q}_1$)

unpredictable way. This chaos, however, does not exhibit any fractal structure which is found in dissipative dynamical systems.

PROBLEMS

3-1   Investigate the properties of the cubic map (see Holmes, 1979)

$$x_{n+1} = y_n$$

$$y_{n+1} = -bx_i + dy_i - y_n^3$$

Find the fixed points and determine their stability as a function of the parameters b, d. Iterate this map for $b = 0.2$, $d = 2.5$, 2.65, 2.77. Can you find a strange attractor?

3-2   Find a 2-D linear transformation that takes a northwest square whose sides are parallel to the x, y-axes and transforms it into a diamond-shaped region.

3-3   Classify the fixed points of a 2-D map for the four cases of local transformation matrices. Sketch a few orbits for each case.

(a) $\begin{bmatrix} \frac{1}{2} & 1 \\ 0 & \frac{1}{3} \end{bmatrix}$ **(b)** $\begin{bmatrix} 0 & -2 \\ 2 & 0 \end{bmatrix}$

(c) $\begin{bmatrix} 1 & 1 \\ 1 & 1 \end{bmatrix}$ **(d)** $\begin{bmatrix} -1 & 1 \\ 1 & -1 \end{bmatrix}$

3-4   Find a transformation that converts the logistic map (3-6.2) into a linear mod 1 map. (*Hint:* Use a trigonometric identity.)

3-5   Show that the Henon map deforms a centered rectangle into a horseshoe-type region as shown in Figure 1-23. What happens to a northwest corner rectangle under one iteration of the map?

3-6   **(a)** Calculate the Jacobian determinant (3-1.4) for the cubic map in Problem 3-1. For what value of b will this map be area preserving?

   (b) Show that the 2-D map (3-1.3) is area-preserving.

3-7   For what values of Ω, κ/Γ will the circle map (3-5.7) have period-1 fixed points?

3-8 A first-order difference equation similar to the logistic map (3-6.2) is the sine map $x_{n+1} = \lambda \sin \pi x_n$, $0 \le x_n \le 1$. Show under what conditions this map has two fixed points in $0 < x < 1$. Also show that the stretching rate at the nonzero fixed point, x_1, is given by $|F'(x)| = \pi(\lambda^2 - x_1^2)^{1/2}$. Estimate the value of λ at which this map period doubles. [*Hint:* When does $|F'(x)| = 1$?]

3-9 Investigate the fixed points and stability as a function of the control parameter of the cubic map

$$x_{n+1} = \lambda x_n(1 - x_n^2)$$

3-10 Derive a 2-D map for a mass rattling between two walls with perfect impact. Assume the two walls move coherently with a small sinusoidal vibration.

3-11 Show that the standard map (3-1.11) has a saddle point on $v = 0$. Derive expressions for the two eigenvectors representing the unstable and stable manifolds. (See also Section 2.4 in Guckenheimer and Holmes, 1983.)

3-12 Calculate the value of the Rayleigh number in the Lorenz equations (3-8.3) at which the origin becomes unstable, and find the two fixed points away from the origin.

3-13 Using a computer, integrate the Lorenz equations (3-8.3) in the chaotic parameter regime and find a Poincaré map as the trajectories penetrate a plane containing the two unstable fixed points.

4

CHAOS IN PHYSICAL SYSTEMS

The world is what it is and I am what I am.... This out there and this in me, all this, everything, the resultant of inexplicable forces. A chaos whose order is beyond comprehension. Beyond human comprehension.
Henry Miller
Black Spring

4.1 NEW PARADIGMS IN DYNAMICS

In his book *The Structure of Scientific Revolutions,* Thomas Kuhn (1962) argues that major changes in science occur not so much when new theories are advanced but when the simple models with which scientists conceptualize a theory are changed. A conceptual model or problem that embodies the major features of a whole class of problems is called a *paradigm.* In vibrations, the spring–mass model represents such a paradigm. In nonlinear dynamics the motion of the pendulum and the three-body problem in celestial mechanics represent classical paradigms.

The theory that new models are precursors for major changes in scientific or mathematical thinking has no better example than the current revolution in nonlinear dynamics. Here the two principal paradigms are the Lorenz attractor [Eq. (1-3.9)] and the logistic equation [Eq. (1-3.6)]. Many of the features of chaotic dynamics are embodied in these two examples, such as divergent trajectories, subharmonic bifurcations, period doubling, Poincaré maps, and fractal dimensions. Just as the novitiate in linear vibrations had to master the subtleties of the spring–mass paradigm to understand vibrations of complex systems, so the budding nonlinear dynamicist of today must understand the phenomena embedded in the Lorenz and logistic models. Other lesser paradigms are also important in dynamical systems, including the forced motions of the Van der Pol equation (1-2.5), the Duffing oscillator models (1-2.4) of Ueda and Holmes (see below, this chapter), the two-dimensional map of Henon (1-3.8), and the circle map (3-5.7).

The implication of this minor revolution in physics means that in this new age of dynamics we will observe dynamical phenomena and conduct dynamics experiments in physical systems from a vastly new perspective. Old experiments will be viewed in a new light while new dynamical phenomena remain to be discovered.

To date, there have appeared many fine books on the subject of chaos. Most of these, however, focus almost entirely on the mathematical analysis of chaotic dynamics. In this chapter, we survey a variety of mathematical and physical models which exhibit chaotic vibrations. An attempt is made to describe the physical nature of the chaos in these examples, as well as to point out the similarities and differences between the physical problems and their more mathematical chaos paradigms mentioned above.

Early Observations of Chaotic Vibrations

Early scholars in the fields of electrical and mechanical vibrations rarely mention nonperiodic, sustained oscillations with the exception of problems relating to fluid turbulence. Yet chaotic motions have always existed. Experimentalists, however, were not trained to recognize them. Inspired by theoreticians, the engineer or scientist was taught to look for resonances and periodic vibrations in physical experiments and to label all other motions as "noise."

Joe Keller, a mathematician at Stanford University, has speculated on the reason for the apparent myopic vision of experimental scientists as regards chaotic phenomena in the last century. He notes that the completeness and beauty of linear differential equations led to its domination of the mathematical training of most scientists and engineers.

Examples of nonperiodic oscillations can be found in the literature, however. Three cases are cited here. First, Van der Pol and Van der Mark (1927), in a paper on oscillations of a vacuum tube circuit, make the following remark at the end of their paper: "Often an irregular noise is heard in the telephone receiver before the frequency jumps." No explanation is offered for these events; and in classical treatises on the Van der Pol oscillator, no further mention is made of "irregular noises."

One of the more interesting stories of observing chaotic oscillations before the age of chaos theory is told by Professor Yoshisuke Ueda of Kyoto University.[1] Professor Ueda was the student of a very famous professor of nonlinear electrical circuits, C. Hayashi, whose well-known treatise is referenced at the end of this book. Ueda tells of collecting data on an analog computer in November of 1961 when he accidently came upon a nonperiodic signal from the simulation of frequency entrainment on a second-order, nonlinear, periodically driven oscillator. At the time, they assumed that nonperiodic motions were always quasiperiodic. But, Ueda's analysis at the time showed that instead of a nice closed Poincaré section, characteristic of quasi-periodicity, he got a ragged picture, what he calls a "shattered egg." However, Ueda's attempts to get his famous professor to acknowledge his observations or publish them met with no success. He attempted to publish some of these results in 1971 in Japan when he was a new professor at Kyoto, but he again met with resistance. Not until 1978 did his famous "Japanese Attractor" paper get published in *Transactions of the Institution of Electrical Engineers in Japan* [reprinted in English in the *International Journal of Non-Linear Mechanics*, **20** (1980), 481–491].

The moral of this story is clear: Even in one of the most advanced laboratories in nonlinear circuits, chaotic dynamics were rejected because they did not fit in with the mathematical theories of the times. (As evidence of the advanced nature of the experiments at Kyoto, Professor Abe Hack invented an automatic Poincaré-map-generating circuit in 1966 to be used with the analog computer.)

In a third example, Tseng and Dugundji (1971) studied the nonlinear vibrations of a buckled beam. The beam was rigidly clamped at both ends and then compressed to buckling. This created an arched structure. When the beam was vibrated transverse to its length and the acceleration forces increased, snap-through occurred. In this regime, intermittent oscillations were observed as well as subharmonic responses. The analysis in the paper, however, only dealt with periodic vibrations.

Many readers may recall similar phenomena in scientific experiments that they have done or have seen in engineering practice. Cha-

[1] This lecture was given by Professor Ueda at the International Symposium "The Impact of Chaos on Science and Society," 15–17 April 1991, organized by the United Nations University and the University of Tokyo.

otic noise has always been around in engineering devices, such as static in old radios and chatter in loose-fitting gears, but until recently we had no models or mathematics to simulate or describe it.

4.2 PARTICLE AND RIGID BODY SYSTEMS

Multiple-Well Potential Problems

A system with a finite number of equilibrium states can often be described by a multiple-well potential energy function $V(\mathbf{x})$ where, for a particle with unit mass, the equation of motion takes the form

$$\ddot{\mathbf{x}} + \nabla V(\mathbf{x}) = \mathbf{F}(\mathbf{x}, \dot{\mathbf{x}}, t) \qquad (4\text{-}2.1)$$

where $\mathbf{x}$ represents the position of the particle in the configuration space, ∇V represents the gradient operator, and $\mathbf{F}(\mathbf{x}, \dot{\mathbf{x}}, t)$ represents additional forcing and dissipation forces. These systems are good candidates for chaotic vibrations because the unforced problem $\mathbf{F} = 0$ has one or more saddle points in the phase space which can

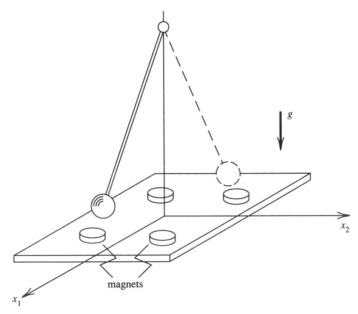

Figure 4-1 Sketch of a pendulum with a ferromagnetic end mass oscillating above four permanent magnets. Model for a two-degree-of-freedom four-well potential oscillator.

lead to horseshoes in the Poincaré map of the system. A sketch of a mechanical system with a four-well potential is shown in Figure 4-1 for a spherical pendulum under gravity with four permanent magnets underneath the bob.

Double-Well Potential Problems. The forced vibrations of a buckled beam were modeled using a Duffing-type equation by Holmes (1979), who showed in analog computer studies that chaotic vibrations were possible. The nondimensional equation derived by Holmes is

$$\ddot{x} + \gamma\dot{x} - \frac{1}{2}x(1 - x^2) = f_0\cos \omega t \qquad (4\text{-}2.2)$$

where x represents the lateral motion of the beam [Here a simple one-mode model is used to represent the beam as in Moon and Holmes (1979).] This equation can also model a particle in a two-well potential (Figure 1-2). This model has been used to study plasma oscillations (e.g., see Mahaffey, 1976). Chaotic solutions obtained from an analog computer are shown in Figure 4-2. An experimental realization of this model was discussed in Chapter 2. A Fourier spectrum based on solutions to this equation (Figure 2-7) shows a continuous spectrum of frequencies which is characteristic of chaotic motions. A Poincaré

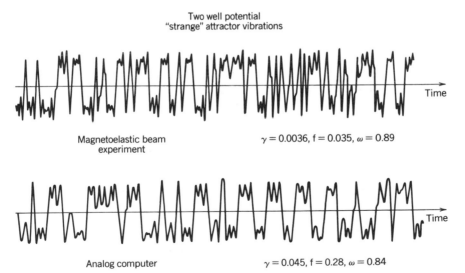

Figure 4-2 Chaotic vibrations of a periodically forced buckled beam: comparison of analog computer simulation and experimental measurements. [From Moon and Holmes (1979).]

map of the strange attractor is shown in Figure 4-3. Fractal dimensions for chaotic solutions are discussed in later chapters. Numerical studies of the double-well problem have also been published by Dowell and Pezeshki (1986), Moon and Li (1985a,b), and Ueda et al. (1986).

In a similar example, Clemens and Wauer (1981) have analyzed the snap-through oscillation of a one-hinged arch. Their equation takes the form

$$m\ddot{y} + \gamma\dot{y} + 2k\left(1 - \frac{1}{(b^2 + y^2)^{1/2}}\right) y = f_0\sin \omega t \qquad (4\text{-}2.3)$$

When only cubic nonlinearities are retained, this equation assumes the form (4-2.2) for the two-well potential Duffing oscillator. In another two-well problem, Shaw and Shaw (1989) has studied the forced vibration of an inverted pendulum with amplitude constraints.

Chaotic motions of an elastoplastic arch have been studied by Poddar et al. (1986). Once the length of the arch is longer than the distance between the pinned ends, there will be two equilibrium positions. Forced excitation can then result in unpredictable jumping from one arched position to another (see also Symonds and Yu, 1985).

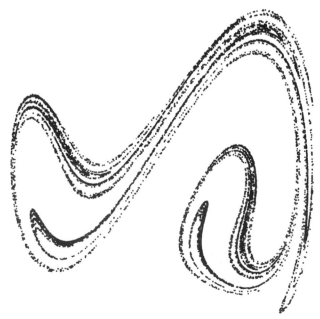

Figure 4-3   Poincaré map of chaotic solutions to the forced two-well potential oscillator; 15,000 points.

Three- and Four-Well Potential Problems. Chaotic dynamics of a particle in three-well and four-well potentials have been studied both experimentally and analytically in M.S. and Ph.D. dissertations of G.-X. Li (see Li and Moon, 1990a,b). A brief discussion of these problems is given in Chapters 6 and 7. A three-well problem can easily be created experimentally by placing three permanent magnets below a cantilevered beam (see Figure 2-2). A four-well potential can also be created by placing four magnets below a spherical pendulum. In this case, the number of degrees-of-freedom is two (Figure 4-1).

Chaotic Dynamics in the Solar System

The first firm test of the Newtonian model of the physical dynamical world was the correct prediction of the motions of the planets. And, for more than two centuries, students of physics have been taught the predictable nature of Newton's orbital dynamics. The time history of the planets in our solar system has been used to measure the history of our world. For over three decades, orbital dynamics has been used to predict with remarkable accuracy the motions of our rockets and satellites. Now, more than three centuries after the publication of the *Principia*, some are challenging the notion of absolute predictability in the motions of some of the objects in our solar system. How can this be?

The law of the gravitational force of attraction of Newton is inversely proportional to the square of the distance between masses, and thus it would appear to be strongly nonlinear. For two masses under mutual attraction, however, the problem can be reduced to a single mass problem moving around a fixed center. Furthermore, a change of variable transforms the nonlinear problem into the linear harmonic oscillator (e.g., see Goldstein, 1980). However, when three or more celestial bodies interact, then stochastic dynamics are possible.

Another departure from the classical Newton orbital problem is the effect of mass distribution. When the mass of the planet or moon has certain symmetrics, then one can reduce the problem to the interaction of point masses. However, for irregularly shaped objects, the angular displacements add complexity similar to that of spinning top dynamics.

In the following, a few examples of rigid body and orbital dynamics are described which may provide clues to chaos in the solar system. An example of chaotic fluid flow on Jupiter may be found in the work of Marcus (1988) as well as Meyers et al. (1989).

Chaotic Tumbling of Hyperion. The NASA mission of Voyager 2 transmitted pictures of an irregularly shaped satellite of Saturn called Hyperion. The pioneering work of J. Wisdom of M.I.T. showed how this nonsymmetric celestial object could exhibit chaotic tumbling in its elliptical orbit around Saturn.

It is well known that an elongated satellite such as a dumbell-shaped object orbiting in a circular orbit could exhibit oscillating planar rotary motions about an axis through the center of mass and normal to the plane of the orbit, with a period $1/\sqrt{2}$ smaller than the orbital period.

When the satellite is asymmetric with three different moments of inertia, $A < B < C$, Wisdom et al. (1984) show that the planar dynamics are described by

$$\frac{d^2\theta}{dt^2} + \frac{\omega_0^2}{2r^3} \sin 2(\theta - f) = 0 \qquad (4\text{-}2.4)$$

where time is normalized by the orbital period $T = 2\pi$ and where $r(t)$ and $f(t)$ are 2π periodic [e.g., see Thompson (1989a for a review of this work]. Here $\omega_0^2 = 3(B - A)/C$, r is the radius to the center of mass, and $\theta(t)$ measures the orientation of the long axes of the satellite. The term $f(t)$ is called the *true anomaly* of the orbit.

This equation is similar to that of a parametrically forced pendulum which has been found to exhibit chaotic dynamics. However complex the planar oscillations may be, Wisdom et al. (1984) show that these planar motions can become unstable with the possibility of three-dimensional tumbling of the satellite in its orbit around Saturn. Imagine living on such a world where the Saturn rise and set are unpredictable and where definitions of east and west, defined on Earth by the fixed axes of rotation, would be hard to determine by intuition.

Chaotic Orbits of Halley's Comet. The dynamics of a celestial object such as Halley's comet about the sun can be approximately described by a two-body problem which is fully integable. However, the motions of several planets, namely, Jupiter and Saturn, can exert perturbations on the orbit of Halley's comet when there is a close encounter with these planets. This can be seen as an analogy to a pendulum that receives a series of short time perturbations similar to that of the kicked rotor problem.

In a recent paper, Chirikov and Vecheslavov (1989) have used this technique of reducing the dynamics of Halley's comet under the influence of Jupiter to a two-dimensional iterated map.

The determination of possible chaotic dynamics requires a large number of observations. Given the limited number of orbital periods in our lifetime of many of the celestial objects in our solar system, the only tool that we have is to perform a simulation backwards and forwards in time. If one uses the differential equations of Newton for each of the relevant celestial bodies, the calculation using digital computers becomes extremely time-consuming. One solution is to construct an electronic Orrery or a dedicated analog or digital computer in which the equations are hard-wired into the electronics. Another solution is to replace the coupled ordinary differential equations with coupled iterated maps.

Sketching only the barest outline of the problem, x_n is chosen to represent the relative phase of the orbit of Jupiter at which Halley's comet reaches its perihelion. Between encounters the comet is assumed to have an energy proportional to ω_n. Using observations of Halley's comet, Chirikov and Vecheslavov (1989), derive a 2-D map of the form

$$\omega_{n+1} = \omega_n + F(x_n)$$
$$x_{n+1} = x_n + \omega_{n+1}^{-3/2}$$

(4-2.5)

where $F(x)$ is the saw-tooth function shown in Figure 4-4a. Iteration of this map for certain initial conditions leads to the stochastic orbit shown in Figure 4-4b.

Pendulum Problems

Forced Single-Degree-of-Freedom Pendulum. The classical pendulum has a restoring force or torque that is proportional to the sine of the angular displacement, θ (Figure 4-5). This implies that there are two equilibrium positions $\theta = 0, \pi$. In the case of zero forcing, there is both a center and a saddle point in the phase plane. As with multiple-potential well problems, saddles often are clues to the existence of horseshoe maps in the Poincaré section when periodic forcing is added to the problem. (See Figures 1-13, 3-5.)

Chaos in a Pendulum. The motion of a particle in both space-periodic and time-periodic force fields serves as a model for several problems in physics. These include the classical pendulum, a charged particle in a moving electric field, synchronous rotors, and Josephson junctions. For example, the equation for the nonlinear dynamics of a

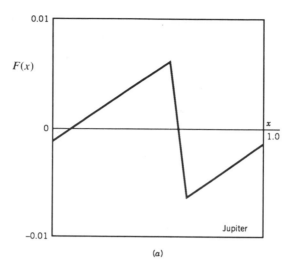

(a)

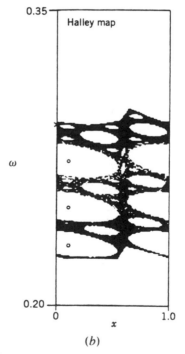

(b)

Figure 4-4 (a) Mapping function (4-2.5) for a model of Halley's comet perturbed by the orbit of Jupiter. (b) Iteration of the map (4-2.5). [From Chirikov and Vecheslavov (1989).]

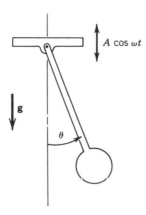

Figure 4-5   Parametrically forced pendulum.

particle in a traveling electric force field takes the form (e.g., see Zaslavsky and Chirikov, 1972)

$$\ddot{x} + \delta\dot{x} + \alpha \sin x - g(kx \quad \omega t) \qquad (4\text{-}2.6)$$

where $g(\ )$ is a periodic function. The study of the forced pendulum problem has revealed complex dynamics and chaotic vibrations (see Hockett and Holmes, 1985; Gwinn and Westervelt, 1985):

$$\ddot{x} + \delta\dot{x} + \alpha \sin x = f \cos \omega t \qquad (4\text{-}2.7)$$

Parametric oscillation is a term used to describe vibration of a system with time-periodic changes in one or more of the parameters of a system. For example, a simply supported elastic beam with small periodic axial compression is often modeled by a one-mode approximation which yields an equation of the form

$$\ddot{x} + \omega_0^2(1 + \beta \cos \Omega t)x = 0 \qquad (4\text{-}2.8)$$

This linear ordinary differential equation is the well-known Mathieu equation. It is known that for certain values of ω_0^2, β, and Ω the equation admits unstable oscillating solutions. When nonlinearities are added, these vibrations result in a limit cycle. A similar example is the pendulum with a vibrating pivot point (Figure 4-5). Chaotic vibrations for this problem have been studied numerically by Levin and Koch (1981) and McLaughlin (1981). The mathematical equation for this problem is similar to (4-2.8):

$$\ddot{\theta} + \beta\dot{\theta} + (1 + A\cos\Omega t)\sin\theta = 0 \qquad (4\text{-}2.9)$$

Period-doubling phenomena have been observed in numerical solutions, and a Feigenbaum number was calculated for the sixth subharmonic bifurcation of $\delta = 4.74$.

Chaotic motion of a double pendulum have been studied by Richter and Scholz (1984).

Spherical Pendulum. The complex dynamics of a spherical pendulum with two degrees of freedom has been examined by Miles (1984a), who found chaotic solutions for this problem in numerical experiments when the suspension point undergoes forced periodic motions. The equation of motion can be derived from a Lagrangian given by

$$L = \frac{1}{2}m\,(\dot{x}^2 + \dot{y}^2 + \dot{z}^2) - m\frac{g}{l}(l - z) \qquad (4\text{-}2.10)$$

where l is the length of the pendulum and the coordinates (x, y, z) satisfy the constraint equation

$$(x - x_0)^2 + y^2 + z^2 = l^2$$

The suspension point is $x_0 = \varepsilon l\cos\omega t$, and gravity acts in the z direction.

Miles (1984a) used a perturbation technique and transformed the resulting equation of motion using

$$\begin{aligned}
x &= [p_1(\tau)\cos\theta + q_1(\tau)\sin\theta]l\varepsilon^{1/3} \\
y &= [p_2(\tau)\cos\theta + q_2(\tau)\sin\theta]l\varepsilon^{1/3}
\end{aligned} \qquad (4\text{-}2.11)$$

where $\theta = \omega t$ and $\tau = \frac{1}{2}\varepsilon^{2/3}\omega t$. The resulting set of four first-order equations for (p_1, p_2, q_1, q_2) with small damping added (represented by α) is found to be

$$\frac{d}{dt}\begin{bmatrix} p_1 \\ p_2 \\ q_1 \\ q_2 \end{bmatrix} = \begin{bmatrix} -\alpha & -\beta & -\delta & 0 \\ \beta & -\alpha & 0 & -\delta \\ \delta & 0 & -\alpha & -\beta \\ 0 & \delta & \beta & -\alpha \end{bmatrix}\begin{bmatrix} p_1 \\ p_2 \\ q_1 \\ q_2 \end{bmatrix} + \begin{bmatrix} 0 \\ 1 \\ 0 \\ 1 \end{bmatrix} \qquad (4\text{-}2.12)$$

where α, β, and δ depend on the variables (p_1, p_2, q_1, q_2). The reader is referred to Miles (1984a) for the definitions of α, β, and δ. The divergence of this flow in the four-dimensional phase space is $\nabla \circ \mathbf{f} = -4\alpha$. Equilibrium points of the set of equations (4-2.12) correspond to either periodic planar or nonplanar motions. Numerical simulation of this set of equations shows a transition from closed orbit trajectories and discrete spectra to complex orbits and broad spectra characteristic of chaotic motions.

Experiments on a Magnetic Pendulum. The pendulum is a classical paradigm in dynamics. To find out if this paragon of deterministic dynamics can exhibit chaotic oscillations, the author and co-workers at Cornell University (Moon et al., 1987) constructed a magnetic dipole rotor with a restoring torque proportional to the sine of the angle between the dipole axis and a fixed magnetic field (Figure 4-6). A time-periodic restoring torque was provided by placing a sinusoidal voltage across two poles transverse to the steady magnetic field. The mathematical model for this forced magnetic pendulum becomes

$$J\ddot{\theta} + c\dot{\theta} + MB_s\sin\theta = MB_d\cos\theta\cos\Omega t \qquad (4\text{-}2.13)$$

where J is the rotational inertia of the rotor, c is a viscous damping constant, M is the magnetic dipole strength of the rotor dipole, and B_s and B_d are the intensities of the steady and dynamic magnetic fields, respectively. Figure 4-7 shows a comparison of periodic and chaotic

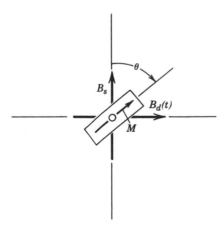

Figure 4-6 Sketch of a magnetic dipole rotor in crossed static and dynamic magnetic fields—a "magnetic pendulum."

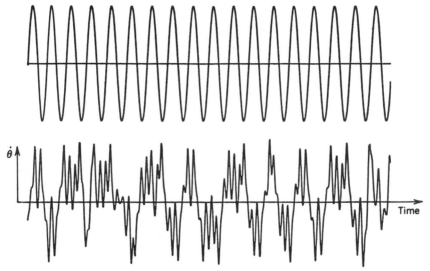

Figure 4-7 *Top:* Periodic motion of a magnetic rotor (Figure 4-6). *Bottom:* Chaotic motion of a magnetic rotor.

rotor speeds under periodic excitation. Additional discussion of this experiment may be found in Chapters 5 and 6.

Chaos theory has also been used to excite nonperiodic vibrations in a multiple-pendulum mobile sculpture by Viet et al. (1983). A brief discussion on chaos and sculptural assemblages of pendulums as in the work of Calder is given in Appendix C, "Chaotic Toys."

Rigid Body Problems

The gyroscopic rotational effects of a free-spinning rigid body such as precession and nutation are well known. Then under what circumstances can rigid bodies exhibit chaotic dynamics? To answer this we review the equations of motion known as *Euler's equations*. Here the components of the rotation vector $\boldsymbol{\omega} = (\omega_1, \omega_2, \omega_3)$ are written with respect to the principal inertial axes centered at the center of mass:

$$\frac{d}{dt} I_1\omega_1 = (I_2 - I_3)\omega_2\omega_3 + M_1$$

$$\frac{d}{dt} I_2\omega_2 = (I_3 - I_1)\omega_1\omega_3 + M_2 \qquad (4\text{-}2.14)$$

$$\frac{d}{dt} I_3\omega_3 = (I_1 - I_2)\omega_1\omega_2 + M_3$$

where $[I_1, I_2, I_3]$ are the principal inertias and (M_1, M_2, M_3) are applied moments. In general, the rotational motion could be coupled to the translational motion, or applied forces could be coupled into the applied moments. We focus here on the force-free case in which the center of mass is stationary. In this case, when the applied moments are zero, the motion is integrable. That is, one can write down the motion in terms of elliptic functions (e.g., see Goldstein, 1980). However, there are several cases where freely rotating rigid bodies can exhibit chaotic behavior.

The first case is when one of the moments M_i varies periodically in time. The second case is where one has parametric excitation through time-periodic changes in the principal inertias, for example, $I_2 = I_0 + B \cos \Omega t$. The third case is where the applied moments are coupled through some feedback mechanism to the rotation velocities, that is,

$$\mathbf{M} = \mathbf{A} \cdot \boldsymbol{\omega}$$

This case has been studied by Leipnik and Newton (1981). Under appropriate choice of constants, they obtained a double strange attractor, each with its own basin of attraction. The choice of constants used by Leipnik and Newton was $\mathbf{I} = [3I_0, 2I_0, I_0]$ and

$$\mathbf{A} = \begin{bmatrix} -1.2 & 0 & -\sqrt{6}/2 \\ 0 & 0.35 & 0 \\ \sqrt{6} & 0 & -0.4 \end{bmatrix}$$

The chaotic dynamics can easily be observed in a three-dimensional phase space.

The general equations (4-2.14) are identical to the equations for the bending of a thin elastic rod or tape. This analogy, discovered by Kirchhoff in 1831, is exploited in Chapter 8 to discuss spatially chaotic bending of a thin elastic tape (Davies and Moon, 1992).

Ship Capsize and Nonlinear Dynamics. Ships and submarines constitute one class of rigid body dynamics under the influence of gravity and hydrodynamic forces. Two groups have done considerable research on ship dynamics using modern methods of nonlinear analysis; J. M. T. Thompson and co-workers at University College, London and A. H. Nayfeh and co-workers at Virginia Polytechnic Institute in Blacksburg, Virginia. In spite of such research, current design criteria for ship stability in naval architecture is largely empirical (Thompson et al., 1990). The simplest models involve the assumption of one-degree-of-freedom rolling subject to a periodic overturning moment in lateral ocean waves sometimes called *regular beam seas* (Figure 4-8):

$$I\ddot{\theta} + B(\dot{\theta}) + C(\theta) = D \sin \omega t \tag{4-2.15}$$

where θ is the angle of roll and I is the moment of inertia. The damping $B(\dot{\theta})$ is usually nonlinear as is the overturning moment $C(\theta)$. The London group has done a lot of analysis on a ship in high winds in which $C(\theta)$ is derived from a one-well potential function (for a review see Thompson et al., 1990; also see Virgin, 1986):

$$C(\theta) = \beta_1\theta - \beta_2\theta^2 \tag{4-2.16}$$

The Virginia Polytechnic Institute group has published papers on the symmetric ship problem using an equation of the form

$$\ddot{\theta} + \omega^2\theta + \alpha_3\theta^3 + 2\mu\dot{\theta} + \mu_3\dot{\theta}^3 = F_0\cos \Omega t \tag{4-2.17}$$

This group has also studied ship rolling oscillations excited by heave-roll coupling with a parametric forcing term proportional to the roll angle: $\theta \cos \Omega t$ (Sanchez and Nayfeh, 1990).

Thompson's work on dynamic stability for ships in a one-well potential has led to safety criteria based on ideas about fractals and basins

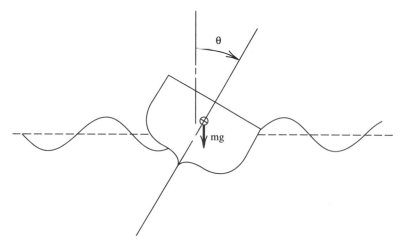

Figure 4-8   Sketch of a model for ship dynamics subject to wind and sea wave forces.

of attraction which are discussed in Chapter 7 (see also Thompson, 1989b and Thompson et al., 1990).

Impact Oscillators

Impact-type problems result in explicit difference equations or maps which often yield chaotic vibration under certain parameter conditions (see also §3.1). A classic impact-type map is the motion of a particle between two walls. When one wall is stationary and the other is oscillatory (Figure 4-9a), the problem is called the Fermi model for cosmic ray acceleration involving charged particles and moving magnetic fields. This model is discussed in great detail by Lichtenberg and Lieberman (1983) in their readable monograph on stochastic motion. Several sets of difference equations have been studied for this model. In one model, the moving wall imparts momentum changes without change of position. The resulting equations are given by

$$v_{n+1} = |v_n + V_0 \sin \omega t_n|$$
$$t_{n+1} = t_n + \frac{2\Delta}{v_{n+1}}$$

$$(4\text{-}2.18)$$

where v_n is the velocity after impact, t_n is the time of impact, V_0 is the maximum momentum per unit mass that the wall can impart, and Δ is the gap between the two walls.

Numerical studies of this and similar equations reveal that stochastic-type solutions exist in which thousands of iterations of the map (4-2.18) fill up regions of the *phase space* (v_n, t_n) as illustrated in Figure 4-9b. In some cases, the trajectory does not penetrate certain "islands" in the (v_n, t_n) plane. In these islands more regular orbits occur. This system can often be analyzed using classical Hamiltonian dynamics. This system is typical of chaos in low- or zero-dissipation problems. In moderate-to-high dissipation, the chaotic Poincaré map becomes localized in a structure with fractal properties as in Figure 3-11. But in low dissipation problems, the Poincaré map fills up large areas of the phase plane with no apparent fractal structure.

The Fermi accelerator model is also similar to one in mechanical devices in which there exists play, as illustrated in Figure 4-10. A mass slides freely on a shaft with viscous damping until it hits stiff springs on either side (see Shaw and Holmes, 1983 and Shaw, 1985). Another mathematical model which is closer to the physics is the bouncing ball on a vibrating surface shown in Figure 4-11. This problem has been

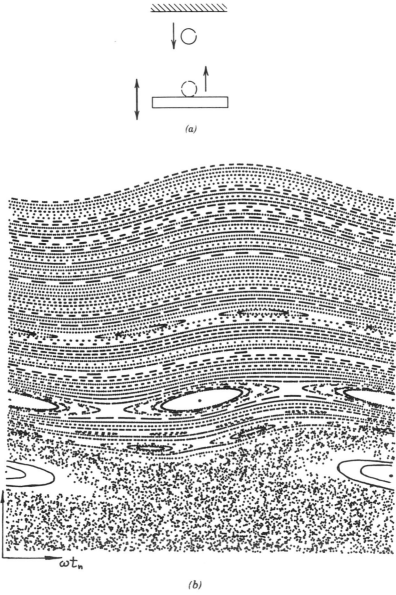

Figure 4-9 (a) Particle impact dynamics model with a periodically vibrating wall. (b) Poincaré map v_n versus ωt_n (mod π) for the impact problem in (a) using Eqs. (4-2.18).

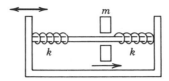

Figure 4-10 Experimental model of mass with a deadband in the restoring force.

studied by Holmes (1982). Using an energy loss assumption for each impact, one can show that the following difference equations result:

$$\phi_{j+1} = \phi_j + v_j$$

$$v_{j+1} = \alpha v_j - \gamma \cos(\phi_j + v_j) \tag{4-2.19}$$

Here ϕ represents a nondimensional impact time, and v represents the velocity after impact. As shown in Figure 4-11a, a steady sinusoidal motion of the table can result in a nonperiodic motion of the ball. A fractal-looking chaotic orbit for this map is shown in Figure 4-11b. This model suffers from the problem of admitting negative velocities at impact. This problem was addressed in a paper by Bapat et al. (1986).

Experiments on the chaotic bouncing ball have been performed by Tufillaro and Albano (1986). Other studies of impact or *bilinear* oscillator problems have been done by Thompson and Ghaffari (1982), Thompson (1983), Isomaki et al. (1985), and Li et al. (1990).

Impact Print Hammer. Impact-type problems have emerged as an obvious class of mechanical examples of chaotic vibrations. The bouncing ball (4-2.19), the Fermi accelerator model (4-2.18), and a beam with nonlinear boundary conditions all fall into this category. A practical realization of impact-induced chaotic vibrations is the impact print hammer experiment studied by Hendriks (1983) (Figure 4-12). In this printing device, a hammer head is accelerated by a magnetic force and the kinetic energy is absorbed in pushing ink from a ribbon onto paper. Hendriks uses an empirical law for the impact force versus relative displacement after impact; u is equal to the ratio of displacement to ribbon-paper thickness:

$$F = \begin{cases} -AE_p u^{2.7}, & \dot{u} > 0 \\ -AE_p \beta u^{11}, & \dot{u} < 0 \end{cases} \tag{4-2.20}$$

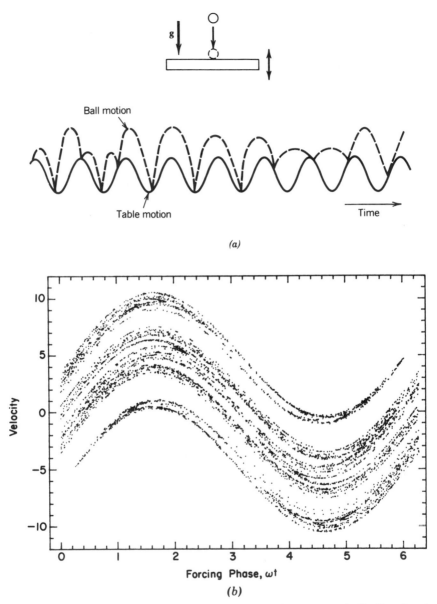

(a)

(b)

Figure 4-11   (a) Chaotic time history of a ball bouncing on a periodically vibrating table. [From Holmes (1982).] (b) Poincaré map of (a); impact velocity versus impact time (mod 2π).

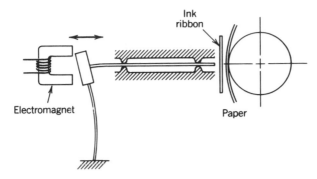

Figure 4-12 Sketch of pin-actuator for a printer mechanism.

where A is the area of hammer-ribbon contact, E_p acts like a ribbon-paper stiffness, and β is a constant that depends on the maximum displacement. The point to be made is that this force is extremely nonlinear.

When the print hammer is excited by a periodic voltage, it will respond periodically as long as the frequency is low. But as the frequency is increased, the hammer has little time to damp or settle out and the impact history becomes chaotic (see Figure 4-13). Thus,

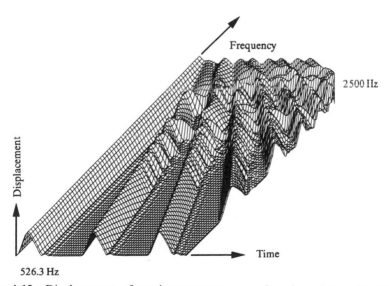

Figure 4-13 Displacement of a printer actuator as a function of time for different input frequencies showing loss of predictable output. [From Hendriks (1983), copyright 1983 by International Business Machines Corporation, reprinted with permission.]

chaotic vibrations restrict the speed at which the printer can work. One potential solution which is under study is the addition of control forces to suppress this chaos. This idea has been explored in the work of Tung and Shaw (1988).

Chaos in Gears and Kinematic Mechanisms

Kinematic mechanisms are generally input–output devices that convert one form of motion into another. For example, a gear transmission converts a rotary input motion into another rotary motion at a different frequency. Or, a slider crank mechanism converts translation into rotary motion. This mechanism is the heart of tens of millions of automobile engines. These devices are called kinematic because, in the ideal mechanism, the relation between input and output depends only on geometry or kinematic relationships; that is, inertia does not determine the mechanical gain. However, in real mechanical devices, linear and nonlinear departures from the ideal mechanism such as elastic members, gaps, play, and friction bring inertial effects into the dynamic behavior of the mechanism. For example, gear transmissions work fine when under load, but they often lead to rattling vibrations when the load becomes small when there is small play in the gears or bearings. The understanding of machine noise in mechanical systems has been a neglected subject (e.g., see Moon and Broschart, 1991). Such noise sometimes leads to fatigue and other material damage as well as creates unwanted acoustic or hydrodynamic noise as in submarines. The modern developments in nonlinear dynamics have given new tools to attack this hitherto unsolved problem.

   A number of papers have appeared which treat the possibility of nonlinear and chaotic vibrations in kinematic mechanisms, including gears [Pfeiffer (1988), Karagiannis and Pfeiffer (1991), Singh et al. (1989)], slider crank mechanism and four-bar and robotic mechanisms (e.g., see Beletzky, 1990).

Gear Rattling Chaos. Two spur gears with diameters d_1, d_2, which ideal geometrics, have a frequency or speed ratio ω_1/ω_2 equal to d_2/d_1. This speed ratio is affected by the meshing of teeth on each gear. However, when elasticity effects in the teeth (or gaps between the teeth) are present, this ideally kinematic problem becomes a dynamic one. Consider, for example, the two gears shown in Figure 4-14 in which a gap of ε exists between the circumferential distance between tooth contacts and the actual width of the tooth. Suppose we assume that the motion of one gear is given, while the motion of the

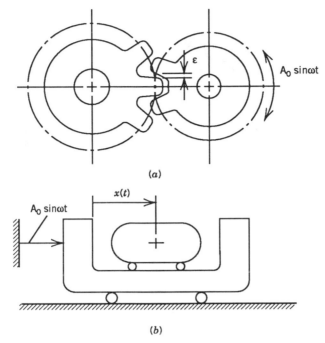

Figure 4-14 (*a*) Sketch of two enmeshed gears with a excessive play Δ. (*b*) Analogous problem of a mass moving between two vibrating constraints.

other is governed by the dynamics. For example, we could imagine the left-hand gear rotating with a small oscillatory motion while the right-hand gear tooth exhibits complex dynamic impacts between the two drive gear teeth. This problem is not unlike the Fermi map problem in (4-2.18) (see Figure 4-14*b*).

A study of this problem and its extension to more complex gear transmission systems has been given by Pfeiffer and co-workers at the Technical University of Munich [e.g., see Pfeiffer (1988), Karagiannis and Pfeiffer (1991)]. This relation between the gear rattling problem and the Fermi map has been studied by Pfeiffer and Kunert (1989). Also, in the United States the gear laboratory of R. Singh at the Ohio State University has looked at various nonlinear vibrations of gear systems including chaotic dynamics (e.g., see Comparin and Singh, 1990).

The Munich group, however, has pioneered in the application of Poincaré map techniques for predicting noise in automative and other gear transmission systems. A typical Poincaré map from two meshed gears with a small gap and periodic excitation on one gear is shown in Figure 4-15. The Munich group has also tried to predict the probability distribution function for the chaotic noise using the Fokker–Planck

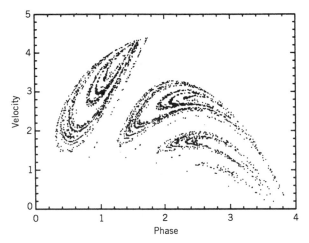

Figure 4-15 Poincaré map for two gears with play where one gear is excited by periodic oscillations. [From Li et al. (1990).]

equation (see Kunert and Pfeiffer, 1991) and to show how different arrangements of gears could possibly reduce gear noise.

Control System Chaos

Imagine a mechanical device with a nonlinear restoring force and suppose a control force is added to move the system from one position to another according to some prescribed reference signal $x_r(t)$. Such a system can be modeled by the following third-order system:

$$m\ddot{x} + \delta\dot{x} + F(x) = -z$$
$$\dot{z} + \alpha z = \Gamma_1[x - x_r(t)] + \Gamma_2\dot{x}$$

(4-2.21)

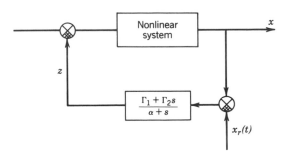

Figure 4-16 Feedback control system: nonlinear plant with linear feedback control.

Here z represents a feedback force, and Γ_1 and Γ_2 represent position and velocity feedback gains, respectively. This system of equations can be represented by the block diagram in Figure 4-16 with a nonlinear mechanical plant and a linear feedback law.

Two types of chaotic vibrations problems can be explored here. First, if the system is autonomous [i.e., the reference signal is zero—$x_r(t) = 0$], one could explore the gain space (Γ_1, Γ_2) for regions of steady, periodic, and chaotic vibrations. The second problem arises if $x_r(t)$ is periodic. That is, we wish to move the mass through a given path over and over again as in some manufacturing robotic device.

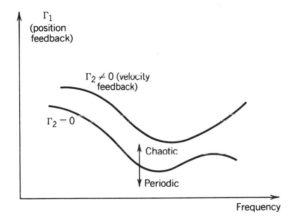

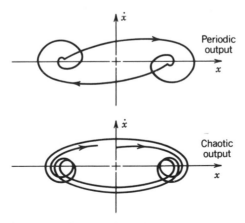

Figure 4-17 *Top:* Chaos boundary as a function of feedback gain and input command frequency. *Bottom:* Trajectories of periodic and chaotic dynamics for a mass with feedback control and nonlinear restoring force with a deadband region (see Figure 4-10). (See Golnaraghi and Moon, 1991.)

One could then explore the parameters of frequency and gain for which the system is periodic or chaotic as in Figure 4-17.

Chaotic vibrations for an autonomous system of the form (4-2.21) were studied by Holmes and Moon (1983) as well as by Holmes (1984). For example, when $F(x) = x(x^2 - 1)(x^2 - B)$, the mechanical system has three stable equilibria. This system has been shown to exhibit both periodic limit cycle oscillation and chaotic motion.

The problem of a forced feedback system has been studied by Golnaraghi and Moon (1991). Also Sparrow (1981) looked into chaotic oscillations in a system with a piecewise linear feedback function. Many other examples of chaotic control systems have since appeared in the literature. See also Baillieul et al. (1980). A discussion of using the chaotic nature of a strange attractor to control the dynamics of a system is presented in Section 4.9.

4.3   CHAOS IN ELASTIC SYSTEMS

Chaos in Elastic Continua

Many experiments on chaotic vibrations in elastic beams have been carried out by the author and co-workers [e.g., see Moon and Holmes (1979, 1985), Moon (1980a,b, 1984b), Moon and Shaw (1983), and Cusumano and Moon (1990)]. Two types of problems have been investigated. In the first problem, the partial differential equation of motion for the beam is essentially linear, but the body forces or boundary conditions are nonlinear. In the second problem, the motions are sufficiently large enough that significant nonlinear terms enter the equations of motion.

The planar equation of motion of an elastic beam with small slopes and deflections is governed by an equation of the form

$$D\frac{\partial^4 v}{\partial x^4} + m\frac{\partial^2 v}{\partial t^2} = f\left(v, \frac{dv}{dt}, t\right) \tag{4-3.1}$$

where v is the transverse displacement of the beam, D represents an elastic stiffness, and m is the mass per unit length. The right-hand term represents the effects of distributed body forces or internal damping. In many of the experiments at Cornell University, we used permanent magnets to create nonlinear body force terms. We also use flow-induced forces to produce self-excited oscillation of elastic beams (see Section 4.4).

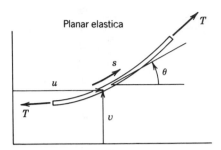

Figure 4-18 Planar deformation of an elastic rod.

When the displacement and slope of the beam centerline are large, we use variables (u, v, θ) to characterize the horizontal and vertical displacements and the slope which are related by (see Figure 4-18)

$$(1 + u')^2 + (v')^2 = 1, \qquad \tan \theta = \frac{v'}{1 + u'} \qquad (4\text{-}3.2)$$

where $(\ )' = \partial/\partial s$ and s is the length along the deformed beam. The balance of momentum equations then take the form (see Moon and Holmes, 1979)

$$m\ddot{v} = f_v - G'$$
$$m\ddot{u} = f_u + H' \qquad (4\text{-}3.3)$$

where

$$G = D\theta''(1 + u') - Tv'$$
$$H = D\theta''v' + T(1 + u')$$

In these equations, (f_u, f_v) represent body force components, while T represents the axial force in the rod. The nonlinearities in these equations are distinguished from those in fluid mechanics by the fact that no convective or kinematic nonlinearities enter the problem. Also, the local stress–strain relations are linear. The nonlinear terms arise from the change in geometric shape and are known as *geometric nonlinearities*. [See Love (1922) for a discussion of nonlinear rod theory. See also Chapter 8.]

Elastic Beam with Nonlinear Boundary Conditions. Multiple equilibrium positions are not needed in a mechanical system to get chaotic

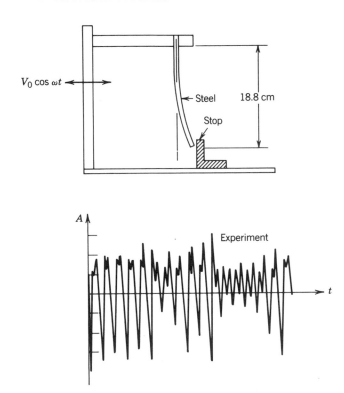

Figure 4-19   Chaotic vibrations of an elastic beam with a nonlinear boundary condition.

vibrations. Any strong nonlinearity will likely produce chaotic noise with periodic inputs. One example of a system with one equilibrium position is an elastic beam with nonlinear boundary conditions (see Moon and Shaw, 1983). Nonlinear boundary conditions are those that depend on the motion. For example, suppose the end is free for one direction of motion and is pinned for the other direction of motion. The chaotic time history of this beam is shown in Figure 4-19. Another variation of this problem is a two-sided constraint with play which gives three different linear regimes for the bending of the beam. Experiments in our laboratory also show chaos for this nonlinear boundary condition. Shaw (1985) has performed an analysis of these mechanical oscillations when play or a dead zone is present. Flow-induced chaotic vibrations have also been observed in a cantilevered pipe with nonlinear end constraints (see Section 4.4).

Magnetoelastic Buckled Beam. In this example, an elastic cantilevered beam is buckled by placing magnets near the free end of the beam [see Chapters 2 and 4 as well as Moon and Holmes (1979) and Moon (1980a,b; 1984b)]. The magnetic forces destabilize the straight unbent position and create multiple equilibrium positions as shown in Figure 4-20. In experiments, we have created up to four stable equilibrium positions with four magnets. In the postbuckled state, the system represents a particle in a two (or more)-well potential (Figure 1-2b). The whole system is placed on a vibration shaker and oscillates with constant amplitude and frequency. For small oscillations, the beam vibrations occur about one of the equilibrium positions. As the amplitude is increased, however, the beam can jump out of the potential well and chaotic motions can occur, with the beam jumping from one well to another (Figure 4-2). A Poincaré map of this phenomenon is shown in Figure 4-21. (We call this map the *Fleur de Poincaré.*)

The equation used to model this system is a modal approximation to the beam equation (4-3.3) with nonlinear magnetic forces acting at the tip.

A one-mode approximation for a damped beam with a free end gives good results. This equation can be rewritten as three first-order equations. Note that here the x variable refers to nondimensional modal amplitude and *not* to the distance along the beam.

$$\dot{x} = y$$
$$\dot{y} - -\gamma\dot{x} + \tfrac{1}{2}x(1 - x^2) - A_0\omega^2\cos z \qquad (4\text{-}3.4)$$
$$\dot{z} = \beta$$

This problem is analogous to a particle in a double-well potential $V = -(x^2 - x^4/2)/4$. This experiment is discussed throughout this book.

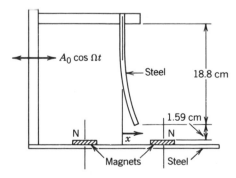

Figure 4-20 Steel elastic beam on a periodically moving support that is buckled by magnetic body forces.

Figure 4-21   Experimental Poincaré map of chaotic motion of the magnetically buck-led beam, Fleur de Poincaré.

The Poincaré section (Figure 4-21) has the character of two-dimensional point mappings. The experiments *do not* always exhibit period doubling before the motion became chaotic. Odd subharmonics were often a precursor to chaos. (Note: A description of the experimental apparatus may be found in Appendix C, "Chaotic Toys.")

Another variation of this experiment is an inverted pendulum with an elastic spring reported in the People's Republic of China by Zhu (1983) from Beijing University. For a weak spring, the inverted pendulum has two stable equilibria similar to the two-well potential problem (see also Shaw and Shaw, 1989).

Three-Dimensional Elastica and Strings

Under certain conditions, the forced planar motion of the nonlinear elastica described by (4-3.3) becomes unstable and three-dimensional motions result. Similar phenomena are known for the planar motion of a stretched string (Miles, 1984b). At Cornell University, we have performed several experiments with very thin flexible steel elastica with rectangular cross section (e.g., 0.25 mm × 10 mm × 20 cm long) known as "Feeler" gauge steel strips (Figure 4-22a). For these beams, small motions in the stiff or lateral direction of the unbent beam are nearly impossible without buckling or twisting of the local cross sections. However, when there is significant bending in the weak direction, lateral displacements are possible accompanied by twisting of the local cross sections. We have shown that planar vibrations of the beam in the weak direction near one of the natural frequencies not only become unstable but can exhibit chaotic motions as well. This is demonstrated in Figure 4-22b, where power spectra (fast Fourier transform; see Chapter 5) show a broad spectrum of frequencies when the driving input has a single-frequency input. Similar phenomena are observed for very thin sheets of paper. In fact, we have shown that chaotic motions of very thin sheets of paper generate a broad spectrum of acoustic noise in the surrounding air. This work is described in the doctoral dissertation of Cusumano (1990). See also Section 7.4 for a discussion of the calculation of fractal dimensions for chaotic attractors in these experiments.

Chaotic ballooning motions of a periodically excited string under tension have been studied both analytically and experimentally by O'Reilly (1991).

Two-Degree-of-Freedom Buckled Beam. To explore the effects of added degrees of freedom, we built an elastic version of a spherical

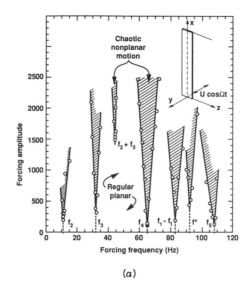

(a)

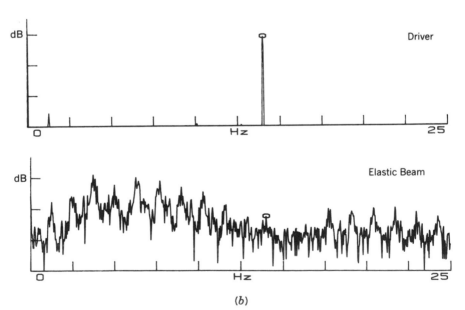

(b)

Figure 4-22 (a) Regions of chaos for a periodically forced thin elastica. (b) Fourier spectra for forced vibrations of a thin elastic beam. Broad-spectrum chaos is the result of out-of-plane vibration. [From Cusumano and Moon (1990).]

pendulum (Figure 4-1) where a beam with circular cross section was used (see Moon, 1980b). Again magnets were used to buckle the beam, but the tip was free to move in two directions. This introduced two incommensurate natural frequencies, and quasiperiodic vibrations occurred which eventually became chaotic (Figure 4-23).

This experimental system can be modeled by equations for two coupled oscillators as given by

$$\ddot{x} + \gamma\dot{x} - \tfrac{1}{2}x(1 - x^2) + \beta xy^2 = f_2 \tag{4-3.5a}$$

$$\ddot{y} + \delta\dot{y} + a(1 + \varepsilon y^2)y + \beta x_2 y = f_0 + f_1\cos \omega t \tag{4-3.5b}$$

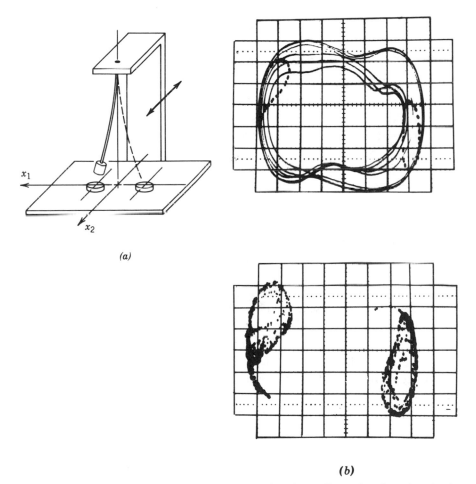

(a)

(b)

Figure 4-23 (*a*) Sketch of an elastic rod undergoing three-dimensional motions in the neighborhood of a double-well potential created by two magnets. (*b*) *Top:* Simultaneous time trace of phase plane motion and Poincaré map of quasiperiodic motion. *Bottom:* Poincaré map of chaotic motion.

The terms f_0 and f_2 account for gravity if the beam is not initially parallel with the earth's gravitational field, and the coupling terms are conservative. If the coupling is small, one can solve for $y(t)$ from Eq. (4-3.5b) and the equation for $x(t)$ looks like a parametric oscillator.

Miles (1984b) has performed numerical experiments on two quadratically coupled, damped oscillators and has found regions of chaotic motions resulting from sinusoidal forcing. He examined the special case when the two linear natural frequencies ω_1 and ω_2 were related by $\omega_2 \simeq 2\omega$.

4.4 FLOW-INDUCED CHAOS IN MECHANICAL SYSTEMS

Flow-Induced Elastic Vibrations

Flow in a Pipe: Fire Hose Chaos. There are many classes of flow-induced vibrations: flow inside flexible bodies such as pipes or rocket fuel tanks, flow around bodies such as wings, or heat exchange tubes and flow over one surface of a body such as over a panel of an aircraft or a rocket. One of the most studied problems has been the steady flow of fluid through flexible tubes or pipes (see Païdoussis, 1980; also see Chen, 1983). This problem has interest not only because of the nonconservative nature of the fluid forces, but also because of the relevance of the problem to flow-induced vibrations in heat exchange systems. This problem has recently received attention using modern methods of nonlinear analysis and experimentation. Although some of the early classical work goes back to dynamicists in the Soviet Union, research using modern methods has been centered in Europe (Steindl and Troger, 1988) and North America (Bajaj and Sethna, 1984; Sethna and Shaw, 1987; Païdoussis and Moon, 1988; Copeland and Moon, 1992; and Tang and Dowell, 1988).

In one study shown in Figure 4-24, fluid flows with constant velocity out of a cantilevered flexible pipe. At a critical flow speed, small limit cycle vibrations appear. The nonlinearity in this problem consists of amplitude constraints near the end of the pipe. When the limit cycle oscillations grow to where the pipe hits the constraints, chaotic vibrations appear (Païdoussis and Moon, 1988). This problem has been modeled as two coupled autonomous nonlinear oscillators, with the dynamics living in a four-dimensional phase space. (See Figure 5-24.)

Aeroelastic Panel Flutter. An example of chaos in autonomous mechanical systems is the flutter resulting from fluid flow over an elastic

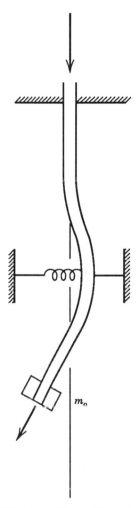

Figure 4-24 Sketch of a flexible tube with nonlinear boundary conditions carrying a steady flow of fluid.

plate. This problem is known as *panel flutter*, and readers are referred to two books by Dowell (1975, 1988) for more discussion of the mechanics of this problem. Panel flutter occurred on the outer skin of the early Saturn rocket flights that put men on the Moon in the early 1970s. Dowell and co-workers have done extensive numerical simulation of panel flutter. In early work, Kobayashi (1962) and Fung (1958) had observed nonperiodic motions in their analyses. In one set of problems, they looked at the combined effects of in-plane compression in the plate and fluid flow. More recent numerical results are given in Figure 4-25, showing stable phase plane trajectories for one set of fluid

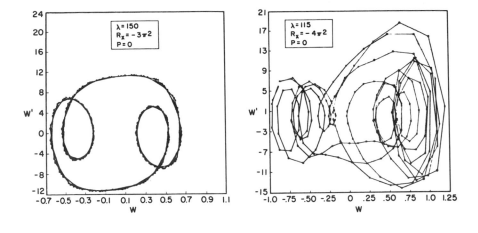

Figure 4-25 Flow over a buckled elastic plate. *Top left:* Periodic aeroelastic vibrations. *Top right:* Chaotic vibrations of the plate. [From Dowell (1982).]

velocity and compressive load conditions and chaotic vibrations for another set of conditions (see also Dowell, 1982, 1984). This example also illustrates a different type of Poincaré map. Because there is no intrinsic time, one must choose a hyperplane in phase space and look at points where the trajectory penetrates that plane. Dowell has done this for thc panel flutter problem and has shown strange attractor-type Poincaré maps.

Supersonic Panel Flutter. An analytic-analog computer study that uncovered chaotic vibrations and predates the Lorenz paper by one year is that of Kobayashi (1962). He analyzed the vibrations of a buckled plate with supersonic flow on one side of the plate. Kobayashi expanded the deflection of the simply supported plate in a Fourier series and studied the coupled motion of the first two modes. Denoting the nondimensional modal amplitudes of these two modes by x and y, the equations studied using an analog computer were of the form

$$\ddot{x} + \delta\dot{x} + [1 - q + x^2 + 4y^2]x - Qy = 0$$
$$\ddot{y} + \delta\dot{y} + 4[4 - q + x^2 + 4y^2]y + Qx = 0$$

$$(4\text{-}4.1)$$

where q is a measure of the in-plane compressive stress in the plate (which can exceed the buckling value) and Q is proportional to the dynamic fluid pressure of the supersonic flow upstream of the plate. In his abstract of this 1962 paper, Kobayashi states, "Moreover the following remarkable results are obtained. (i) In some unstable region of a moderately buckled plate, only an *irregular vibration* is observed" [italics added]. He also refers to earlier experimental studies in 1957 at the NACA in the United States which was the pre-Sputnik ancestor of NASA (see also Fung, 1958).

4.5 INELASTIC AND GEOMECHANICAL SYSTEMS

Nonlinear Dynamics of Friction Oscillators

Early models of chaotic dynamics, as presented in the first edition of this book, included mainly polynomial or trigonometric nonlinearities. As the field matures, more realistic physical nonlinearities are being studied. One of these is the dynamics of vibrating systems with dry friction (Popp and Steltzer, 1990; Feeny and Moon, 1989, 1992). The study of friction has a long history (e.g., see Den Hartog, 1940; Stoker, 1950), and we cannot begin to mention all the literature on the subject based on classical dynamical methods. The skidding of a car on a dry pavement, the screeching of chalk on a blackboard, and other experiences of technical devices with friction have always suggested that more complex dynamics are involved than simple periodic or steady motions. To date there is still debate between materials scientists and mechanicians about the nature of the friction force between two solid objects. We cannot resolve them here. What we can say is that certain classical models can lead to chaotic dynamics and that the global character of this chaotic attraction in phase space is similar to that measured in experiments.

In the classical friction problem, the friction force in the direction tangential to the surface depends on the force that is applied normal to the surface as shown in Figure 4-26. The equation of motion for a harmonically forced friction oscillator is given by

$$\ddot{x} + 2\delta\dot{x} + x + \eta(x)f(\dot{x}) = A\cos\Omega t$$

Here the time and distance are normalized by the mass and linear spring constant. We also allow the normal force effect $\eta(x)$ to depend on the motion. In many models the tangential friction force is written as a function of velocity. In a study by Feeny and Moon (1992),

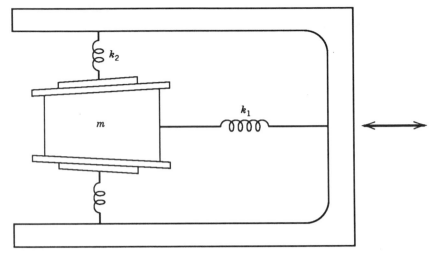

Figure 4-26   Sketch of a linear oscillator with nonlinear friction force.

two models were explored: the classical Coulomb friction law with a discontinuity at $\dot{x} = 0$, and a continuous approximation to the Coulomb law:

$$f(\dot{x}) = \mu \, \text{sgn}(\dot{x})$$

or

$$f(\dot{x}) = [\mu_d + (\mu_s + \mu_d)\text{sech}(\beta\dot{x})] \tanh \alpha\dot{x}$$

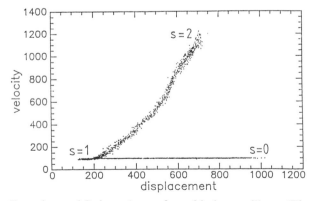

Figure 4-27   Experimental Poincaré map for a friction oscillator (Figure 4-26). The horizontal part is the sticking region. [From Feeny and Moon (1989) with permission of Elsevier Science Publishers, copyright 1989.]

The tanh term models the jump from positive friction to negative friction and approaches a discontinuity as $\alpha \to \infty$. The sech term represents a transition from the static friction $\dot{x} \simeq 0$ to the dynamic friction. A comparison of numerical integration of the continuous function model with experimental observations shows good agreement. In the experiment, the moving mass was designed so that the normal force varied linearly with the displacement (see Feeny, 1990). Color Plate 2 shows a three-dimensional phase space $(x, v = \dot{x}, \Omega t \pmod{2\pi})$.

The chaotic attractor is composed of three sections: positive and negative velocities and a sticking region. A Poincaré map is shown in Figure 4-27. This shows a nearly one-dimensional structure with two branches, namely, a positive velocity branch and a sticking branch (lower curve). This 2-D map can be reduced to a 1-D map (see Chapter 5). Also one can use bimodal symbol dynamics (+1 for slipping, −1 for sticking) from which a Lyapunov exponent can be calculated (see also Table 3-1 and Chapter 5, Figure 5-17).

Chaos, Fractals, and Surface Machining

It has been observed that all surfaces of solid objects are created by dynamic processes, be they mechanical, chemical, or thermal. There is also evidence from measurements of surface topography that the apparently random displacements from the mean obey a scaling law (Feder, 1988), and experiments on fractured surfaces appear to show fractal scaling (Mandelbrot et al., 1984). These *static* properties of fractal surface topography have led some to propose that nonlinear dynamics may play a role in the machining or surface creation process (Scott, 1989). In a series of papers, Grabec (1986, 1988) has studied a nonlinear model of the cutting dynamics for an elastic tool bit using a friction law between the workpiece and the tool. The equations for this model are given in terms of the two-degree-of-freedom displacement (x, y) of the tip of the tool bit:

$$m\ddot{x} + r_x\dot{x} + k_x x = F_x$$
$$m\ddot{y} + r_y\dot{y} + k_y y = F_y$$

where $F_y = kF_x$ and

$$F_x = F_0\left[1 + C_1\left(1 - \frac{v}{v_0}\right)^2\right]$$

$$k = k_0 \left[1 + C_2 \left(1 - \frac{v}{v_0} \right)^2 \right] \left[1 + C_3 \left(\frac{h}{h_0} - 1 \right)^2 \right]$$

$$h(t) = h_0 - y(t), \qquad v = v_1 - \dot{x}$$

In this empirical model, the friction force decreases as the velocity increases. This effect is believed necessary to model the tool chatter noise that one often hears in machining. Numerical simulation by Grabec (1988) using these equations seems to lead to chaotic tool chatter and a fractal dimension of the attractor between 2 and 3.

These studies are still speculative, but experiments in our laboratory do show chaotic dynamics in machining. Thus to produce quality machine surfaces the study of nonlinear processes in cutting, grinding, polishing, and other manufacturing processes, must be important.

Fracture, Fatigue, and Chaos

The failure of solid objects by fracture and fatigue remains one of the unsolved problems of classical physics. Experimental data in this field has always had a large scatter, even when material specimens are carefully prepared. The physics involves nonlinear processes at the molecular scale, micron scale (e.g., dislocations), and at the macro or machine-structure scale. A few papers are beginning to appear in the literature in which the methods of nonlinear dynamics and fractal mathematics are explored. The relation between fracture and fractals has been studied by Mandelbrot et al. (1984), and Lung (1986). The nonlinear dynamics of a running crack is explored in the papers of Fineberg et al. (1991), and Markworth et al. (1988). And, a discussion of a fatigue law based on nonlinear dynamics has been presented by Russell et al. (1991). This is an area of study that should see increasing attention in the next few years. A further discussion of fractals and fracture may be found in Section 7.8.

Earthquakes and Chaos

If unpredictabiliy of weather is the ultimate paradigm of fluid chaos, then is the uncertainty surrounding earthquakes the comparable paradigm of chaos in solid mechanics? This is the question a number of researchers are trying to answer (Levi, 1990). Current theories of earthquakes are based on the relative motion of the global tectonic plates of the Earth's crust. Thus it is natural that models would be based on the combination of stored elastic energy in the plates and

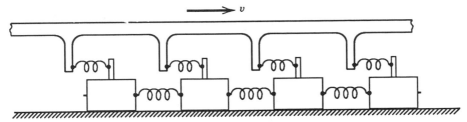

Figure 4-28   Earthquake model for the chaotic motions between two tectonic plates. [From Carlson and Langer (1989).]

some effective friction forces along the fault zones where the deformation occurs. One such model, proposed by Carlson and Langer (1989), is shown in Figure 4-28. In this and other models of earthquakes, the energy source is the slow but steady velocity of one of the plates relative to the other. The buildup of elastic energy is then released when the sticking force exceeds some critical value. The unpredictable nature of this stick–slip motion is thought to be a paradigm for the unpredictable nature of earthquakes.

A two-block model with friction proposed by Nussbaum and Ruina (1987) at first produced only time-periodic behavior, but a recent adaptation of this model by Huang and Turcotte (1990) using unequal friction forces on each block seems to result in chaotic dynamics of the blocks. In a more physics-based model which looks at both the spatial and temporal deformations between two elastic plates with friction contact along these common edges (i.e., the fault line), Horowitz and Ruina (1989) show through calculations that complex spatial patterns of slip can develop along the fault line.

From a more general view of geomechanics and nonlinear dynamics, Turcotte (1992) has published a monograph which tries to relate scaling law behavior in geology to the theorem of fractals and chaotic dynamics.

4.6   CHAOS IN ELECTRICAL AND MAGNETIC SYSTEMS

Nonlinear Electrical Circuits—Ueda Attractor

One of the first discoveries of chaos in electrical circuits was that of a periodically excited nonlinear inductor studied by Ueda (1979, 1991). The equation for a circuit with nonlinear inductance and linear resistor, driven by a harmonic voltage, can be written in a nondimensional form

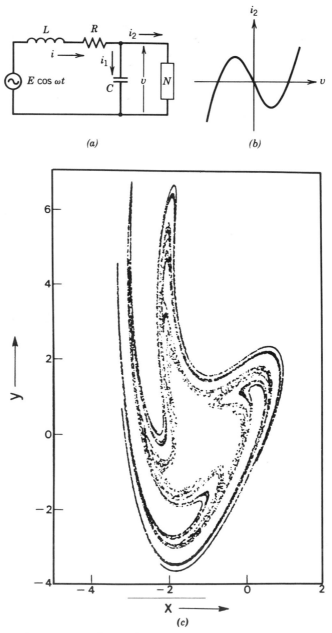

Figure 4-29 Poincaré map of chaotic analog computer simulation of a forced Van der Pol-type circuit. [From Ueda and Akamatsu (1981).]

as follows:

$$\ddot{x} + k\dot{x} + x^3 = B \cos t \qquad (4\text{-}6.1)$$

which is a special case of Duffing's equation (1-2.4). Professor Y. Ueda of Kyoto University in Japan has obtained beautiful Poincaré maps of the chaotic dynamics of this equation using analog and digital simulation (Figure 3-33).

Ueda has also modeled a negative resistor oscillator, shown in Figure 4-29. The equation for this system is a modified Van der Pol equation (1-2.5):

$$\ddot{x} + (x^2 - 1)\dot{x} + x^3 = B \cos \omega t \qquad (4\text{-}6.2)$$

It is interesting to note that both the Duffing and Van der Pol equations have been studied for decades, yet nowhere in any of the standard references on nonlinear vibrations are chaotic solutions reported. Other nonlinear chaotic circuits are discussed in the next section.

Nonlinear Circuits

Periodically Excited Circuits: Chaos in a Diode Circuit. The idealized diode is a circuit element that either conducts or does not. Such on–off behavior represents a strong nonlinearity. A number of experiments in chaotic oscillations have been performed using a particular diode element called a *varactor diode* (Linsay, 1981; Testa et al., 1982; Rollins and Hunt, 1982) using a circuit similar to the one in Figure 4-30. Both period-doubling and chaotic behavior were reported. The

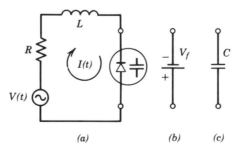

(a)   (b)   (c)

Figure 4-30 (a) Model for a varactor diode circuit. (b) Circuit element when the diode is conducting. (c) Circuit element when the diode is off. [From Rollins and Hunt (1982) with permission of the American Physical Society, copyright 1982.]

period doubling suggests that an underlying mathematical model is a one-dimensional map in which the absolute value of the maximum current value in the circuit during the $(n + 1)$st cycle depends on that in the nth cycle:

$$|I_{max}|_{n+1} = F(|I_{max}|_n) \tag{4-6.3}$$

One of the interesting questions regarding this system was the physical origin of the nonlinearity. In the earlier work of Linsay, it was proposed that the diode could be modeled as a highly nonlinear capacitance, where

$$c = c_0(1 - \alpha V)^{-\gamma}$$

$$\frac{d}{dt} c(V)V = I \tag{4-6.4}$$

$$L\frac{dI}{dt} = -RI - V + V_0\sin \omega t$$

where $\gamma \simeq 0.44$. Rollins and Hunt (1982), however, have proposed an entirely different model in which the circuit acts as either one of two linear circuits, shown in Figure 4-30b,c. Each cycle consists of a conducting and a nonconducting phase. The nonlinearity arises in determining when to switch from the conducting circuit with bias voltage V_f to the nonconducting circuit with constant capacitance. The switching time is a function of the maximum current value $|I_{max}|$. In this model, exact solutions of the circuit differential equations are known in each interval, with unknown constants to be determined using continuity of current and voltage at the switching times. Rollins and Hunt used this technique to calculate numerically the mapping function shown in Figure 4-31. Later experiments showed that this model accounted for more of the physics than the earlier version using nonlinear capacitance. See also Hunt and Rollins (1984).

Another study with a varactor diode was reported by Bucko et al. (1984), who looked at a series circuit with a diode, inductor, and resistor driven by a sinusoidal voltage. They assumed a mathematical model of the form

$$L\frac{dI}{dt} + RI + f\left(I, \int I\, dt\right) = V_0\cos \omega t \tag{4-6.5}$$

where the properties of the nonlinear diode $f(I)$ were discussed in the

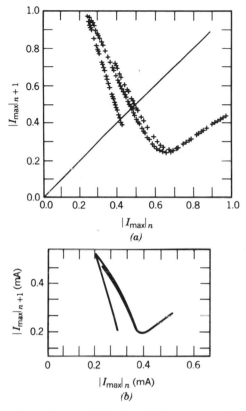

Figure 4-31 Comparison of (*a*) calculated and (*b*) measured one-dimensional maps for the varactor diode circuit of Figure 4-32. [From Rollins and Hunt (1982) with permission of the American Physical Society, copyright 1982.]

previous section. Bucko et al. explored the parameter plane (V_0, ω) and outlined regions of subharmonic and chaotic response. These results are shown in Figure 4-32. Figure 4-32*a* shows a driving frequency range $0.5 < \omega/2\pi < 4.0$ MHz. These data show that one can choose a parameter path that results in a period-doubling route to chaos. However, one can also follow paths that apparently do not follow this route. Figure 4-32*a* also shows chaotic islands which, when expanded in Figure 4-23*b*, exhibit further islands of chaos. This example shows that when the basic equations (4-6.5) are three differential equations, the Poincaré map of the dynamics is *two*-dimensional and the period-doubling properties of the one-dimensional map may not hold in such systems.

   For certain parameter regimes, however, the two-dimensional map may look one-dimensional and the dynamics are likely to behave as a

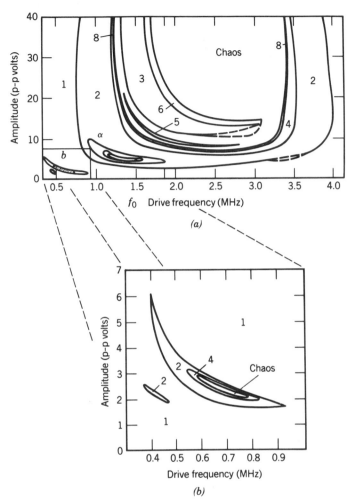

Figure 4-32 (*a*) Subharmonic and chaotic oscillation regions in the driver voltage amplitude–frequency plane for an inductor–resistor–diode series circuit. (*b*) Enlargement of diagram in (*a*). [From Bucko et al. (1984) with permission of Elsevier Science Publishers, copyright 1984.]

one-dimensional noninvertible map. The experimental moral of this is the following: When there is more than one essential nondimensional group in a physical problem, one should explore a region of parameter space to uncover the full range of possibilities in the nonlinear dynamics.

Nonlinear Inductor. Bryant and Jeffries (1984a) have studied a sinusoidally driven circuit with a linear negative resistor and a nonlinear inductor with hysteresis. In this work, they looked at four circuit elements in parallel: a voltage generator, negative resistor, capacitor, and a coil around a toroidal magnetic core, with typical values of $C \approx 7.5~\mu\text{F}$, $R = -500~\Omega$, and a forcing frequency around 200 Hz or higher. The negative resistor was created by an operational amplifier circuit. If N is the number of turns around the inductor, A the effect core cross section, and l the magnetic path length, the equation for the flux density B in the core is given by

$$NAC\ddot{B} + \frac{NA}{R}\dot{B} + \frac{l}{N}H(B) = I(t) \tag{4-6.6}$$

where $H(B)$ is the nonlinear magnetic field constitutive relation of the core material. In their experiments, they used $N = 100$ turns, $A \approx 1.5$ $10^{-5}~\text{m}^2$, and $l \approx 0.1$ m.

Using this circuit, they observed quasiperiodic vibrations, phase-locked motions, period doubling, and chaotic oscillations.

Autonomous Nonlinear Circuits—Chua Attractor. Autonomous chaotic oscillations in a tunnel diode circuit have been observed by Gollub et al. (1980) for the circuit shown in Figure 4-33*a*.

The nonlinear elements in this circuit are two tunnel diodes. The current–voltage relation shown in Figure 4-33*b* is obviously nonlinear and exhibits a hysteresis loop for cyclic variations in current I_D. In this work, the authors use return maps to construct pseudo-phase-

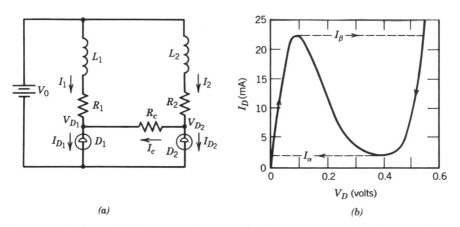

(a)   *(b)*

Figure 4-33 Tunnel diode circuit which admits autonomous chaotic oscillations. [From Gollub et al. (1980) with permission of Plenum Publishing Corp., copyright 1980.]

plane Poincaré maps. That is, they time sample the current

$$x_n \equiv I_{D_2} (t_0 + n\tau) \tag{4-6.7}$$

where n is an integer, and then plot x_n versus x_{n+1}. The data were sampled when the voltage V_{D_1} passed through a value of 0.42 V in the decreasing sense. The authors also used Fourier spectra and calculation of Lyapunov constants to measure the divergence rate of nearby trajectories.

As noted above, Ueda (1979) studied chaos in a circuit with negative resistance. A novel way to achieve negative resistance in the laboratory is with an operational amplifier. Two examples of experiments on chaotic oscillations in nonlinear circuits using this technique are those by Matsumoto et al. (1984, 1985) and Bryant and Jeffries (1984a,b).

The circuit studied by Matsumoto et al. is shown in Figure 4-34a and consists of three coupled current circuits with a nonlinear resistor. This circuit is autonomous; that is, there is no driving voltage. Thus, the system can produce oscillations only if the nonlinear resistor has negative resistance over some voltage range. In their model, Matsumoto et al. (1984) chose a trilinear current–voltage relation

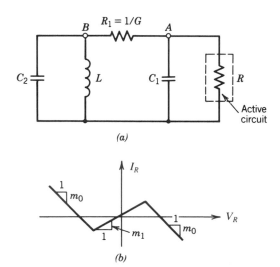

(a)

(b)

Figure 4-34 Circuit with trilinear active circuit elements which leads to autonomous chaotic oscillations. [From Matsumoto et al. (1985), copyright 1985 Institute of Electrical and Electronic Engineers.]

shown in Figure 4-34b which has the form

$$g(V_1) = m_0 V_1 + \tfrac{1}{2}(m - m_0)|V_1 + b| + \tfrac{1}{2}(m_0 - m_1)|V_1 - b| \quad (4\text{-}6.8)$$

The resulting circuit equations are obtained by summing currents at nodes A and B in Figure 4-36a and summing voltages in the left-hand circuit loop:

$$C_1 \dot{V}_1 = \frac{1}{R}(V_2 - V_1) - g(V_1)$$

$$C_2 \dot{V}_2 = \frac{1}{R}(V_1 - V_2) - 1 \quad\quad\quad (4\text{-}6.9)$$

$$L\dot{I} = -V_2$$

where V_1 and V_2 are the voltages across the capacitors C_1 and C_2 and I is the current through the inductor. Chua and co-workers created the trilinear resistor (4-6.8) by using an operational amplifier with diodes. For small voltages, the nonlinear resistance is negative, and the equilibrium position $(V_1, V_2, I) = (0, 0, 0)$ is unstable and oscillations occur. Chaotic oscillations were found for $1/C_1 = 9$, $1/C_2 = 1$, $1/L - 7$, $G = 0.7$, $m_0 = -0.5$, $m_1 = -0.8$, and $b = 1$ in a set of consistent units. A chaotic time history is shown in Figure 4-35, which has the same character as the Lorenz attractor (Figure 1-27). See also Chua et al. (1986).

Magnetomechanical Models

Dynamo Models. A physical model that has received considerable attention is the rotating disk in a magnetic field. This system is of interest to geophysicists as a potential model to explain reversals of the earth's magnetic field. A single-disk dynamo is shown in Figure 4-36. The equation governing the rotation Ω and the currents I_1 and I_2 are of the form (see Robbins, 1977).

$$J\dot{\Omega} = -k\Omega - \mu_2 I_1 (I_1 + I_2) + T$$

$$L_1 \dot{I}_1 = -RI_1 - R_3 I_2 + \mu_1 \Omega I_1 \quad\quad (4\text{-}6.10)$$

$$L_1 \dot{I}_2 = -R_2 I_2 + \mu_2 \Omega I_1$$

where T is an applied constant torque. The time traces in Figure 4-36 show that the current (and hence the magnetic field) can reverse in an

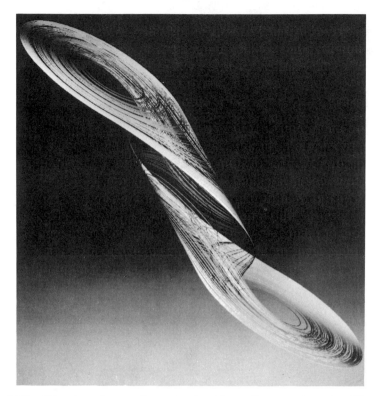

Figure 4-35   Chaotic trajectory for a circuit with a trilinear resistor (see Figure 4-34) numerical simulation. This attractor, based on Chua's circuit, is called the *double scroll*. [From Matsumoto et al. (1985), copyright 1985 Institute of Electrical and Electronic Engineers.]

apparently random manner. [See also Jackson (1990) for a lengthy discussion of an analysis of Eqs. (4-6.10).]

Magnetically Levitated Vehicles.   Suspension systems for land-based vehicles must provide vertical and lateral restoring forces when the vehicle departs from its straight path. Conventional suspension systems such as pneumatic tires and steel wheels on steel rails, as well as the futuristic systems of air cushion or magnetic levitation, all exhibit nonlinear stiffness and damping behavior and are thus candidates for chaotic vibrations. As an illustration, some experiments performed at Cornell University on a magnetically levitated vehicle are described. [See the book by Moon (1984a, 1993) which describes magnetic levitation transportation mechanics.]

In this experiment, permanent magnets were attached to a rigid platform and a continuous L-shaped aluminum guideway was moved

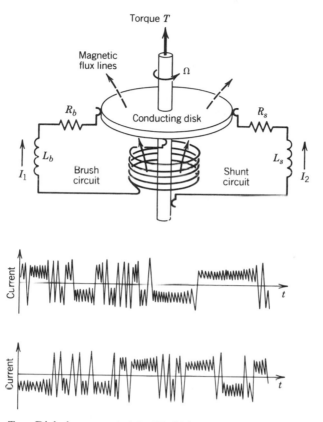

Figure 4-36 *Top:* Disk dynamo model of Robbins (1977) for reversals of the Earth's magnetic field. *Bottom:* Chaotic current reversals from numerical solutions of disk dynamo equations (4-6.10).

past the model using a 1.2-meter-diameter rotating wheel (Figure 4-37). The induced eddy currents in the aluminum guideways interact with the magnetic field of the magnets to produce lift, drag, and lateral guidance forces. The magnetic drag force is nonconservative and can pump energy into the vibrations of the model. Thus, under certain conditions, the model can undergo limit cycle oscillations. As the speed is increased, damped vibrations change to growing oscillations (see bottom of Figure 4-37). The nonlinearities in the suspension forces limit the vibration and a limit cycle motions results. [This bifurcation in stability is known in mathematics as a *Hopf bifurcation* (Chapter 1). In mechanics it is called a *flutter* oscillation.]

In addition to flutter or limit cycle oscillations, the levitated model can undergo static bifurcations. Thus, at certain speeds, the equilibrium state can change from vertical to two stable tilted positions as

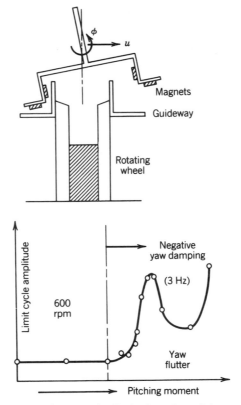

Figure 4-37   *Top:* Sketch of magnetically levitated model on a rotating aluminum guideway. *Bottom:* Limit cycle bifurcation of levitated model.

shown in Figure 4-37. This latter instability is known in aircraft dynamics as *divergence* and is analogous to buckling of an elastic column. In our experiments, chaotic vibrations occurred when the system exhibited both divergence (multiple equilibrium states) and flutter. The flutter provides a mechanism to throw the model from one side

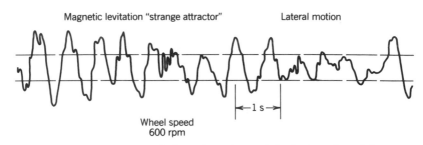

Figure 4-38   Chaotic lateral motions of the levitated model.

of the guideway to another, similar to what occurred in the buckled beam problem discussed in Chapter 2. The mathematical model for this instability, however, has two degrees of freedom. Lateral and roll dynamics were measured from films of the chaotic vibrations (Figure 4-38). These vibrations were quite violent and if they occurred in an actual vehicle traveling at 400–500 km/h, the vehicle would probably derail and be destroyed.

Optical Systems

We have seen that multiple-well potential problems are a natural source of chaotic oscillations. The creation of coherent light using devices known as *lasers* involves the stimulation of electrons between two or more atomic energy levels. Thus, it is not surprising that chaotic and complex dynamical behavior may be found in laser systems.

Many papers have been published in the physics literature on chaotic behavior of laser systems as well as on the chaotic propagation of light through nonlinear optical devices. An extensive review of chaos in light systems has been written by Harrison and Biswas (1986), and a very readable introduction to nonlinear dynamics of lasers may be found in Haken (1985). In elementary laser systems the nonlinearity originates from the fact that the system oscillates between at least two discrete energy levels. The simplest mathematical model for such systems involves three first-order equations for the electric field in the laser cavity, the population inversion, and the atomic polarization. These equations, known as the *Maxwell–Block equations*, are similar in structure to the Lorenz equations discussed in Chapters 1 and 3 [see Eq. (3-8.3)]. Chaotic phenomena in lasers have been observed in both the autonomous mode and the modulation mode.

The simplest model for laser dynamics is derived from Maxwell's equations of electromagnetics and the semiclassical theory of quantum mechanics. At one level of the theory, electrons are assumed to reside in one of two states, each governed by wave functions φ_1, φ_2. The complete wave function is then assumed to be a superposition of these two with time-varying amplitudes $c_1(t)$, $c_2(t)$. Here $|c_i|^2$ is the probability of finding the electron in the state "i". The difference $d = |c_2|^2 - |c_1|^2$ is called the *electron occupation difference*, and its macroscopic measure is denoted by $D(t)$ and is called the *inversion density*. When there are more electrons in one state than the other, the material has an atomic electric dipole density whose macroscopic measure is denoted by the polarization density, $P(x)$.

The third dynamical variable in the laser problem is the macroscopic electric field $E(t)$. In the dynamical equations, only the slowly varying or modulation parts of these variables are of concern (i.e., we filter out the high-frequency dynamics of the light wave). The basic dynamical equations in these "slow" variables take the form as described in Haken (1985):

$$\dot{P} = \gamma_1 P + \gamma ED$$
$$\dot{D} = -\gamma_2 D + \gamma_2 (\Lambda + 1) - \gamma_2 \Lambda EP \qquad (4\text{-}6.11)$$
$$\dot{E} = -\kappa E + \kappa P$$

It is left to the reader as an exercise to show that these equations can be related to the Lorenz equations (1-3.9), or see Section 8.3 of Haken's book. Many experimental observations of period doubling and other chaotic phenomena have been reported [e.g., see Milonni et al. (1987) for a review].

The other class of problems discussed in Harrison and Biswas (1986) involves passive nonlinear optics. Here the index of refraction (speed of light in the medium) depends on the intensity of the light, for example, through the Kerr effect.

4.7   FLUID AND ACOUSTIC SYSTEMS

Chaotic Dynamics in Fluid Systems

Although the primary focus of this book is on low-order mechanical and electrical systems, the major impact of the new dynamics on fluid mechanics warrants mention of at least a few fluid experiments in chaotic motions. We recall from Chapter 1 that the major nonlinearity in fluid problems is a convective acceleration term $\mathbf{v} \circ \nabla \mathbf{v}$ in the equations of motion (1-1.3). However, other nonlinearities may also play a role such as free surface or interface conditions and non-Newtonian viscous effects. We can classify five types of fluid experiments in which chaotic motions have been observed:

1. Closed-flow systems: Rayleigh–Benard convection, Taylor–Couette flow between cylinders
2. Open-flow systems: pipe flow, boundary layers, jets
3. Fluid particles: dripping faucet
4. Waves on fluid surfaces: gravity waves
5. Reacting fluids: chemical stirred tank reactor, flame jets

Another set of fluid problems has been the collapse of fluid bubbles which can create acoustic chaos (Lauterborn and Cramer, 1981).

One reason for the intense interest in chaotic dynamics and fluids is its potential for unlocking the secrets of turbulence. [For example, see Swinney (1983) for a review and see the edited volume by Tatsumi (1984) for a collection of papers on fluids and chaos.] Some feel that this may be too ambitious a goal for a theory based on a few ordinary differential equations and maps. One view is that dynamical systems theory will provide a good model for the transition to turbulence, but will require major breakthroughs to solve the more difficult problem of fully developed spatial and temporal turbulence (strong turbulence). However, a group at Cornell University has recently studied the dynamics of coherent structures in a turbulent boundary layer for open flow over a wall (Aubry et al., 1988) using modern global bifurcation theory. This work was one of the first to seek to examine both spatial and temporal complexities in fluid problems. Whatever the ultimate progress, nonlinear dynamical theory has added new tools to the study of experimental fluid mechanics.

Closed-Flow Systems: Rayleigh–Benard Thermal Convection.   We recall from Chapter 1 that a thermal gradient in a fluid under gravity produces a buoyancy force that leads to a vortex-type instability with resulting chaotic and turbulence motions (Figure 4-39). By far the most studied experimental system is the thermal convection of fluid in a closed box. This is the system that Lorenz tried to model with his famous equations (3-8.3).

Experimental studies of Rayleigh–Benard thermal convection in a box have shown period-doubling sequences as precursors to the chaotic state. They have been carried out in helium, water, and mercury for a wide range of nondimensional Prandtl numbers and Rayleigh

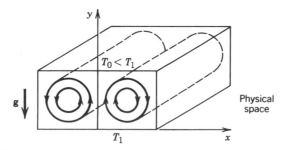

Figure 4-39   Sketch of thermofluid convection rolls.

numbers. These experiments emerged in the late 1970s. For example, Libchaber and Maurer (1978) observed period-doubling convection oscillation in helium. A number of experimental papers have emerged from a group at the French National Laboratory at Saclay, France, associated with Bergé and co-workers (1980, 1982, 1985); see also Dubois et al. (1982). The experiment is similar to that pictured in Figure 4-39 with a fluid of silicone oil in a rectangular cell with dimensions 2 cm × 2.4 cm × 4 cm. These authors have observed both the quasiperiodic route to chaos (Newhouse et al., 1978) and intermittent chaos. In the former, they observe the following sequence of dynamic events as the temperature gradient is increased:

$$\begin{array}{ccccccccc} \text{steady} & \to & \text{monofrequency} & \to & \text{quasiperiodic} & \to & \text{chaotic} & \to & \text{thermal} & \to \\ \text{state} & & \text{motion} & & \text{motion} & & \text{motion} & & \text{gradient} \end{array}$$

The frequency range observed in their experiments is very low, for example, $9\text{–}30 \times 10^{-3}$ Hz. They were one of the first groups to obtain Poincaré maps in fluid experiments. This was facilitated by their discovery of regions in the flow where one frequency or oscillator was predominant. Thus, they could use one frequency to synchronize the Poincaré maps. Two maps are shown in Figure 2-19. The first is quasiperiodic and the frequency ratio is close to 3. The second is based on 1500 Poincaré points and shows a breakup of the toroidal attractor before chaos sets in. The techniques used to measure the motion included laser Doppler anemometry and a differential interferometric method. More recent work involving mode-locking and chaos in convection problems has been done by Haucke and Ecke (1987).

Lorenz–Saltzman Model. Perhaps the most famous model to date is the Lorenz equations which attempt to model atmospheric dynamics. In this model, one imagines a fluid layer, under gravity, which is heated from below so that a temperature difference is maintained across the layer (Figure 4-39). When this temperature difference becomes large enough, circulatory, vortex-like motion of the fluid results in which the warm air rises and the cool air falls. The tops of parallel rows of convection rolls can sometimes be seen when flying above a cloud layer. The two-dimensional convective flow is assumed to be governed by the classic Navier–Stokes equations (1-1.3). These equations are expanded in the two spatial directions in Fourier modes with fixed boundary conditions on the top and bottom of the fluid layer. For a small temperature difference ΔT, no fluid motion takes place, but at a critical ΔT, convective or circulation flow occurs. This motion is referred to as *Rayleigh–Benard convection*.

Truncation of the Fourier expansion in three modes was studied by Lorenz (1963). An earlier study by Saltzman (1962) used a five-mode truncation. In this simplification, the velocity in the fluid (v_x, v_y) is written in terms of a stream function ψ:

$$v_x = \frac{\partial \psi}{\partial y}, \qquad v_y = \frac{-\partial \psi}{\partial x} \qquad (4\text{-}7.1)$$

In the Lorenz model, the nondimensional stream function and perturbed temperature are written in the form [see Lichtenberg and Lieberman (1983, pp. 443–446) for a derivation]

$$\psi = \sqrt{2}\, x(t)\sin \pi a x \sin \pi y \qquad (4\text{-}7.2)$$

$$\theta = \sqrt{2}\, y(t)\cos \pi a x \sin \pi y - z(t)\sin 2\pi y \qquad (4\text{-}7.3)$$

where the fluid layer is taken as a unit length. The resulting equations for (x, y, z) are then given by

$$\dot{x} = \sigma(y - x)$$
$$\dot{y} = \rho x - y - xz \qquad (4\text{-}7.4)$$
$$\dot{z} = -\beta z + xy$$

The parameter σ is a nondimensional ratio of viscosity to thermal conductivity (Prandtl number), ρ is a nondimensional temperature gradient (related to the Rayleigh number), and $\beta = 4(1 + \alpha^2)^{-1}$ is a geometric factor, with $\alpha^2 = \frac{1}{2}$.

For the parameter values $\sigma = 10$, $\rho = 28$, and $\beta = \frac{8}{3}$ (studied by Lorenz), there are three equilibrium points, all of them unstable. The origin is a saddle point, whereas the other two are unstable foci or spiral equilibrium points (see Figure 1-26). However, globally, one can show that the motion is bounded. Thus, the trajectories have no home but remain confined to an ellipsoidal region of phase space. A numerical example of one of these wandering trajectories is shown in Figure 1-27. A discussion of the bifurcation sequence as the thermal gradient is increased is given in Section 3.8.

Thermal Convection Model of Moore and Spiegel. It is often the case that discoveries of major significance are not singular and that different people in different places observe new phenomena at about the same time. Such appears to be the case regarding the discovery of low-order models for thermal convection dynamics. Above we discussed the now famous Lorenz (1963) equations, (4-7.4), which later received tremendous attention from mathematicians. Yet around the same time, Moore and Spiegel (1966) of the Goddard Institute and New York University, respectively, proposed a model for unstable oscillations in fluids which rotate, have magnetic fields or are compressible, and have thermal dissipation. The equations derived in their paper, like Lorenz's, are equivalent to three first-order differential equations. If z represents the vertical displacement of a compressible fluid mass in a horizontally stratified fluid (Figure 4-40a), restoring forces in the fluid are represented by a spring force and a buoyancy force resulting from gravity. Also, the fluid element can exchange heat with the surrounding fluid. Thus, the dynamics are modeled by a second-order equation (Newton's law) coupled to a first-order equation for heat transfer, leading to a third-order equation. In nondimensional form,

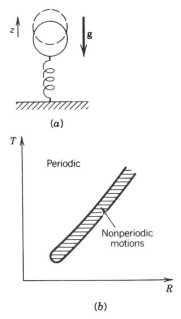

Figure 4-40   (a) Spring–mass model for thermal convection of Moore and Spiegel (1966). (b) Region of nonperiodic motions in the nondimensional parameter space for the thermal convection model of Moore and Spiegel (1966), Eq. (4-7.5).

this equation becomes

$$\ddot{z} + \dot{z} + (T - R + Rz^2)\dot{z} + Tz = 0 \qquad (4\text{-}7.5)$$

where a nonlinear temperature profile of the form

$$\theta = \theta_0 \left[1 - \left(\frac{z}{L}\right)^2 \right]$$

is assumed. In Eq. (4-7.5), T and R are nondimensional groups:

$$T = \left(\frac{\text{thermal relaxation time}}{\text{free oscillation time}}\right)^2$$

$$R = \left(\frac{\text{thermal relaxation time}}{\text{free fall time}}\right)^2$$

In their numerical studies, Moore and Spiegel discovered an entire region of aperiodic motion as shown in Figure 4-40b. In a follow-up paper, Baker et al. (1971) analyzed the stability of periodic solutions in the aperiodic regime.

They showed that Eq. (4-7.5) can be put into the form

$$\ddot{s} = -(1 - \delta)s + \theta$$
$$\dot{\theta} = -R^{-1/2}\theta + (1 - \delta s^2)\dot{s} \qquad (4\text{-}7.6)$$

The limit of $R \rightarrow \infty$ is the zero dissipative case. In this limit (R large), Baker et al. showed that in the range of periodicity, the periodic solutions of (4-7.6) become unstable locally. This property of global stability and local instability seems to be characteristic of chaotic differential equations. In a more recent paper, Marzec and Spiegel (1980) studied a more general class of third-order equations of the form

$$\dot{x} = y$$
$$\dot{y} = -\frac{dV(x, \lambda)}{dx} - \varepsilon\mu y \qquad (4\text{-}7.7)$$
$$\dot{\lambda} = -\varepsilon[\lambda + g(x)]$$

where $V(x, y)$ is thought of as a potential function. They show that both the Moore–Spiegel oscillator (4-7.5) and the Lorenz system (4-7.4) (with a change of variables) can be put into the above form (4-7.6). Strange attractor solutions to specific examples of (4-7.6) were found numerically. The above set of equations also models a second-order oscillator with feedback control λ similar to (4-2.21).

It will be an interesting study for historians of science to answer why the Lorenz system received so much study and the Moore–Spiegel model was virtually ignored by mathematicians. Both purported to model convection. Lorenz published his article in the *Journal of Atmospheric Sciences*, whereas Moore and Spiegel published theirs in the *Astrophysics Journal*.

Closed-Loop Thermosiphon. It is curious, given the great amount of attention to the Lorenz attractor as a paradigm for convective flow chaos, that only a few attempts were made to design an experiment that incorporated the assumptions of the Lorenz model. One of these is the flow of fluid in a circular channel under gravity, called a *thermosiphon*. The relevance of this experiment to the Lorenz model was pointed out by Hart (1984). Convectively driven flows are of interest as models for geophysical flows such as warm springs or groundwater flow through permeable layers in the Earth's crust, and they are also of interest as applications for solar heating systems or reactor core cooling.

Early experiments by Bau and Torrance (1981) were performed in a rectangular loop thermosiphon. They derived equations that describe flow in a closed circular tube with gravity acting in the vertical plane, as shown in Figure 4-41. Essentially, all variables are assumed to be independent of the radial direction. The principal dependent variables are the circumferential velocity $v(t)$ and the temperature $T(\theta, t)$. A

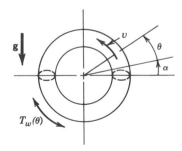

Figure 4-41   Thermal convection in a vertical one-dimensional fluid circuit. A model for a thermosiphon.

viscous wall stress is assumed to act in the fluid. Also, a prescribed wall temperature $T_w(\theta)$ is assumed with a linear cooling law proportional to $T - T_w$.

The basic equations are the balance of angular momentum for the fluid mass and a partial differential equation for the energy or heat balance law.

A buoyancy force or moment is introduced by assuming that the fluid density depends on the temperature,

$$\rho = \rho_0[1 - \beta(T - T_0)] \tag{4-7.8}$$

so that a net torque acts on the fluid proportional to

$$g\beta a \int_{-\pi}^{\pi} T(\theta)\cos(\theta + \alpha)\, d\theta \tag{4-7.9}$$

where θ is as defined in Figure 4-42.

In a method similar to that used in deriving the Lorenz equations (4-7.4), the temperature is expanded in a Fourier series. In this way, the partial differential equation for the heat balance is reduced to a set of ordinary differential equations.

Following Hart (1984), one writes

$$T(\theta) = \sum C_n(t)\cos n\theta + S_n(t)\sin n\theta \tag{4-7.10}$$

He shows that only the $n = 1$ thermal modes determine the dynamics. By redefining variables $x = v$, $y = C_1$, and $z = S_1 + R_a$, where R_a is similar to the Reynolds number, the resulting coupled first-order equations take the form

$$\dot{x} = P_r[-F(x) + (\cos\alpha)y - (\sin\alpha)(z - R_a)]$$
$$\dot{y} = -xz - y + R_a x \tag{4-7.11}$$
$$\dot{z} = xy - z$$

where $F(x)$ is a nonlinear friction law. To obtain the Lorenz equations, one sets $\alpha = 0$ and $F(x) = Cx$. The Lorenz limit corresponds to antisymmetric heating about the vertical. In their experiments, Bau and Torrance (1981) investigated the stability of flow but did not explore the chaotic regime. Given the close correspondence between Eqs. (4-7.11) and the Lorenz equation (4-7.4), it would appear natural that experimental exploration of the chaotic regime of the thermosiphon would be attempted. Another analysis of the relation between the Lorenz equations and fluid in a heated loop has been reported by Yorke et al. (1985).

Earlier experiments with a fluid convection loop by Creveling et al. (1975) did not report chaotic motions. However, recent experiments by Gorman et al. (1984, 1986) have reproduced some of the features of the Lorenz attractor. The working fluid was water and the apparatus consisted of a 75-cm-diameter loop of 2.5-cm-diameter Pyrex (glass) tubing. The bottom half was heated with electrical resistance tape while the top half was kept in a constant-temperature bath.

Taylor–Couette Flow Between Cylinders. A classic fluid mechanics system which exhibits perturbulent chaos is the flow between two rotating cylinders (called *Taylor–Couette flow*) shown in Figure 4-42. Much work has been done on this system [e.g., see Swinney (1983) for a review]. This flow is sensitive to the Reynolds number $R = (b - a)a\Omega_i/v$ and the ratios b/a and Ω_o/Ω_i, where the latter is the quotient of the outer cylinder rotation rate to the inner as well as the boundary conditions on the ends. This system exhibits a prechaos behavior of quasiperiodic oscillations before broad-band chaotic noise sets in. Other work includes that of Brandstater and co-workers (1983, 1984, 1987). Taylor–Couette flow also exhibits complex spatiotemporal dynamics that are now under study by the group at the University of Texas at Austin under Professor H. Swinney.

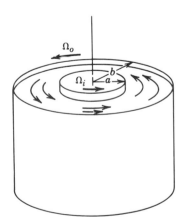

Figure 4-42   Sketch of flow between two rotating cylinders known as *Taylor–Couette flow*.

Pipe Flow Chaos. While closed-flow problems have captured the bulk of the attention vis-à-vis dynamical systems theory, open-flow problems are of great importance to engineering design. These include flows over airfoils, boundary layers, jets, and pipe flow. Recently, increased attention has been focused on applying the theory of chaotic dynamics to the laminar–turbulent transition problem in open-flow systems. One example is the experiment of Sreenivasan (1986) of Yale University who is studying intermittency in pipe flows. In this problem, low-velocity flow is laminar and steady, whereas for sufficiently high mean flow velocity the flow field becomes turbulent. At some critical velocity, the transition from laminar to turbulent appears to occur in intermittent bursts of turbulence. As the velocity increases, the fraction of time spent in the chaotic state increases until the flow is completely turbulent. Some observations of this phenomenon go back to Reynolds in 1883. The current focus of attention is to try to relate features of the intermittency, such as distribution of burst times, to dynamical theories of intermittency (e.g., see Pomeau and Manneville, 1980).

Fluid Drop Chaos. A simple system with which the reader can observe chaotic dynamics in one's home is the dripping faucet. This experiment is described by R. Shaw of the University of California—Santa Cruz in a monograph on chaos and information theory (1984). The experiment and sketch of experimental data are shown in Figure 4-43. The observable variable is the time between drops as measured with a light source and photocell, and the control variable is the flow rate from the nozzle. In Shaw's experiment, he measures a sequence of time intervals $\{T_n, T_{n+1}, T_{n=2}\}$ but does not measure the drop size or other physical properties of the drop such as shape. He and his students obtained periodic motion and period-doubling phenomena as well as chaotic behavior. Different maps of T_{n+1} versus T_n are obtained for different flow rates. The map in Figure 4-43 shows a classic one-dimensional parabolic map similar to the logistic map of Feigenbaum (1978). They also observed a more complicated map which is best represented in a three-dimensional phase space T_n versus T_{n+1} versus T_{n+2}. This is an example of using discrete data to construct a pseudo-phase-space and suggests that another dynamics variable should be observed (such as drop size).

Surface Wave Chaos. It is well known that waves can propagate on the interface between two immiscible fluids under gravity (e.g., air on

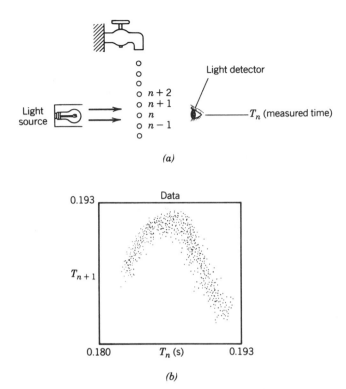

(a)

(b)

Figure 4-43 Experimental one-dimensional map for the time between drops in a dripping faucet. [From Shaw (1984) with permission of Ariel Press, copyright 1984.]

water). Such waves can be excited by vibrating a liquid in the vertical direction in the same way that one can parametrically excite vibrations in a pendulum. Subharmonic excitation of shallow water waves goes back to Faraday in 1831. An analysis of this phenomenon in the context of period doubling has been performed by a group at UCLA (Keolian et al., 1981). They looked at saltwater waves in an annulus of 4.8-cm mean radius with a cross section of 0.8×2.5 cm. The system is driven in the vertical direction by placing the annulus on an acoustic loudspeaker. By measuring the wave height versus time at several locations around the annulus, the UCLA group measured a subharmonic sequence before chaos that does not follow the classic period-doubling sequence; for example, they observe resonant frequencies pf/m, where f is the driving frequency, for $m = 1, 2, 4, 12, 13, 16, 18,$ 20, 24, 28, 36, which differs from the 2^n sequence of the logistic equation.

In another study of forced surface waves, Ciliberto and Gollub (1985) looked at a cylindrical dish of water with radius 6.35 cm with a depth of about 1 cm. They also used a loudspeaker excitation to explore regions of periodic and chaotic motion of the fluid height. In the region around 16 Hz, for example, they obtained chaotic wave motion for a vertical driving height of around 0.15 mm. The tried to interpret the results in terms of nonlinear interaction between two linear spatial modes (see Figure 6-8, Section 6.2). A theoretical analysis of this problem has been done by Holmes (1986).

Acoustic Chaos

At first look, the propagation of sound waves in air or water would appear to be a linear phenomenon, since they are usually modeled by the linear wave equation. But, as anyone who has tried to blow into a trumpet or a wind instrument has found out, it is not difficult to create noisy, chaotic-like acoustic effects. The origin of chaos in acoustics has at least three sources: the nonlinearities in the medium itself; the acoustic generator; and the reflection, impedance, or reception of the acoustic waves. A review of chaotic dynamics in some acoustics problems has been given by Lauterborn and Holzfuss (1991) of the Technical University of Darmstadt, Federal Republic of Germany. This group has pioneered in the study of chaotic noise from bubbles and cavitation in liquids. In this class of problems, a high-intensity source creates bubbles in the fluid. The nonlinear behavior of the bubbles is then believed to be the source of period doubling and chaotic acoustic phenomena in the fluid (e.g., see Lauterborn and Holzfuss, 1989 and Lauterborn and Cramer, 1981).

Two papers on musical chaos have been published by Maganza et al. (1986), who studied period doubling and chaos in clarinet-like instruments, and by Gibiat (1988), who studied a similar system. Embedding space or pseudo-phase-space techniques were used to look for qualitative behavior of the clarinet-like resonator.

In a study in our laboratory, chaotic modulation in an organ-pipe generator of sound was obtained when a nonlinear mechanical impedance was placed at the open end of the meter-long pipe (Moon, 1986).

Other examples of acoustic chaos are discussed in the review by Lauterborn and Holzfuss (1991).

4.8 CHEMICAL AND BIOLOGICAL SYSTEMS

Chaos in Chemical Reactions

Rössler (1976a,b) and Hudson et al. (1984) have observed chaotic dynamics in a small reaction–diffusion system. Also, Schrieber et al. (1980) have observed similar behavior in two coupled stirred-cell reactors. If (x_1, y_1) represents the chemical concentration in one cell and (x_2, y_2) represents the concentration in another cell, a set of equations can be derived to model the dynamic behavior:

$$\dot{x}_1 = A - (B + 1)x_1 + x_1^2 y_1 + D_1(x_2 - x_1)$$

$$\dot{y}_1 = B_1 x_1 - x_1^2 y_1 + D_2(y_2 - y_1)$$

$$\dot{x}_2 = A - (B + 1)x_2 + x_2^2 y_2 + D_1(x_1 - x_2) \tag{4-8.1}$$

$$\dot{y}_2 = B x_2 - x_2^2 y_2 + D_2(y_1 - y_2)$$

A now classic example of chemical chaos is the Belousov–Zhabotinski reaction in a stirred-flow reactor. Subharmonic oscillations and period doubling have been observed by a group under Professor H. Swinney of the University of Texas at Austin (Simoyi et al., 1982). With the input chemical concentrations held fixed, the time history of the concentration of the bromide ion, one of the reaction chemicals, shows complex subharmonic behavior of different flow rates. See Argoul et al. (1987b) for a review. In a more recent study, this group has developed a model to explain spatially induced chaos in Belousov–Zhabotinski-type reaction–diffusion systems (Vastano et al., 1990).

Biological Chaos

One of the exciting aspects of the new mathematical models in nonlinear dynamics is the wide applicability of these paradigms to many different fields of science. Thus, it is no surprise that dynamic phenomena in biological systems have been explained by some of the very same equations used to describe chaos in the electrical and mechanical sciences. A collection of papers concerning chaos in biological systems may be found in the book edited by Degn et al. (1987). A readable book on this subject has been published by Glass and Mackey (1988). A few other examples are described here.

Chaotic Heart Beats. Glass et al. (1983) have performed dynamic experiments on spontaneous beating in groups of cells from embryonic chick hearts. Without external stimuli, these oscillations have a period between 0.4 and 1.3 s. However, when periodic current pulses are sent into the group using microelectrodes, entrainment, quasiperiodicity, and chaotic motions have been observed (see Fig. 2-22). The circle map has been used as a model to explain some of these phenomena [e.g., see Guevara et al. (1990), Glass (1991), and Arnold (1991)].

A discussion of the relevance of nonlinear dynamics and chaotic models to ventricular fibrillation was given by Goldberger et al. (1986) and Goldberger and West (1987a). These papers contain a number of references on cardiac dynamics. In another paper, Rigney and Goldberger (1989) used a parametrically excited pendulum equation to model a period doubling of the heart rate observed in electrocardiograph experiment.

Another work, that of Lewis and Guevara (1990), described the dynamic modeling of ventricular muscle due to periodic excitation in the sinoatrial node of the heart. The authors started with the telegraph equation, a partial differential equation used in electrical engineering, and reduced the dynamics to a one-dimensional map.

Nerve Cells. In a similar type of experiment, sinusoidal stimulation of a giant neuron in a marine mollusk by Hayashi et al. (1982) also showed evidence of chaotic behavior.

Biological Membranes

Biological membranes control the flow of ions into a cell such as potassium and sodium. Time history measurements of the ion current through a channel in a cell of the cornea as shown in Figure 4-44 seem to indicate a chaotic opening and closing of the channel. Earlier models of this phenomenon were based on intrinsic random processes. However, a dynamic model of the ion kinetics using an iterated one-

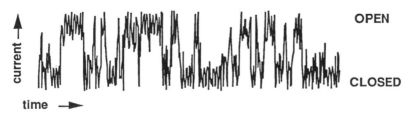

Figure 4-44 Experimental time history of the current through a single-ion-channel protein from a cell in the cornea. [From Leibovitch and Czegledy (1991).]

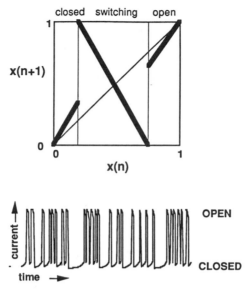

Figure 4-45   *Top:* One-dimensional map of chaotic ion channel kinetics. *Bottom:* Iteration of map. [From Leibovitch and Toth (1991) with permission of Academic Press, Ltd.]

dimensional map has been proposed by Liebovitch and Toth (1991), shown in Figure 4-45. The apparent chaotic switching from open to closed is similar to the particle in a two-well potential described in Chapters 1 and 2. In another paper, Liebovitch and Czegledy (1991) used a two-well Duffing oscillator as a possible model for the ion dynamics.

4.9   NONLINEAR DESIGN AND CONTROLLING CHAOS

Is chaos good for anything? Can engineers use chaos theory to invent new technology? Can one design a chaotic controller? In a time of debate between the role of basic research in enhancing technology and productivity, these questions are being raised if not by dynamicists themselves, then by their funding sponsors. Certainly the role of chaotic dynamics in the mixing of fluids and chemicals has drawn interest in chemical engineering (see Section 8.4 and Ottino, 1989b). But, in electrical and mechanical systems, chaotic behavior is more often avoided in design. However, recent progress has been made in using nonlinear behavior to advantage in the design of electronic circuits, and of even using the chaotic nature of a strange attractor to design a control system.

Much of this recent work stems from the work of a group at the University of Maryland. Research conducted in 1992 points the way toward future applications of what I would call "nonlinear thinking" in design of dynamic systems. Perhaps the most clever set of papers are those inspired by recent research from the group at the University of Maryland entitled "Controlling Chaos." In Ott, Grebogi, and Yorke (1990) they proposed using the stochastic nature of a chaotically behaved system to lock onto one of the many unstable periodic motions embedded in the strange attractor.

Two papers which apply this nonlinear thinking to control chaos in different experimental systems are Spano and Ditto (1991) and Hunt (1991). Spano and Ditto describe how the use of small perturbations can be used to switch between different orbits embedded in a strange attractor using a control scheme derived from the underlying map for the attractor. They use a flexible beam made of an amorphous ferroelastic material called Metglas ($Fe_{81}B_{13.5}Si_{3.5}C_2$) whose effective Young's modulus is sensitive to small magnetic fields. When an applied field is periodically perturbed, the beam undergoes chaotic motions. Embedded in this attractor are an infinite set of unstable periodic orbits. Additional perturbations to the field were then used to control the motion to switch from one formerly unstable periodic orbit to another.

In the paper by Hunt (1991) of Ohio University the system consists of a rectifier-type diode in series with an inductor driven by a sinusoidal voltage. The resulting map generated by the strange attractor is then used to produce a control signal. Using this method the author was able to stabilize almost all uncontrolled unstable orbits up to period 23. These two papers essentially employ digital control schemes.

Extending this idea of using chaotic systems to build useful oscillators, Pecora and Carroll (1991) describe how one might "paste together" nonlinear subsystems to build useful chaotically driven systems. For example, one might build a circuit (the receiver) that can be synchronized with the chaotic output of another subsystem (the transmitter). In this way, one hopes to be able to introduce modulation on the chaotic carrier for secure transmission of information.

In another use of the chaotic attractor for control, Shinbrot, et al. (1990) of the University of Maryland, described how one can devise control algorithms to move a system from one chaotic orbit to another at will in the phase space. This idea uses the exponential sensitivity of a chaotic system and small perturbations (control impulses) to direct a system from one point on an attractor to a target point on the attractor.

There are two aspects to the oxymoron "controlling chaos." In one set of problems one examines how a control parameter can either quench a chaotically dynamic system or produce chaotic dynamics in an otherwise quiet or periodically behaved system. One example of this type of "Controlled Chaos" is a paper by Golnaraghi and Moon (1990) using a servo-controlled "pick and place" robotic device.

However, the new concepts of "controlling chaos" inspired by the Maryland group illustrate how one might constructively use the exponentially divergent nature of chaotic orbits and the extreme sensitivity of these systems to small perturbations to design useful systems. These problems might also become paradigms in the education of engineers into the new methods of "nonlinear design." This design philosophy recognizes that nonlinear systems exhibit multiple basins of dynamic behavior which can lead to more creative solutions than those based on linear dynamic systems.

PROBLEMS

4-1 Consider an inverted pendulum: a spherical mass at the end of a massless rod of length L. The pendulum is constrained by two rigid walls on each side. At equilibrium the pendulum mass will rest on one of the two walls. Assume that the rest angles are small and show that for undamped free vibrations the dynamics are governed by (see Shaw and Rand, 1989)

$$\ddot{x} - x = 0, \qquad |x| < 1$$
$$x \to -x, \qquad |x| = 1$$

Show that a saddle point exists at the origin of the phase space $(x, \dot{x})$. Sketch a few trajectories.

4-2 A mass particle is constrained to move under a constant gravity force in a circular path which lies in a vertical plane. Assume that the plane rotates with frequency Ω_0 about the vertical axis through the center of the circle. Find the value of Ω_0 for which the number of equilibria changes from one to three. Show that this problem is similar to a particle in a two-well potential.

4-3 Consider the two-degree-of-freedom system of a linear spring–

mass oscillator confined to the diameter of a circular disk with spring constant, k_r. Assume also that the disk can rotate about its axis with a linear torsional spring restraint, k_ϕ. Neglecting gravity and dissipation, show that the equations of motion take the form

$$m\ddot{r} - mr\dot{\varphi}^2 + k_r r = 0$$

$$J\ddot{\varphi} + m(r^2\ddot{\varphi} + 2r\dot{r}\dot{\varphi}) + k_\phi\varphi = 0$$

[*Hint:* Note that the kinetic energy is given by $\frac{1}{2}m(\dot{r}^2 + r\dot{\varphi}^2) + \frac{1}{2}J\dot{\varphi}^2$.] Show that energy is conserved in this problem, and also show that the problem can be posed in terms of Hamiltonian dynamics.

4-4 For the computationally ambitious, use the conservation of energy in Problem 4-3 to obtain a Poincaré map and show that integration of the equations in Problem 4-3 leads to periodic, quasiperiodic, and stochastic orbits as in Figure 1-13. (See also Chapter 3, Figures 3-34, 3-35.)

4-5 *Chaos in Gears.* Consider two planar spur gears where one tooth on one gear is enmeshed between two teeth on the second gear. Assume that the gear diameters at the contact circle are d_1, d_2 and that play, Δ, exists between the tooth in gear #1 and the two teeth in gear #2 (see Figure 4-14). Derive an equation of motion for gear #1 assuming that gear #2 oscillates sinusoidally with a small angle. Assume a coefficient of restitution at each gear tooth impact. How is this problem similar to or different from the Fermi map problem (4-2.18) shown in Figure 4-9*a*?

4-6 *The Kicked Rotor.* Equations for the kicked rotor in Chapter 3 [Eqs. (3-5.4)] were derived for a damped system. Derive the two-dimensional map for zero damping ($c = 0$). Iterate this map on a small computer. Do you expect fractal structure?

4-7 *Chaotic Comet.* The discussion leading to the difference equations (4-2.5) describes a simple map for the dynamics of Halley's comet as perturbed by the orbit of Jupiter. In one model the perturbation function $F(x)$ in (4-2.5) is represented by the sawtooth curve in Figure 4-4*a*. Assume that the positive and negative amplitudes are equal and are given by $|F_{max}| = 6.35 \times 10^{-3}$ with positions $x_+ = 0.552$, $x_- = 0.640$. Also, $F(0.083) = 0$. Use the mod 1 property of x to find the equations of the piecewise linear curves and iterate the map to obtain Figure 4.4b. Can you show whether this map is area preserving or not? [See Chirikov and Vecheslavov (1989) for a discussion of the physics.]

4-8 *Tumbling of Hyperion.* The rigid body tumbling of the irregularly shaped satellite of Saturn called Hyperion has been modeled by the equation (4-2.4) by Wisdom et al. (1984), where the perturbing functions $f(t)$ and $r(t)$ are periodic in time. In the original paper, $f(t)$ and $r(t)$ are found from the elliptic orbit of the satellite using an eccentricity of the orbit of $e = 0.1$. However, Eq. (4-2.4) is analogous to a periodically forced pendulum. As an approximation, assume that $r = 1$ (i.e., not time-dependent), $\omega_0 = 0.2$, and $f = \theta_0\cos t$. Numerically integrate these equations on a computer and plot θ versus $\dot\theta$ when $f(t) = 0$ (i.e., a Poincaré map). Choose several values of $\theta_0 < \pi/2$ and several different initial conditions and compare your results with Figures 1 and 2 of Wisdom et al. (1984).

4-9 When $f(t)$ and $r(t)$ are periodic in Problem 4-8, show that the dynamical system (4-2.4) is a Hamiltonian problem (e.g., chaotic orbits will be stochastic and not exhibit fractal structure).

4-10 *Chaotic Scatterer.* Imagine two semi-infinite one-dimensional mechanical wave guides in which the displacement field $u(x, t)$ satisfies the classical wave equation $u_{tt} = c_0^2 u_{xx}$. The first wave guide lies along the negative x-axis, whereas the second lies along the positive x-axis. The two wave guides are connected by two springs to a scatterer of mass m. The left-hand spring is assumed to be linear of strength k, whereas the right-hand spring force has a cubic nonlinearity of the form $\kappa[u(0, t) - U(t)]^3$, where $U(t)$ is the displacement of the scatterer. Assuming that the force in the wave guide is given by $\kappa\partial u/\partial x$, derive a set of ordinary differential equations for the motion of the scatterer when periodic right moving waves are generated at $x \to -\infty$. [*Answer:* $\dot U = V$, $\dot V = -F_1(U - g - f) - F_2(U - h)$, $\dot g = \alpha_1 F_1(U - g - f) + \dot f$, $\dot h = \alpha_2 F_2(U - h)$, $F_1(x) = kx$, $F_2(x) = \kappa x^3$, $f = f_0\cos \omega t$.]

4-11 In the case of the nonlinear scatterer of Problem 4-10, show that elements of the phase space contract under the action of the dynamics. Where does the damping come from?

4-12 A magnetic compass needle is assumed to be pivoted at its center and subjected to a rotating magnetic field of intensity B_0.

Assume that the needle carries a magnetic moment **M** and that the torque about the axis is given by the cross-product of the moment **M** and the magnetic field. Derive the equation of motion (Croquette and Poitou, 1981)

$$J\ddot{\theta} = -\mu[\sin(\theta - \omega t) + \sin(\theta + \omega t)] - \gamma\dot{\theta}$$

When $\gamma = 0$, is this a Hamiltonian system? Show either analytically or computationally that either clockwise or counterclockwise motions are solutions.

4-13 Consider the magnetic pendulum described by (4-2.13). Derive an equation of motion for the case when the magnetic field is uniform, but whose direction oscillates with a small angle about the vertical. [Set $B_d = 0$ in (4-2.13).] Show that the equation is similar to the equation (4-2.4) for the tumbling of the Saturnian satellite Hyperion when $r(t) = $ constant.

4-14 Consider a simple serial circuit with a nonlinear iron core inductor, a linear resistor, and a periodic voltage source. Assume that the flux in the inductor depends on the cube of the current. Derive the equation of motion and show how it is related to the equation for the Japanese attractor (4-6.1).

4-15 The simplest model of a laser is written as a system of three first-order differential equations with a quadratic nonlinearity (4-6.11). Show how these equations can be transformed under certain assumptions into the Lorenz equations for thermal convection (1-3.9) (Haken, 1985).

5

EXPERIMENTAL METHODS IN CHAOTIC VIBRATIONS

Perfect logic and faultless deduction make a pleasant theoretical struc-
ture, but it may be right or wrong; The experimenter is the only one to
decide, and he is always right.

L. Brillouin
Scientific Uncertainty and Information, 1964

5.1 INTRODUCTION: EXPERIMENTAL GOALS

A review of physical systems which exhibit chaotic vibrations was
presented in Chapter 4. In this chapter, we discuss some of the experimental techniques that have been used successfully to observe and
characterize chaotic vibrations and strange attractors. To a great extent, these techniques are specific to the physical medium—for example, rigid body, elastic solid, fluid, optical, or reacting medium. However, many of those measurements which are unique to chaotic
phenomena, such as Poincaré maps or Lyapunov exponents, are applicable to a wide spectrum of problems. Since publication of the first
edition, some researchers have turned their attention to the spatial as
well as to the temporal aspects of chaos. In this chapter we shall focus
only on temporal chaos. An introduction to spatial dynamics is given
in Chapter 8 along with a few experimental examples.

A diagram outlining the major components of an experiment is
shown in Figure 5-1. The source of the vibration is either (a) an external
energy source such as an electromagnetic shaker or (b) an internal
source of self-excitation. In the case of an autonomous system, such

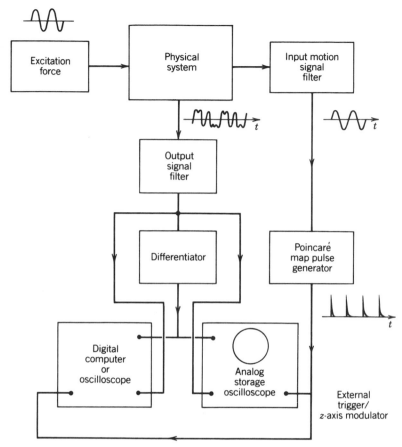

Figure 5-1 Diagram showing components of an experimental system to measure the Poincaré map of a chaotic physical system.

as the Rayleigh–Benard convection cell, the source of instability is a prescribed temperature difference across the cell, and the nonlinearities reside in the convective terms in the acceleration of each fluid element.

The other major elements include *transducers* to convert physical variables into electronic voltages, a *data acquisition* and storage system, *graphical display* (such as an oscilloscope), and data analysis computer.

The techniques that must be mastered for experiments in chaotic vibrations depend on some extent on the goals that one sets up for the experimental study. These goals could include the following:

1. Establish existence of chaotic vibration in a particular physical system.

2. Determine critical parameters for bifurcations.
3. Determine criteria for chaos.
4. Map out chaotic regimes.
5. Measure qualitative features of chaotic attractor—for example, Poincaré maps.
6. Measure quantitative properties of attractor—for example, Fourier spectrum, Lyapunov exponent, probability density function.
7. Seek methods to quench, control, prevent or exploit chaos in a technical system.

5.2 NONLINEAR ELEMENTS IN DYNAMICAL SYSTEMS

The phenomena of chaotic vibrations cannot occur if the system is linear. Thus, in performing experiments in chaotic dynamics, one should understand the nature of the nonlinearities in the system. To refresh one's memory, a linear system is one in which the principle of superposition is valid. Thus if $x_1(t)$ and $x_2(t)$ are each possible motions of a given system, then the system is linear if the sum $c_1x_1(t) + c_2x_2(t)$ is also a possible motion. Another form of the superposition principle is more easily described in mathematical terms. Suppose the dynamics of a given system can be modeled by a set of differential or integral equations of the form

$$L[\mathbf{X}] = \mathbf{f}(t) \qquad (5\text{-}2.1)$$

where $\mathbf{X} = (x_1, x_2, \ldots, x_k(t), \ldots, x_n)$ represents a set of independent dynamical variables that describe the system. Suppose the system is forced by two different input functions $\mathbf{f}_1(t)$ and $\mathbf{f}_2(t)$ with outputs $\mathbf{X}_1(t)$ and $\mathbf{X}_2(t)$. Then if the system is linear, the effect of two simultaneous inputs can be easily found:

$$L[c_1\mathbf{X}_1 + c_2\mathbf{X}_2] = c_1\mathbf{f}_1(t) + c_2\mathbf{f}_2(t) \qquad (5\text{-}2.2)$$

The only way that this property can hold is for the terms in the differential equations (5-2.1) to be to the first power $\mathbf{X}$ or $\dot{\mathbf{X}}_1$, and so on—hence the term *linear system*. Nonlinear systems involve the unknown functions in forms other than to the first power, that is, x^2, x^3, $\sin x$, x^a, $1/(x^2 + b)$, or similar forms for the derivatives or integrals of the function, for example, $\dot{x}^2$, $[\int x\, dt]^2$.

Experimental nonlinearities can be created in many ways, some of them quite subtle. In mechanical or electromagnetic systems, nonlinearities can occur in the following forms:

(a) Nonlinear material or contitutive properties (stress versus strain, voltage versus current)
(b) Nonlinear acceleration or kinematic terms (e.g., centripetal or Coriolis acceleration terms)
(c) Nonlinear body forces
(d) Geometric nonlinearities

Material Nonlinearities

Examples of material nonlinearities in mechanical and electrical systems include the following:

Solid Materials. Nonlinear stress versus strain: (1) elastic (e.g., rubber) and (2) inelastic (e.g., steel stressed beyond the yield point; also plasticity, creep).

Magnetic Materials. Nonlinear magnetic field intensity **H** versus flux density **B**:

$$\mathbf{B} = \mathbf{f(H)}$$

(e.g., ferromagnetic material iron, nickel cobalt—hysteretic in nature).

Dielectric Materials. Nonlinear electric displacement **D** versus electric field intensity **E**:

$$\mathbf{D} = \mathbf{f(E)}$$

(e.g., ferroelectric materials).

Electric Circuit Elements. Nonlinear voltage versus current:

$$V = f(I)$$

[e.g., Zener and tunnel diodes, nonlinear resistors, field effect transistors (FET), metal oxide semiconductors (MOSFET)]. Nonlinear voltage versus charge:

$$V = g(Q)$$

(e.g., capacitors). Other material nonlinearities include nonlinear optical materials (e.g., lasers), heat-flux–temperature gradient properties, nonlinear viscosity properties in fluids, voltage–current relations in electric arcs, and dry friction.

Kinematic Nonlinearities

This type of nonlinearity occurs in fluid mechanics in the Navier–Stokes equations, where the acceleration term includes a nonlinear velocity operator

$$v\frac{\partial v}{\partial x} \quad \text{or} \quad \mathbf{v} \cdot \nabla \mathbf{v}$$

which represents convective effects. (See Eq. (1-1.3).)

In particle dynamics, one often uses local coordinate systems to describe motion relative to some inertial reference frame. When the local frame rotates with angular velocity Ω relative to the large frame, the absolute acceleration is given by

$$\mathbf{A} = \mathbf{a} + \mathbf{A}_0 + \dot{\mathbf{\Omega}} \times \boldsymbol{\rho} + \mathbf{\Omega} \times \mathbf{\Omega} \times \boldsymbol{\rho} + 2\mathbf{\Omega} \times \mathbf{v} \quad (5\text{-}2.3)$$

where $\mathbf{A}_0$ is the acceleration of the origin of the small frame relative to the reference, $\boldsymbol{\rho}$ and $\mathbf{v}$ are local position vector and velocity, respectively, of the particle. The last two terms are called the *centripetal* and *Coriolis* acceleration terms. The last three terms are *nonlinear* in the variables $\boldsymbol{\rho}$, $\mathbf{v}$, Ω.

For a rigid body in pure rotation, these terms appear in Euler's equations for the rotation dynamics [see Eq. (4-2.14)]:

$$M_x = I_x\frac{d\omega_x}{dt} - (I_z - I_y)\omega_y\omega_z$$

$$M_y = I_y\frac{d\omega_y}{dt} - (I_x - I_z)\omega_z\omega_x \quad (5\text{-}2.4)$$

$$M_z = I_z\frac{d\omega_z}{dt} - (I_y - I_x)\omega_x\omega_y$$

where (M_x, M_y, M_z) are applied force moments and (I_x, I_y, I_z) are principal second moments of mass about the center of mass.

Nonlinear Body Forces

Electromagnetic forces are represented as follows:

$$\text{Currents:} \qquad F = \alpha I_1 I_2 \quad \text{or} \quad \beta IB$$

$$\text{Magnetization:} \quad \mathbf{F} = \mathbf{M} \cdot \nabla \mathbf{B}$$

$$\text{Moving Media:} \quad \mathbf{F} = q\mathbf{v} \times \mathbf{B}$$

(Here I is current, $\mathbf{B}$ is the magnetic field, $\mathbf{M}$ is the magnetization, q represents charge, and $\mathbf{v}$ is the velocity of a moving charge.)

Geometric Nonlinearities

Geometric nonlinearities in mechanics involve materials with linear stress–strain behavior, but the geometry changes with deformation. A classic example of a geometric nonlinearity is the elastica shown in Figure 5-2. In this problem, the material is linearly elastic but the large deformations produce a nonlinear force–displacement or moment–angle relation of the form

$$M = A\kappa$$

$$\kappa = \frac{u''}{[1 + (u')^2]^{3/2}} \tag{5-2.5}$$

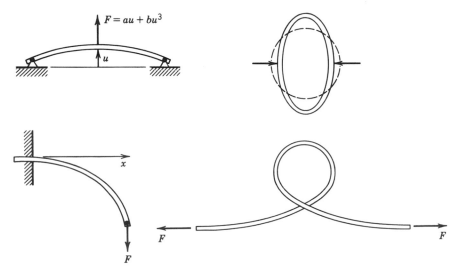

Figure 5-2   Examples of geometric nonlinearities in elastic structures.

where M is the bending moment, κ is the curvature of the neutral axis of the beam, and $u(x)$ is the transverse displacement of the beam. This problem is an interesting one for study of chaotic vibrations because the elastica can exhibit multiple equilibrium solutions (see Chapter 8). Cylindrical and spherical shells also exhibit geometric elastic nonlinearities (see Evenson, 1967).

5.3 EXPERIMENTAL CONTROLS

First and foremost, the experimenter in chaotic vibrations should have control over noise, both mechanical and electronic. If one is to establish chaotic behavior for a deterministic system, the noise inputs to the system must be minimized.

For example, mechanical experiments such as vibration of structures or autonomous fluid convection problems should be isolated from external laboratory or building vibrations. This can be accomplished by using a large-mass table with low-frequency air bearings. A low-cost solution is to work at night when building noise is at a minimum.

Second, one should build in the ability to control significant physical parameters in the experiments, such as forcing amplitude or temperature gradient. This is especially important if one wishes to observe bifurcation sequences such as period-doubling phenomena. Where possible, one should use continuous element controls and avoid devices with incremental or step changes in the parameters. In some problems, there is more than one dynamic motion for the same parameters. Thus, control over the initial state variables may also be important.

Another factor is the number of significant figures required for accurate measurement. For example, to plot Poincaré maps from digitally sampled data, an 8-bit system may not be sensitive enough and one may have to go with 12-bit electronics or better in order to resolve the fine fractal structure in the maps.

Frequency Bandwidth

Most experiments in fluid, solid, or reacting systems may be viewed as infinite-dimensional continua. However, one often tries to develop a mathematical model with a few degrees of freedom to explain the major features of the chaotic or turbulent motions of the system. This is usually done by making measurements at a few spatial locations in the continuous system and by limiting the frequency bandwidth over which one observes the chaos. This is especially important if velocity measurements for phase-plane plots are to be made from deformation histories. Electronic differentiation will amplify higher-frequency signals, which may not be of interest in the experiment. Thus, extremely good electronic filters are often required, especially ones that have little or no phase shift in the frequency band of interest.

5.4   PHASE-SPACE MEASUREMENTS

It was pointed out in Chapter 2 that chaotic dynamics are most easily unraveled and understood when viewed from a phase-space perspective. In particle dynamics, this means a space with coordinates composed of the position and velocity for each independent degree of freedom. In forced problems, time becomes another dimension. Thus, the periodic forcing of a two-degree-of-freedom oscillator with generalized positions $(q_1(t), q_2(t))$ has a phase-space representation with coordinates $(q_1, \dot{q}_1, q_2, \dot{q}_2, \omega t)$, where ω is the forcing frequency. (The phase variable ωt is usually plotted modulo 2π.)

If one measures displacement $q(t)$, a differentiation circuit is required. If velocity is measured, the phase space may be spanned by $(v, \int v \, dt)$, which calls for an integrator circuit. As noted above, in building integrator or differentiator circuits, care should be taken that the phase as well as the amplitude is not disturbed within the frequency band of interest.

In electronic or electrical circuit problems, the current and voltage can be used as state variables. In fluid convection problems, temperature and velocity variables are important.

Pseudo-Phase-Space Measurements

In many experiments, one has access to only one measured variable $\{x(t_1), x(t_2), \ldots\}$ (where t_1 and t_2 are sampling times, not to be confused with Poincaré maps). When the time increment is uniform, that is, $t_2 = t_1 + \tau$ and so on, then a pseudo-phase-space plot can be made using $x(t)$ and its past (or future) values:

$(x(t), x(t - \tau))$ or $(x(t), x(t + \tau))$

(two-dimensional phase space)

$(x(t), x(t - \tau), x(t - 2\tau))$

(three-dimensional phase space)

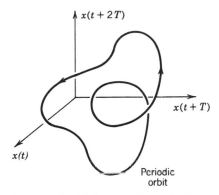

Figure 5-3 Periodic trajectory of a third-order dynamical system using pseudo-phase-space coordinates.

One can show that a closed trajectory in a phase space in $(x, \dot{x})$ variables will be closed in the $(x(t), x(t - \tau))$ variables (one must connect the points when the system is digitally sampled) as shown in Figure 5-3. Likewise, chaotic trajectories in $(x, \dot{x})$ look chaotic in $(x(t), x(t - \tau))$ variables. The plots can be carried out after the experiment by a computer, or one may perform on-line pseudo-phase-plane plots using a *sample-and-hold circuit.*

The one difficulty with pseudo-phase-space variables is taking a Poincaré map. For example, when there is a natural time scale, such as in forced periodic motion of a system with frequency ω, the sample time τ is usually chosen much smaller than the driving period, that is, $\tau \ll 2\pi/\omega \equiv T$. If τ is not an integer of T, Poincaré maps may lose some of their fine fractal structure.

5.5 BIFURCATION DIAGRAMS

As discussed in Chapter 2, one of the signs of impending chaotic behavior in dynamical systems is a series of changes in the nature of the periodic motions as some parameter is varied. Typically, in a single-degree-of-freedom oscillator, as the control parameter approaches a critical value for chaotic motion, subharmonic oscillations appear. In the now classic "logistic equation," a series of period-2 oscillations appear [Eq. (1-3.6)]. The phenomenon of sudden change in the motion as a parameter is varied is called a *bifurcation.* A sample experimental bifurcation diagram is shown in Figure 5-4. Such diagrams can be obtained experimentally by time sampling the motion as in a Poincaré map and displaying the output on an oscilloscope as

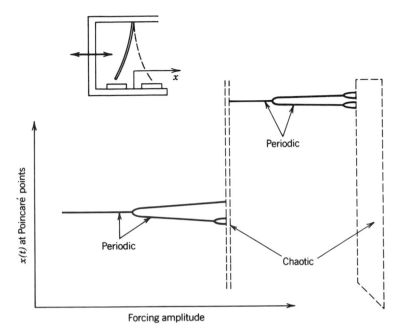

Figure 5-4 Experimental bifurcation diagram for the vibration of a buckled beam: Poincaré map samples of bending displacement versus amplitude of forcing vibrations.

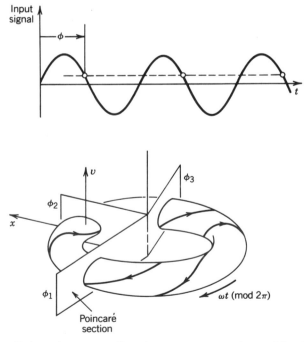

Figure 5-5 *Top:* Poincaré map sampling times at constant phase of forcing function. *Bottom:* Geometric interpretation of Poincaré sections in the three-dimensional phase space.

shown in Figure 5-4. Here the value of the control parameter—for example, a forcing amplitude or frequency—is plotted on the horizontal axis and the time-sampled values of the motion are plotted on the vertical axis. This diagram actually represents a series of experiments, where each value of the control parameter is an experiment. When the control parameter can be varied automatically, such as by a computer and digital-to-analog device, the diagram can be obtained quite rapidly. Care must be taken, however, to make sure transients have died out after each change in the control parameter.

In the bifurcation diagram of Figure 5-4, the continuous horizontal lines represent periodic motions of various subharmonics. The values in the dashed-line areas represent chaotic regions. The boundary between chaotic and periodic motions can clearly be seen in this diagram.

When this is automated, one must be careful not to mistake a quasiperiodic motion for a chaotic motion. A phase-plane Poincaré map is still very useful for distinguishing between quasiperiodic and chaotic motions.

5.6   EXPERIMENTAL POINCARÉ MAPS

Poincaré maps are one of the principal ways of recognizing chaotic vibrations in low-degree-of-freedom problems (see Table 2-2). We recall that the dynamics of a one-degree-of-freedom forced mechanical oscillator or $L–R–C$ circuit may be described in a three-dimensional phase space. Thus, if $x(t)$ is the displacement, $(x, \dot{x}, \omega t)$ represents a point in a cylindrical phase space where $\phi = \omega t$ represents the phase of the periodic forcing function. A Poincaré map for this problem consists of digitally sampled points in this three-dimensional space, for example, $(x(t_n), \dot{x}(t_n), \omega t_n = 2\pi n)$. As discussed in Chapter 2, this map can be thought of as slicing a torus (see Figure 5-5).

Experimentally, this can be done in several ways. If one has a storage oscilloscope, the Poincaré map is obtained by intensifying the image on the screen at a certain phase of the forcing voltage (sometimes called *z-axis modulation*) (Figure 5-1). In our laboratory, we were able to generate a 5 to 10 V pulse of 1 to 2 μs duration when the forcing function reached a certain phase:

$$\omega t_n = \phi_0 + 2\pi n \qquad (5\text{-}6.1)$$

This pulse was then used to intensify a phase-plane image, $(x(t), \dot{x}(t))$, using two vertical amplifiers as in Figure 5-6.

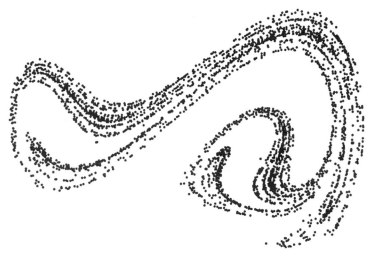

Figure 5-6 Example of an experimental Poincaré map for periodic forcing of a buckled beam.

One can also use a digital oscilloscope in an external sampling rate mode with the same narrow pulse signal used for the analog oscilloscope. A similar technique can be employed using an analog-to-digital (A-D) signal converter by storing the sampled data in a computer for display at a later time. The important point here is that the sampling trigger signal must be exactly synchronous with the forcing function.

Poincaré Maps—Change of Phase. As noted in Chapter 2, chaotic phase-plane trajectories can often be unraveled using the Poincaré map by taking a set of pictures for different phases ϕ_0 in (5-6.1) (see Figure 5-7). This is tantamount to sweeping the Poincaré plane in Figure 5-5. While one Poincaré map can be used to expose the fractal nature of the attractor, a complete set of maps varying ϕ_0 from 0 to 2π is sometimes needed to obtain a complete picture of the attractor on which the motion is riding.

A series of pictures of various cross sections of a chaotic torus motion in a three-dimensional phase space is shown in Figure 5-7. Note the symmetry in the $\varphi = 0°$ and 180° maps for the special case of the buckled beam.

Poincaré Maps—Effect of Damping. If a system does not have suffi-cient damping, then the chaotic attractor will tend to uniformly fill up a section of phase space and the Cantor set structure which is

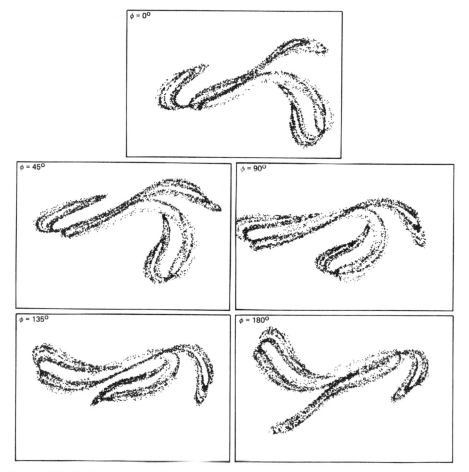

Figure 5-7 Poincaré maps of a chaotic attractor for a buckled beam for different phases of the forcing function.

characteristic of strange attractors will not be evident. An example of this is shown in Figure 2-11 for the vibration of a buckled beam. A comparison of low- and high-damping Poincaré maps shows that adding damping to the system can sometimes bring out the fractal structure. On the other hand, if the damping is too high, the Cantor set sheets can appear to collapse onto one curve as in Figure 3-14.

Poincaré Maps—Quasiperiodic Oscillations. Often what appears to be chaotic may very simply be a superposition of two incommensurate harmonic motions, for example,

$$x(t) = A \cos(\omega_1 t + \phi) + B \cos(\omega_2 t + \phi_2) \qquad (5\text{-}6.2)$$

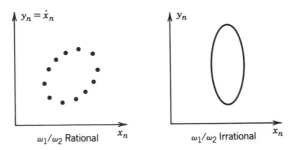

Figure 5-8 Poincaré map of a motion with two harmonic signals with different frequencies.

where ω_1/ω_2 is irrational. One can use one frequency to sample a Poincaré map, for example, $\omega_1 t_n = 2\pi n$. Then the phase-plane points $(x(t_n), \dot{x}(t_n))$ will fill in an elliptically shaped closed curve (Figure 5-8). If ω_1/ω_2 is rational, a finite set of points will be seen in the Poincaré map. The case of ω_1/ω_2 irrational can be thought of as motion on a torus or doughnut-shaped figure in a three-dimensional phase space.

When three or more incommensurate frequencies are present, one may not see a nice closed curve in the Poincaré map and the Fourier spectrum should be used. The difference between chaotic and quasiperiodic motion can also be detected by taking the Fourier spectrum of the signal. A quasiperiodic motion will have a few well-pronounced

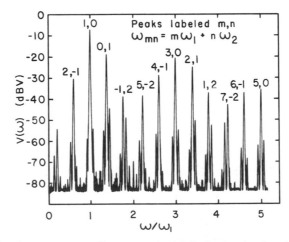

Figure 5-9 Fourier spectrum of an experimental electronic signal from a circuit with a nonlinear inductor. Frequency components are linear combinations of two frequencies. [From Bryant and Jeffries (1984a) with permission of the American Physical Society, copyright 1984.]

peaks as shown in Figure 5-9. Chaotic signals often have a broad spectrum of Fourier components as in Figure 2-7.

Position-Triggered Poincaré Maps

When one does not have a natural time clock, such as a periodic forcing function, then more sophisticated techniques must be used to get a Poincaré map (see also Henon, 1982).

Suppose we imagine the motion as a trajectory in a three-dimensional space with coordinates (x, y, z). To construct a Poincaré map, we intercept a trajectory with a plane defined by

$$ax + by + cz = d \qquad (5\text{-}6.3)$$

as shown in Figure 5-10. The Poincaré map consists of those points in the plane for which the trajectory penetrates the plane with the same sense [i.e., if we define a front and back to the plane (5-6.3), then we collect points only on trajectories that penetrate the plane from front to back or back to front, but not both ways.]

Experimentally, one can do this by using a mechanical or electronic *lever detector*. Examples of position-triggered Poincaré maps are discussed below.

In the impact oscillator shown in Figure 5-11, there are three convenient state variables: the position x, velocity v, and phase of the driving signal $\phi = \omega t$. If one triggers on the position when the mass hits the elastic constraint, the Poincaré map becomes a set of values $(v_n^{\pm}, \omega t_n)$

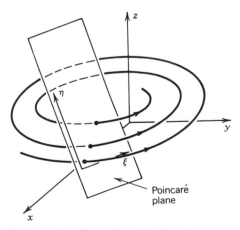

Figure 5-10   General Poincaré surface of section in phase space for the motion of a third-order dynamical system.

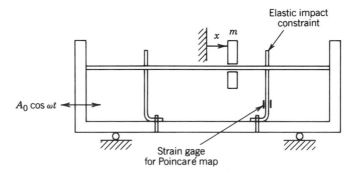

Figure 5-11   Sketch of experimental setup for a position-triggered Poincaré section.

where $v_n^{\pm}$ is the velocity before or after impact and t_n is the time of impact. Here the points in the map can be plotted in a cylindrical space where $0 < \omega t_n < 2\pi$.

An example of the experimental technique to obtain a (v_n, ϕ_n) Poincaré map is shown in Figure 5-11. When the mass hits the position constraint, a sharp signal is obtained from a strain gauge or accelerom-

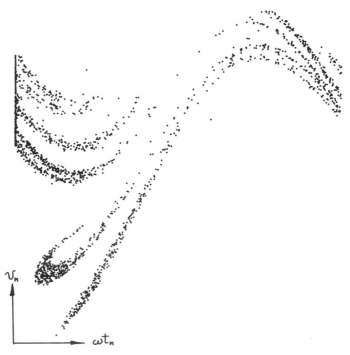

Figure 5-12   Position-triggered Poincaré map for an oscillating mass with impact constraints (Figure 5-11).

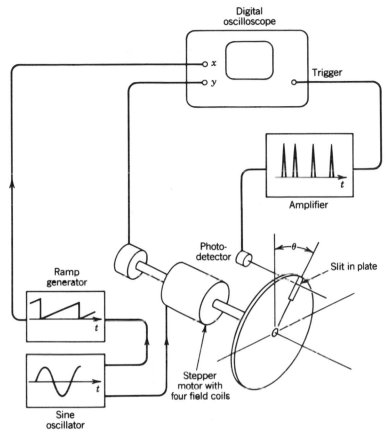

Figure 5-13   Diagram of experimental apparatus to obtain position-triggered Poincaré maps for a periodically forced rotor with a nonlinear torque–angle relation.

eter. This sharp signal can be used to trigger a data storage device (such as a storage or digital oscilloscope) to store the value of the velocity of the particle. [In the case shown in Figure 5-11, a linear variable differential transformer (LVDT) is used to measure position, and this signal is electronically differentiated to get the velocity.] To obtain the phase ϕ_n mod 2π, we generate a periodic ramp signal in phase with the driving signal where the minimum value of zero corresponds to $\phi = 0$, and the maximum voltage of the ramp corresponds to the phase $\phi = 2\pi$. The impact-generated sharp spike voltage is used to trigger the data storage device and store the value of the ramp voltage along with the velocity signal before or after impact. A Poincaré map for a mass bouncing between two elastic walls using this (v_n, ϕ_n) technique is shown in Figure 5-12. (See also Figure 4-14.)

Figure 5-14   Position-triggered Poincaré map for chaos in a nonlinear rotor (see Figure 5-13).

Another example of this kind of Poincaré map is shown in Figure 5-13 for the chaotic vibrations of a motor. In this problem, the motor has a nonlinear torque–angle relation created by a dc current in one of the stator poles, and the permanent magnet rotor is driven by a sinusoidal torque created by an ac current in an adjacent coil. The equation of motion for this problem is [see Section 4.2, Eq. (4-2.13)]

$$J\ddot{\theta} + \gamma\dot{\theta} + \kappa \sin \theta = F_0 \cos \theta \cos \omega t \qquad (5\text{-}6.4)$$

To obtain a Poincaré map, we choose a plane in the three-dimensional space $(\theta, \dot{\theta}, \omega t)$, where $\theta = 0$ (Figure 5-13). This is done experimentally by using a slit in a thin disk attached to the rotor and using a light-emitting diode and detector to generate a voltage pulse every time the rotor passes through $\theta = 0$ (see Figure 5-13). This pulse is then used to sample the velocity and measure the time. The data can be directly

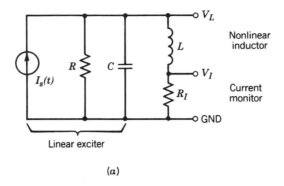

(a)

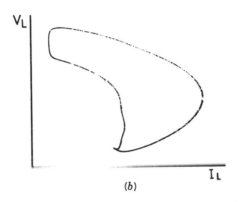

(b)

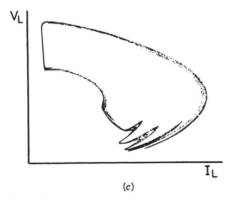

(c)

Figure 5-15   Peak amplitude-generated Poincaré maps for a circuit with nonlinear inductance [from Bryant and Jeffries (1984a) with permission of the Americal Physical Society, copyright 1984].

displayed on a storage oscilloscope or, using a computer, can be replotted in polar coordinates as shown in Figure 5-14.

Another variation of the method of Poincaré sections is to sample data when some variable attains a *peak* value. This has been used by Bryant and Jefferies (1984b) of the University of California—Berkeley. They examined the dynamics of a circuit with a nonlinear hysteretic iron core inductor shown in Figure 5-15. (The nonlinear properties are related to the ferromagnetic material in the inductor.) They sampled the current in the inductor $I_L(t)$ as well as the driving voltage $V_s(t)$, when $V_L = 0$. This is tantamount to measuring *peak* value of the flux in the inductor φ. This is because $V_L = -\dot{\varphi}$, where φ is the magnetic flux in the inductor, and $\varphi = \hat{\varphi}(I)$, so that when $\dot{\varphi} = 0$, the flux is at a maximum or minimum. The Poincaré map is then a collection of pairs of points (V_{sn}, I_{Ln}) which can be displayed on a storage or digital oscilloscope.

Construction of One-Dimensional Maps from Multidimensional Attractors

There are a number of physical and numerical examples where the attracting set appears to have a sheetlike behavior in some three-dimensional phase space as illustrated in Figure 5-16. [The Lorenz equations (1-3.9) are such an example. See also Section 3.8.] This often means that a Poincaré section, obtained by measuring the sequence of

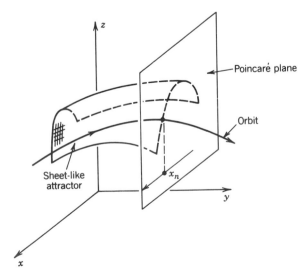

Figure 5-16   Construction of a one-dimensional return map in a three-dimensional phase space.

points which pierce a plane transverse to the attractor, will appear as a set of points along some one-dimensional line. This suggests that if one could parameterize these points along the line by a variable x, it would be possible that a function exists which relates x_{n+1} and x_n:

$$x_{n+1} = f(x_n)$$

The function (called a *return map*) may be found by simply plotting x_{n+1} versus x_n. One example of this is the experiments of Shaw (1984) on the dripping faucet shown in Figure 4-43 or the nonlinear circuit in Figure 4-31 (see also Simoyi et al., 1982). The existence of such a function $f(x)$ implies that the mathematical results for one-dimensional maps, such as period doubling and Feigenbaum scaling (Section 3.6), may be applicable to the more complex physical problems in explaining, predicting, or organizing experimental observations.

For some problems, the function $f(x)$, when it exists, appears to cross itself or is tangled. This may suggest that the mapping function can be untangled by plotting the dynamics in a higher-dimensional embedding space using three successive values of the Poincaré sampled data $[x(t_n), x(t_{n+1}), \text{ and } x(t_{n+2})]$. The three-dimensional nature of the relationship can sometimes be perceived by changing the projection of the three-dimensional curve onto the plane of a graphics computer monitor. This may suggest a special two-dimensional map of the form

$$x_{n+2} = f(x_{n+1}, x_n)$$

$$x_{n+1} = f(x_n, y_n) \tag{5-6.5}$$

$$y_{n+1} = Ax_n$$

This form is similar to the Henon map (1-3.8). This method has been used successfully by Van Buskirk and Jeffries (1985) in their study of circuits with p–n junctions and by Brorson et al. (1983), who studied a sinusoidally driven resistor–inductor circuit with a varactor diode.

Example: 1-D Map for a Friction Oscillator. In Chapter 4, we described experiments with a dry friction oscillator (Figure 4-26) which was modeled as a forced oscillator of the form (see Feeny, 1990)

$$\ddot{x} + F(x, \dot{x}) + \omega_0^2 x = f_0 \sin \Omega t \tag{5-6.6}$$

The natural phase space is thus three-dimensional $(x, \dot{x}, \Omega t)$, and a Poincaré map triggered on the forcing phase would lead to a

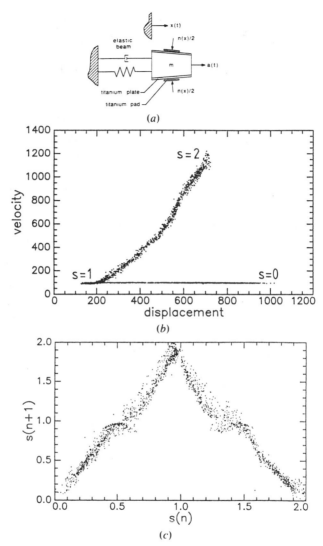

Figure 5-17   (a) Sketch of a one-degree-of-freedom oscillator with dry-friction force. (b) Poincaré map of chaotic dynamics; $0 \leq s < 1$ represents "sticking" motions, whereas $1 \leq s \leq 2$ represents slipping motions. (c) Return map based on the above Poincaré map. [From Feeny and Moon (1989).]

2-D map, $(x_n, \dot{x}_n)$. A sample of such a Poincaré map is shown in Figure 5-17a, It shows an almost 1-D structure and exhibits no obvious fractal patterns. This suggests trying a 1-D map by replotting the 2-D map by projecting the points onto a one-dimensional manifold $0 \leq S \leq 2$. In the experiment, this resulted in a tent- or hump-type map as shown in Figure 5-17b:

$$S_{n+1} = F(S_n) \qquad (5\text{-}6.7)$$

where $0 \le S < 1$ corresponds to the sticking regime and $1 \le S \le 2$ corresponds to the slipping regime.

Double Poincaré Maps

So far we have only talked of Poincaré maps for third-order systems, such as a single-degree-of-freedom oscillator with external forcing. But what about higher-order systems with motion in a four- or five-dimensional phase space? For example, a two-degree-of-freedom autonomous aeroelastic problem would have motion in a four-dimensional phase space (x_1, v_1, x_2, v_2). A Poincaré map triggered on one of the state variables would result in a set of points in a three-dimensional space. The fractal nature of this map, if it exists, might not be evident in three dimensions and certainly not if one projects this three-dimensional map onto a plane in two of the remaining variables.

A technique to observe the fractal nature of three-dimensional Poincaré map of a fourth-order system has been developed in our laboratory which we call a *double Poincaré section* (see Figure 5-18). This technique enables one to slice a finite-width section of the three-dimensional map in order to uncover fractal properties of the attractor and hence determine if it is "strange" (see Moon and Holmes, 1985).

We illustrate this technique with an example derived from the forced motion of a buckled beam. In this case we examine a system with two incommensurate driving frequencies. The mathematical model has the form (See also Wiggins (1988), Section 4.2e)

$$
\begin{aligned}
\dot{x} &= y \\
\dot{y} &= -\gamma y + F(x) + f_1 \cos \theta_1 + f_2 \cos(\theta_2 - \phi_0) \\
\dot{\theta}_1 &= \omega_1 \\
\dot{\theta}_2 &= \omega_2
\end{aligned}
\qquad (5\text{-}6.8)
$$

The experimental apparatus for a double Poincaré section is shown in Figure 5-19. The driving signals were produced by identical signal generators and were added electronically. The resulting quasiperiodic signal was then sent to a power amplifier which drove the electromagnetic shaker.

The first Poincaré map was generated by a 1-μs trigger pulse synchronous with one of the harmonic signals. The Poincaré map (x_n, v_n)

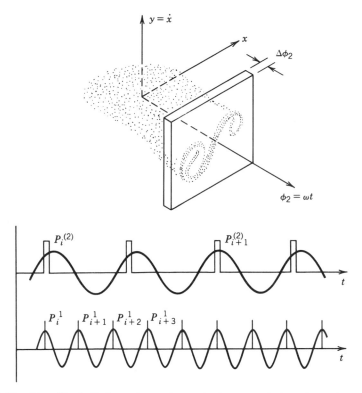

Figure 5-18   *Top:* Single Poincaré map dynamical system; finite width slice of second Poincaré section. *Bottom:* Poincaré sampling voltages for a second-order oscillator with two harmonic driving functions.

using one trigger results in a fuzzy picture with no structure as shown in Figure 5-20*a*. To obtain the second Poincaré section, we trigger on the second phase of the driving signal. However, if the pulse width is too narrow, the probability of finding points coincident with the first trigger is very small. Thus, we set the second pulse width 1000 times the first, at 1 ms. The second pulse width represents less than 1% of the second frequency phase of 2π. The (x, v) points were only stored when the first pulse was *coincident* with the second as shown in Figure 5-18. This was accomplished using a digital circuit with a logical NAND gate. Because of the infrequency of the simultaneity of both events, a map of 4000 points took upwards of 10 h compared to 8–10 min to obtain a conventional Poincaré map for driving frequencies less than 10 Hz.

The experimental results using this technique are shown in Figure 5-20 which compares a single with a double Poincaré map for the two-frequency forced problem. The single map is fuzzy, whereas

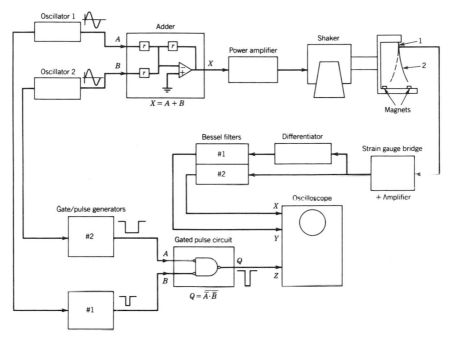

Figure 5-19 Sketch of experimental apparatus to obtain a Poincaré map for an oscillator with two driving frequencies. *Note:* Strain gauges—1; steel beam—2. [From Moon and Holmes (1985) with permission of Elsevier Science Publishers, copyright 1985.]

the double section reveals a fractal-like structure characteristic of a strange attractor.

One can of course generalize this technique to five- or higher-dimensional phase-space problems. However, the probability of three or more simultaneous events will be very small unless the frequency is order of magnitudes higher than 1–10 Hz. Such higher-dimensional maps may be useful in nonlinear circuit problems.

This technique can of course be used in numerical simulation and has been employed by Lorenz (1984) to examine a strange attractor in a fourth-order system of ordinary differential equations. Kostelich and Yorke (1985) have also employed this method to study the dynamics of a kicked or pulsed double rotor. They call the method "Lorenz cross sections" (see also Kostelich et al., 1987).

Experimentally Measured Circle Maps: Quasiperiodicity and Mode-Locking

In Chapter 2 we mentioned briefly the phenomena of quasiperiodicity and mode-locking when two oscillators interact. This can occur when

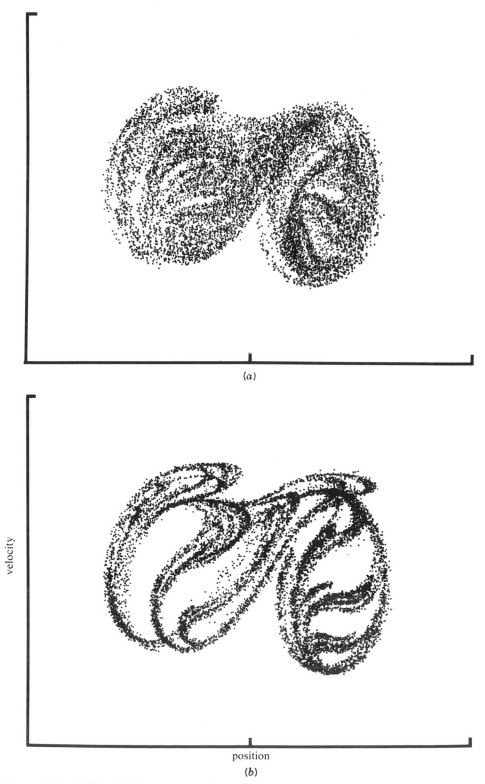

Figure 5-20 (*a*) Single Poincaré map of a nonlinear oscillator with two driving frequencies. (*b*) Double Poincaré map showing fractal structure characteristic of a strange attractor.

one tries to periodically force a nonlinear oscillator with frequency ω_2 which is initially in a limit cycle with frequency ω_1 (e.g., the forced Van der Pol oscillator; see Chapter 1). In many problems, the dynamics can be reduced to a circle map

$$\theta_{n+1} = \theta_n + \Omega + \frac{k}{2\pi} \sin 2\pi\theta_n, \quad (\text{mod } 1) \qquad (5\text{-}6.9)$$

An excellent review of mode-locking, quasiperiodicity, and the circle map from an experimentalist's point of view has been given by Glazier and Libchaber (1988). We shall not attempt to reproduce all the material in this paper, but we shall discuss a few techniques from this paper. These authors also survey the successful application of the circle map to experimental problems, including periodically forced Rayleigh–Benard convection (Jensen et al., 1985), and solid-state systems [e.g., electrically forced germanium (Held and Jeffries, 1986), electrically driven barium sodium niobate (Martin and Martienssen, 1986), and a periodically forced semiconductor laser (Winful et al., 1986)]. In all of these studies, Arnold-tongue mode-locking regimes were found (Chapter 2, Figure 2-23) with dynamic observations qualitatively similar to those of the circle map.

If the measured data are sampled at a Poincaré section synchronous with the driving frequency ω_2, a set of data is generated $\{\cdots x_{i-1}, x_i, x_{i+1} \cdots\}$. The problem for the experimentalist is to find a mapping from x_i to θ_i.

In an approximate method, the Poincaré points $(x_n, \dot{x}_n)$ are plotted. If the closed curve for the quasiperiodic motion is symmetric and elliptic in shape, the points are projected onto a circle centered at the ellipse and the angle is measured on this circle. An example of a periodically excited flexible tube with steady fluid flow is shown in Figure 5-21a. However, if the closed curve of the Poincaré map is twisted or badly distorted, another method may be used.

Following Glazier and Libchaber (1988), one plots the x_i variable versus a fictitious time variable $\theta = Wt_i$ (mod 1). In general, a random choice of W will not reveal any pattern to the x_i versus Wt_i curve. (In practice, one starts close to the uncoupled frequency ratio ω_1/ω_2.) If there is an underlying one-dimensional generalized circle map, a unique value of W will reveal a periodic function. For values of W close to the critical value, the curve will drift. The procedure is best done in an interactive mode with a computer terminal screen. The experimenter replots the data $\{x_i\}$ for different values of W until a unique curve is achieved as in Figure 5-21b.

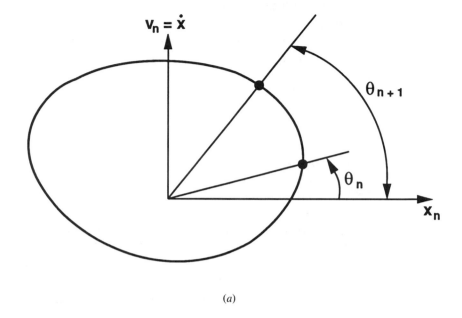

(a)

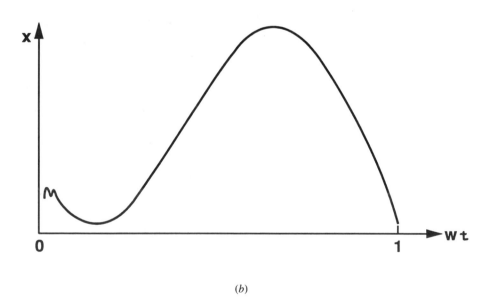

(b)

Figure 5-21 (a) Sketch of procedure to determine existence of a circle map $\theta_{n+1} = \theta_n + F(\theta_n)$. (b) Determination of rotation number or winding number from data generated in a flow induced vibration experiment shown in Figure 5-21c.

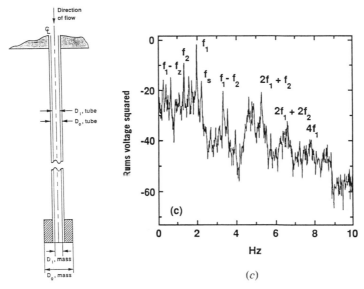

(c)

Figure 5-21 (c) Sketch of a flexible tube with an end mass carrying a steady flow. Lateral periodic force leads to quasiperiodic motion (Copeland and Moon, 1992).

The critical value W is called the *winding number*. If one suspends the "mod" operation on θ_i, then W is the rate at which the angle θ_i changes with the discrete Poincaré time "i," that is,

$$W = \lim_{n \to \infty} \frac{\theta_i - \theta_0}{i} \qquad (5\text{-}6.10)$$

For two uncoupled oscillators, $W = \omega_1/\omega_2 \equiv \Omega$ (the frequency ratio). For the circle map (5-6.9) with $k = 0$, $\theta_{n+1} = \theta_n + \Omega$, so that $W = \Omega$ is a uniform rate of change of angle. For $k \neq 0$ in (5-6.9), we have a measure of the nonlinear coupling between the oscillators.

Once the function $x_i(Wt_i)$ is found from the above procedure, then one can parameterize the curve from say 0 to 2π and assign corresponding angle values θ_i. A plot of a return map should reveal a periodic circle type map

$$\theta_{n+1} = f(\theta_n) \qquad (5\text{-}6.11)$$

In addition to ascertaining if the experimental data can be modeled by a circle map, one can plot mode-locked regions as the control frequency is varied (see also Chapter 2, Figure 2-23). In some problems, one can also determine multifractal properties of the attractor as described in Chapter 7 (see also Glazier and Libchaber, 1988).

5.7   QUANTITATIVE MEASURES OF CHAOTIC VIBRATIONS

Poincaré maps and phase-plane portraits, when they can be obtained, often provide graphic evidence for chaotic behavior and the fractal properties of strange attractors. However, quantitative measures of chaotic dynamics are also important and in many cases are the only hard evidence for chaos. The latter is especially true for systems with extreme frequencies 10^6–10^9 (as in laser systems) in which Poincaré maps may be difficult or impossible to capture. In addition, there are systems with many degrees of freedom where the Poincaré map will not reveal the fractal structure of the attractor (see §5.6 on double or multiple Poincaré maps) or where the damping is so low that the Poincaré map shows no structure but looks like a cloud of points.

At this time in the development of the field there are three principal measures of chaos and another of emerging importance:

(a) Fourier distribution of frequency spectra
(b) Fractal dimension of chaotic attractor
(c) Lyapunov exponents
(d) Invariant probability distribution of attractor

It should be pointed out that while phase-plane pictures and Poincaré maps can be obtained directly from electronic laboratory equipment, the above measures of chaos require a computer to analyze the data, with the possible exception of the frequency spectrum measurement. Electronic spectrum analyzers can be obtained but are often expensive, and one might be better off investing in a laboratory micro- or minicomputer which has the capability to perform other data analysis besides Fourier transforms.

If one is to digitally analyze the data from chaotic motions, then usually an *A-D converter* will be required as well as some means of storing the data. For example, the digitized data can be stored in a buffer in the electronic A-D device and then transmitted directly or over phone lines to a computer. Another option is a *digital oscilloscope* which performs the A-D conversion, displays the data graphically on the oscilloscope, and stores the data on a floppy or hard disk.

Finally, if one has the funds, one can store the output from the A-D converter directly into a hard disk for direct transfer to a laboratory computer.

Frequency Spectra: Fast Fourier Transform

This is by far the most popular measure mainly because the idea of decomposing a nonperiodic signal into a set of sinusoidal or harmonic signals is widely known among scientists and engineers. The assumption made in this method is that the periodic or nonperiodic signal can be represented as a synthesis of sine or cosine signals:

$$f(t) = \frac{1}{2\pi} \int_\Gamma F(\omega) e^{i\omega t} \, d\omega \qquad (5\text{-}7.1)$$

where $e^{i\omega t} = \cos \omega t + i \sin \omega t$.

Because $F(\omega)$ is often complex, the absolute value $|F(\omega)|$ is used in graphical displays. In practice, one uses an electronic device or computer to calculate $|F(\omega)|$ from input data from the experiment while varying some parameter in the experiment (see section in Chapter 2 entitled "Bifurcations: Routes to Chaos"). When the motion is periodic or quasiperiodic, $|F(\omega)|$ shows a set of narrow spikes or lines indicating that the signal can be represented by a discrete set of harmonic functions $\{e^{\pm i\omega_k t}\}$, where $k = 1, 2, \ldots$. Near the onset of chaos, however, a continuous distribution of frequency appears (as shown in Figure 5-22a), and in the fully chaotic regime the continuous spectrum may dominate the discrete spikes.

Numerical calculation of $F(\omega)$, given $f(t)$, can often be very time-consuming even on a fast computer. However, most modern spectrum analyzers use a discrete version of (5-7.1) along with an efficient algorithm called the *fast Fourier transform* (FFT). Given a set of data sampled at discrete even time intervals $\{f(t_k) = f_0, f_1, f_2, \ldots, f_k, \ldots, f_N\}$, the discrete time FFT is defined by the formula

$$T(J) = \sum_{I=1}^{N} f(I) e^{-2\pi i (I-1)(J-1)/N} \qquad (5\text{-}7.2)$$

where I and J are integers.

Several points should be made here which may appear obvious. First, the signal $f(t)$ is time sampled at a fixed time interval τ_0; thus, information is lost for frequencies above $\frac{1}{2}\tau_0$. Second, only a finite set of points are used in the calculation, usually $N = 2^n$, and some built-in FFT electronics only do $N = 512$ or 1028 points. Thus, information is lost about very low frequencies below $1/N\tau_0$. Finally the representation (5-7.2) having no information about $F(t)$ before $t = t_0$ or after

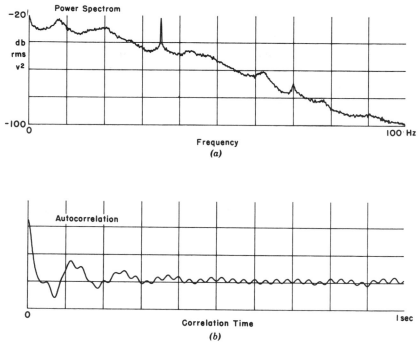

Figure 5-22   (*a*) Fourier spectrum of a chaotic signal. (*b*) Autocorrelation function of a chaotic signal.

$t = t_N$ essentially treats $f(t)$ as a periodic function. In general, this is not the case and because $f(t_0) \neq f(t_N)$, the Fourier representation treats this as a discontinuity which adds spurious information into $F(\omega)$. This is called *aliasing error*, and methods exist to minimize its effect on $F(\omega)$. The reader using the FFT should be aware of this, however, when interpreting Fourier spectra about nonperiodic signals and should consult a signal processing reference for more information about FFTs.

Autocorrelation Function.   Another signal processing tool that is related to the Fourier transform is the autocorrelation function given by

$$A(\tau) = \int_0^\infty x(t)x(t + \tau)\, dt \qquad (5\text{-}7.3)$$

When a signal is chaotic, information about its past origins is lost. This means that $A(\tau) \rightarrow 0$ as $\tau \rightarrow \infty$, or the signal is only correlated with its recent past. This is illustrated in Figure 5-22*b* for the chaotic vibrations of a buckled beam. The Fourier spectrum shows a broad band of frequencies, whereas the autocorrelation function has a peak at the origin $\tau = 0$ and drops off rapidly with time.

Autocorrelation Function Using Symbol Dynamics. In this book we have talked a great deal about iterated maps and differential equations, but not much about symbol dynamics. However, modern dynamical systems theory points out the equivalence between chaotic dynamics with sequences of symbols. This equivalence is more than conceptual, however, and can be used to obtain some quantitative measures of the dynamics. A case in point is the dry friction oscillator discussed above and in Chapter 4 (see Figure 4-26).

Lyapunov exponents are very difficult to measure in physical experimental systems. An alternative to this measure of chaos is the use of an autocorrelation function. There has been speculation that the time for zero autocorrelation would be inversely proportional to the Lyapunov exponent (e.g., see Singh and Joseph, 1989).

To test these ideas, an autocorrelation function based on symbol dynamics has been applied to a chaotic dry-friction oscillator to estimate the largest Lyapunov exponent (Feeny and Moon, 1989). The friction problem is well-suited for symbol dynamics because two distinct states of motion can be identified: sticking and slipping. The experiment consisted of a mass attached to the end of a cantilevered elastic beam, as shown in Figure 5-17a. The mass had titanium plates on both sides, providing surfaces for sliding friction. Spring-loaded titanium pads rested against the titanium plates. The titanium plates were not parallel in the direction of sliding, and thus a displacement of the mass caused a change in the force on the spring-loaded pads and caused a change in the normal load and the friction force. The device was excited by periodic acceleration using an electromagnetic shaker. The displacement of the mass was measured with a strain gage attached to the cantilevered beam.

The dynamics of the oscillator can be reduced to a noninvertible 1-D map (Figure 5-17c), which has been studied in terms of binary symbol sequences. The 2-D map in Figure 5-17b is reduced to a 1-D map in Figure 5-17c by defining a variable "s" along the sticking and slipping curves of the Poincaré map. To obtain Figure 5-17c, one plots s_{n+1} versus s_n. The 3-D picture of the experimental attractor is shown in Color Plate 2.

Singh and Joseph (1989) have proposed a technique of extracting quantitative information from a binary symbol sequence. First, it is necessary to represent the symbol sequence $u(k)$ as a string of 1's and

− 1's. These values are chosen so that the expected mean of a random sequence of equally likely symbols is zero. As a trajectory passes through the Poincaré section for the kth time, if it is not sticking, we set $u(k) = 1$. If it is sticking, we set $u(k) = -1$. An autocorrelation on such a symbol sequence is defined as

$$r(n) = \frac{1}{N} \sum_{k=1}^{N} u(k + n)u(k) \qquad (n = 0, 1, 2, \ldots ; \quad N \gg n) \quad (5\text{-}7.4)$$

If the sequence is chaotic, the autocorrelation should have the property $r(n) \rightarrow 0$ as $n \rightarrow \infty$.

If the sequence becomes uncorrelated, an estimate of the largest Lyapunov exponent can be obtained using the binary autocorrelation function. The macroscopic Lyapunov exponent, λ_m, is rewritten via a derivation in Singh and Joseph (1989) as

$$\lambda_m = \tfrac{1}{2} \alpha[1 - r(1)^2] \qquad (5\text{-}7.5)$$

Application of the equations to a symbol sequence derived from the tent map yields a rapidly decaying autocorrelation and a Lyapunov exponent $\lambda_t = 0.787$ for a string of 100,000 symbols, and an exponent of $\lambda_t = 0.787$ for a string of 2048 symbols, compared to its exact value, calculated using $\log_2$, $\lambda_{tc} = 1$.

The binary autocorrelation function for an experimental sequence of length 2048 was obtained (5-7.4) as shown in Figure 5-23. Applying

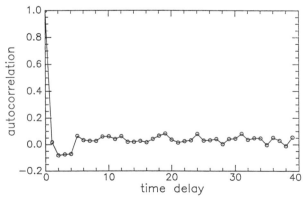

Figure 5-23   Autocorrelation based on stick–slip symbol dynamics of a friction oscillator. [From Feeny and Moon (1989) with permission of Elsevier Publishers, copyright 1989).]

Eqs. (5-7.4) and (5-7.5), the resulting Lyapunov exponent is $\lambda_{exp} = 0.790$. Using Eqs. (5-7.4) and (5-7.5) on numerical smooth-friction law data (2048 symbols) yields an autocorrelation similar to that in Figure 5-23 and a Lyapunov exponent of $\lambda_{s1} = 0.792$.

Wavelet Transform

To characterize the fractal nature of dynamic data, a new signal processing technique has been developed, called the *wavelet transform*, which generalizes the Fourier transform (see e.g., Argoul et al. (1989), and Pezeshlki et al. (1992) for applications). This transform introduces a spectrum of scales to unfold the self-similar nature of the fractallike data.

Fractal Dimension

I will not go into too many technical details about fractal dimensions because Chapter 7 is entirely devoted to this topic. However, the basic idea is to characterize the "strangeness" of the chaotic attractor. If one looks at a Poincaré map of a typical low-dimensional strange attractor, as in Figure 5-6, one sees sets of points arranged along parallel lines. This structure persists when one enlarges a small region of the attractor. As noted in Chapter 2, this structure of the strange attractor differs from periodic motions (just a finite set of Poincaré points) or quasiperiodic motion which in the Poincaré map becomes a closed curve. In the Poincaré map, one can say that the dimension of the periodic map is zero and that the dimension of the quasiperiodic map is one. The idea of the fractal dimension calculation is to attach a measure of dimension to the Cantor-like set of points in the strange attractor. If the points uniformly cover some area on the plane, we might say its dimension was close to two. Because the chaotic map in Figure 5-6 has an infinite set of gaps, its dimension is between one and two—thus the term *fractal dimension*.

Another use for the fractal dimension calculation is to determine the lowest-order phase space for which the motion can be described. For example, in the case of some preturbulent convective flows in a Rayleigh–Benard cell (see Figure 4-39), the fractal dimension of the chaotic attractor can be calculated from some measure of the motion $\{x(t_n) \equiv x_n\}$ (see Malraison et al., 1983). From $\{x_n\}$, pseudo-phase-spaces of different dimension can be constructed (see Section 5.4). Using a computer algorithm, the fractal dimension d was found to reach an asymptotic $d = 3.5$ when the dimension of the pseudo-phase-space was four or larger. This suggests that a low-order approximation of the Navier–Stokes equation may be used to model this motion. The reader is referred to Chapter 7 for further details.

Although there are questions about the ability to calculate fractal dimensions for attractors of dimensions greater than four or five, this technique has gained increasing acceptance among experimentalists especially for low-dimensional chaotic attractors. If this trend continues in the future, it is likely that electronic computing instruments will be available commercially that automatically calculate fractal dimension in the same way as FFTs are done at present.

Lyapunov Exponents

Chaos in dynamics implies a sensitivity of the outcome of a dynamical process to changes in initial conditions. If one imagines a set of initial conditions within a sphere or radius ε in phase space, then for chaotic motions trajectories originating in the sphere will map the sphere into an ellipsoid whose major axis grows as $d = \varepsilon e^{\lambda t}$, where $\lambda > 0$ is known as a *Lyapunov exponent*. [Lyapunov (1857–1918) was a great Russian mathematician and mechanician.]

A number of experimenters in chaotic dynamics have developed algorithms to calculate the Lyapunov exponent λ. For regular motions $\lambda \leq 0$, but for chaotic motion $\lambda > 0$. Thus, *the sign of λ is a criterion for chaos*. The measurement involves the use of a computer to process the data. Algorithms have been developed to calculate λ from the measurement of a single dynamical variable $x(t)$ by constructing a pseudo-phase-space (e.g., see Wolf, 1984). Another experimental technique is the use of the autocorrelation function discussed above (Singh and Joseph, 1989; Feeny and Moon, 1989).

A more precise definition of Lyapunov exponents and techniques for measuring them are given in Chapter 6.

Probability or Invariant Distributions

If a nonlinear dynamical system is in a chaotic state, precise prediction of the time history of the motion is impossible because small uncertainties in the initial conditions lead to divergent orbits in the phase space. If damping is present, we know that the chaotic orbit lies somewhere on the strange attractor. Because we lack specific knowledge about the whereabouts of the orbit, there is increasing interest in knowing the probability of finding the orbit somewhere on the attractor. One suggestion is to see if one can use a probability density in phase space to provide a statistical measure of the chaotic dynamics [see Section

3.7, Eq. (3.7-9)]. There is some mathematical and experimental evidence that such a distribution does exist and that it does not vary with time. (See e.g., Figure 3-27.)

Increasingly, measurement of the probability distribution function is being used as a diagnostic tool in chaotic vibrations, especially in periodically forced systems. In general, the dynamic attractor in a three-dimensional phase space would have a probability measure with three variables $P(x, y, z)$, one for each of the state variables. For a chaotic attractor with fractal properties, however, the distribution function would also have fractal properties. Experimentally, and even computationally, a small amount of noise will smooth out the distribution function. However, another smoothing operation is to integrate $P(x, y, z)$ over one or more of the state variables. For example, for a forced single-degree-of-freedom oscillator, integration over the forcing phase (e.g., $0 \leq \Omega t \leq 2\pi$) and the velocity variable will yield an experimental probability density function $P(x)$ which is piecewise smooth and gives the probability that the motion at any time will lie between x_1 and x_2:

$$\mathcal{P}(x_1; x_2) = \int_{x_2}^{x_1} P(x)\, dx, \qquad \mathcal{P}(-\infty, \infty) = 1 \qquad (5\text{-}7.6)$$

Also, modern signal processing systems often have a function (sometimes called a *histogram*) that will partition an interval $x_1 \leq x \leq x_2$ into N bins and that will count the number of times the digitalized signal lies in each bin in a given finite length data record. With suitable normalization, this procedure will yield an approximation to $P(x)$.

To function as a good diagnostic tool, a signal processing algorithm must provide qualitatively different patterns for periodic and chaotic signals. In the case of forced systems, this is usually the case. A periodic motion usually has an elliptic shape in the phase plane $(x, v = \dot{x})$. If the points on the orbit are subdivided and protected onto the x-axis, then the probability density function $P(x)$ is continuous over a finite interval with singularities at the edges; that is,

$$P(x) = \frac{1}{\pi} \frac{1}{\sqrt{A^2 - x^2}}$$

for a harmonic orbit centered at the origin. (A is the maximum amplitude of the limit cycle.) For chaotic signals, the singularities often disappear and $P(x)$ looks more Gaussian and often non-Gaussian. Two examples are shown in Figures 5-24 and 5-25.

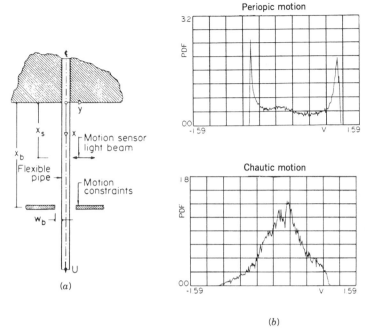

Figure 5-24 Probability density function for flow-induced chaotic vibrations of a flexible tube carrying steady fluid flow. [From Païdoussis and Moon (1988).]

The first case is for the flow-induced vibration os an elastic tube carrying a steady flow of fluid (see Paidoussis and Moon, 1988). A comparison of the probability density function (PDF) for periodic and chaotic states shows a clear distinction (Figure 5-24).

The second example is the experimental two-well potential problem using a buckled elastic beam (Moon, 1980a). Here we show the PDF for both displacement and velocity, that is, $P(x)$ and $P(\dot{x})$ (Figure 5-25). The PDF for velocity looks Gaussian, whereas the PDF for the displacement has a double peak.

A few attempts have been made to analytically calculate the PDF for chaotic attractors. The solution for a one-dimensional map was outlined in Chapter 3. A special case of the standard map was presented by Tsang and Lieberman (1984) using the Fokker–Planck equation. For applied scientists or engineers, one should consult either Soong (1973) or Gardiner (1985) for a discussion of this equation. Consider the forced oscillator with some random noise $W(t)$:

$$\ddot{x} + g(x, \dot{x}) + f(x) = F_0\cos \Omega t + W(t) \qquad (5\text{-}7.7)$$

The Fokker–Planck equation for the PDF $P(x, y = \dot{x}, t)$ is given by

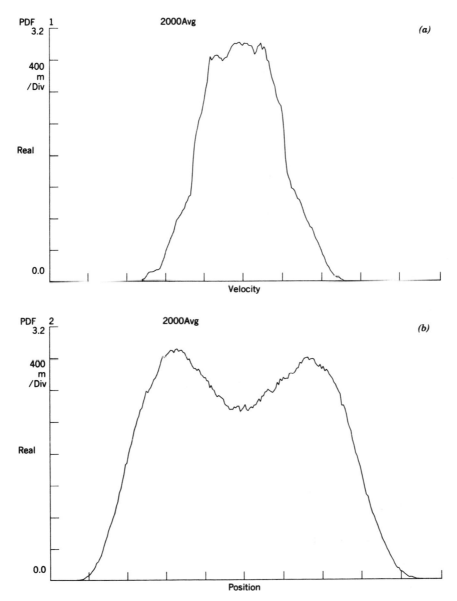

Figure 5-25   Experimental probability density function for chaotic vibration of a buckled beam averaged in time over many thousands of forcing periods. (*a*) Distribution of velocities at the beam tip. (*b*) Distribution of position of the beam tip.

$$\frac{\partial P}{\partial t} + \frac{\partial}{\partial x}(yP) + \frac{\partial}{\partial y}(-g(x, y) - f(x) + f_0\cos \Omega t)P = \frac{1}{2}S_0\frac{\partial^2 P}{\partial y^2}$$

$$(5\text{-}7.8)$$

where the constant S_0 is a measure of the strength of the Gaussian white noise. For $f_0 = 0$, one can find an explicit stationary solution, that is, $\partial/\partial t = 0$. For linear damping, $g(x, y) = \gamma y$, the solution has the form

$$P(x, y) = ce^{-\gamma y^2/S_0} e^{-\{\gamma \int_0^x f(x)\, dx\}/S_0}$$

$$(5\text{-}7.9)$$

For a two-well potential with $f(x) = -x + x^3$, one has

$$P(x, y) = ce^{-\gamma y^2/S_0} e^{\gamma(x^2 - x^4/2)/S_0}$$

$$(5\text{-}7.10)$$

A similar problem has been studied by Kunert and Pfeiffer (1991). What is remarkable is that the analytical solution (5-7.10) for Gaussian white noise also produces a double-hump PDF for the x variable which is qualitatively similar to the deterministic periodically excited problem (i.e., $W(t) = 0$) as shown in the experiments (Figure 5-25).

Thus, there is some suggestion that an approximate PDF for a deterministic chaotic attractor may be related to the PDF for a randomly excited oscillator. These ideas, however, are speculative at this time and remain an area of potential fruitful research.

The use of the probability density function to calculate Lyapunov exponents is discussed in Chapters 3 and 6.

The usefulness of probability distribution for chaotic vibrations is similar to that for random vibrations (e.g., see Soong, 1973 or Lin, 1976). If the probability distribution can be determined for a chaotic system, then one can calculate the mean square amplitude, mean zero crossing times, and probability of displacements, voltages, or stresses exceeding some critical value. However, much remains to be done in this subject at both the mathematical and experimental levels.

Cell Mapping Methods

The use of probabilistic methods of analysis in chaotic vibrations has been developed by C. S. Hsu and co-workers at the University of California at Berkeley (Hsu, 1981, 1987; Hsu and Kim, 1985; Kreuzer, 1985 Tongue, 1987). This method, called the *Cell Mapping Method*, which divides the phase space into many cells, uses ideas from the theory of Markov processes. This computer based method may be useful for low order systems in obtaining a global picture of the possible dynamic attractors. New algorithms have been developed to improve numerical efficiency in this method (see e.g., Tongue, 1987).

PROBLEMS

5-1 Show that the autocorrelation function of a periodic function is periodic.

5-2 Suppose a signal has a subharmonic

$$x(t) = A_1 \cos \omega t + A_2 \cos \frac{\omega}{2} t$$

What does the autocorrelation function look like?

5-3 Consider a signal with quasiperiodic sinusoidal components. What does the FFT and autocorrelation function look like?

5-4 Sketch the PDF, $P(x)$, of a two-frequency quasiperiodic signal.

5-5 Consider a periodic signal $x(t)$ with a single frequency. Derive the PDF, $P(x)$. Sketch the PDF in the phase plane, that is, $P(x, \dot{x})$.

5-6 Suppose a signal has a subharmonic

$$x(t) = A_1 \cos \omega t + A_2 \cos \left(\frac{\omega}{2} t + \varphi_0 \right)$$

Sketch the PDF, $P(x)$. (Hint, see Figure 3-17.)

5-7 Consider the problem of a linear harmonic oscillator linearly coupled to an auto-oscillatory system such as the Van der Pol oscillator (Chapter 1). Show that this represents a fourth-order dynamical system. Numerically find a chaotic parameter regime. How would you define a Poincaré map? Write a program to calculate a double Poincaré map for this system. How would you define the second map?

5-8 Write a program to plot the Henon map (Chapter 1) in the chaotic regime. What is the effect of adding increasing amounts of random noise?

5-9 Design an analog computer circuit to simulate the dynamics of a dynamical system studied by Rössler (1976b) (e.g., see Appendix C):

$$\dot{x} = -y - z, \qquad \dot{y} = x + ay, \qquad \dot{z} = b + z(x - c)$$

(A popular set of parameters to use is $a = 0.398$, $b = 2$, $c = 4$.) Rössler named this a spiral chaos attractor, and it is one of the simplest that exhibits folding action in the three-dimensional phase space.)

6

CRITERIA FOR
CHAOTIC VIBRATIONS

*But you will ask, how could a uniform chaos coagulate at first irregularly
in heterogeneous veins or masses to cause hills—Tell me the cause of
this, and the answer will perhaps serve for the chaos.*

Isaac Newton
On Creation—from a letter circa 1681

6.1 INTRODUCTION

In this chapter, we study how the parameters of a dynamical system
determine whether the motion will be chaotic or regular. This is analogous to finding the critical velocity in viscous flow of fluids above
which steady flow becomes turbulent. This velocity, when normalized
by a characteristic length and by the kinematic viscosity of the fluid,
is known as the critical *Reynolds number,* Re. A reliable theoretical
value for the critical Re has eluded engineers and physicists for over
a century, and for most fluid problems experimental determination of
$(Re)_{crit}$ is necessary. In like manner, the determination of criteria for
chaos in mechanical or electrical systems in most cases must be found
by experiment or computer simulation. For such systems the search
for critical parameters for deterministic chaos is a ripe subject for
experimentalists and theoreticians alike.

Despite the paucity of experimentally verified theories for the onset
of chaotic vibrations, there are some notable theoretical successes
and some general theoretical guidelines.

We distinguish between two kinds of criteria for chaos in physical systems: a predictive rule and a diagnostic tool. A *predictive* rule for chaotic vibrations is one that determines the set of input or control parameters that will lead to chaos. The ability to predict chaos in a physical system implies either that one has some approximate mathematical model of the system from which a criterion may be derived or that one has some empirical data based on many tests.

A *diagnostic* criterion for chaotic vibrations is a test that reveals if a particular system was or is in fact in a state of chaotic dynamics based on measurements or signal processing of data from the time history of the system.

We begin with a review of empirically determined criteria for specific physical systems and mathematical models which exhibit chaotic oscillations (Section 6.2). These criteria were determined by both physical and numerical experiments. We examine such cases for two reasons. First, it is of value for the novice in this field to explore a few particular chaotic systems in detail and to become familiar with the conditions under which chaos occurs. Such cases may give clues to when chaos occurs in more complex systems. Second, in the development of theoretical criteria, it is important to have some test case with which to compare theory with experiment.

In Section 6.3 we present a review of the principal, predictive models for determining when chaos occurs. These include the period-doubling criterion, homoclinic orbit criterion, Shil'nikov criterion, and the overlap criterion of Chirikov for conservative chaos, as well as intermittency and transient chaos. We also review several ad hoc criteria that have been developed for specific classes of problems.

Finally, in Section 6.4 we discuss an important diagnostic tool, namely, the Lyapunov exponent. Another diagnostic concept, the fractal dimension, is described in Chapter 7.

6.2 EMPIRICAL CRITERIA FOR CHAOS

In the many introductory lectures the author has given on chaos, the following question has surfaced time and time again: *Are chaotic motions singular cases in real physical problems or do they occur for a wide range of parameters?* For engineers this question is very important. To design, one needs to predict system behavior. If the engineer chooses parameters that produce chaotic output, then he or she loses predictability. In the past, many designs in structural engineering, electrical circuits, and control systems were kept within the realm of linear system dynamics. However, the needs of modern technology have pushed devices into nonlinear regimes (e.g., large deformations and deflections in structural mechanics) that increase the possibility of encountering chaotic dynamic phenomena.

To address the question of whether chaotic dynamics are singular events in real systems, we examine the range of parameters for which chaos occurs in seven different problems. A cursory scan of the figures accompanying each discussion will lead one to the conclusion that chaotic dynamics are not a singular class of motions and that *chaotic oscillations occur in many nonlinear systems for a wide range of parameter values.*

We examine the critical parameters for chaos in the following problems?:

(a) Circuit with nonlinear inductor: Duffing's equation
(b) Particle in a two-well potential or buckling of an elastic beam: Duffing's equation
(c) Experimental convection loop: a model for Lorenz's equation
(d) Vibrations of nonlinear coupled pendulums
(e) Rotating magnetic dipole: pendulum equation
(f) Circuit with nonlinear capacitance
(g) Surface waves on a fluid

Forced Oscillations of a Nonlinear Inductor: Duffing's Equation

In Chapter 4, we examined the chaotic dynamics of a circuit with a nonlinear inductor (see also Figure 3-33). Extensive analog and digital simulation for this system was peformed by Y. Ueda (1979, 1980) of Kyoto University. The nondimensional equation, where x represents the flux in the inductor, takes the form

$$\ddot{x} + k\dot{x} + x^3 = B \cos t \qquad (6\text{-}2.1)$$

The time has been nondimensionalized by the forcing frequency so that the entire dynamics can be determined by the two parameters k and B and the initial conditions $(x(0), \dot{x}(0))$. Here k is a measure of the resistance of the circuit, while B is a measure of the driving voltage. Ueda found that by varying these two parameters one could obtain a wide variety of periodic, subharmonic, ultrasubharmonic, and chaotic motions. The regions of chaotic behavior in the (k, B) plane are plotted in Figure 6-1. The regions of subharmonic and harmonic motions are quite complex, and only a few are shown for illustration. The two

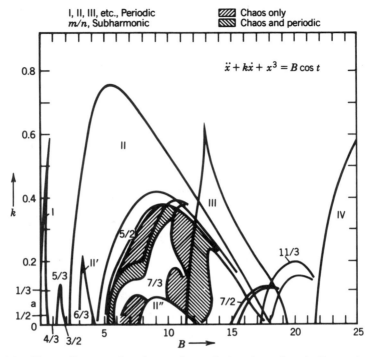

Figure 6-1 Chaos diagram showing regions of chaotic and periodic motions for a nonlinear circuit as functions of nondimensionalized damping and forcing amplitude. [From Ueda (1980).]

different hatched areas indicate either (a) regions of only chaos or (b) regions with both chaotic and periodic motion, depending on initial conditions. A theoretical criterion for this relatively simple equation has been suggested by Szemplinska–Stupnika and Bajkowski (1986). (See also the Color Plate 1 for solutions of (1-2.4).)

Forced Oscillations of a Particle in a Two-Well Potential

This example was discussed in great detail in Chapters 2 and 3. It was first studied by Holmes (1979) and was later studied in a series of papers by the author and co-workers. The mathematical equation describes the forced motion of a particle between two states of equilibrium, which can be described by a two-well potential:

$$\ddot{x} + \delta x - \tfrac{1}{2}x(1 - x^2) = f \cos \omega t \tag{6-2.2}$$

This equation can represent a particle in a plasma, a defect in a solid, and, on a larger scale, the dynamics of a buckled elastic beam (see Chapter 3). The dynamics are controlled by three nondimensional groups (δ, f, ω), where δ represents nondimensional damping and ω is the driving frequency nondimensionalized by the small-amplitude natural frequency of the system in one of the potential wells.

Regions of chaos from two studies are shown in Figures 6-2 and 6-3. The first represents experimental data for a buckled cantilevered beam (Chapter 2). The ragged boundary is the experimental data, whereas the smooth curve represents a theoretical criterion (see Section 6.3). Recently, an upper boundary has been measured beyond which the motion becomes periodic. The experimental criterion was determined by looking at Poincaré maps of the motion (see Chapters 2 and 5).

Results from numerical simulation of Eq. (6-2.2) are shown in Figure 6-3. The diagnostic tool used to determine if chaos was present was the Lyapunov exponent using a computer algorithm developed by Wolf et al. (1985) (see Section 6.4). This diagram shows that there are complex regions of chaotic vibrations in the plane (f, ω) for fixed damping δ. For very large forcing $f \gg 1$, one expects the behvior to emulate the previous problem studied by Ueda.

The theoretical boundary found by Holmes (1979) is discussed in the next section. It has special significance because below this boundary

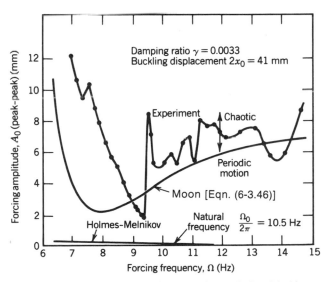

Figure 6-2 Experimental chaos diagram for vibrations of a buckled beam for different values of forcing frequency and amplitude. [From Moon (1980b), reprinted with permission from *New Approaches to Nonlinear Problems in Dynamics*, edited by P. J. Holmes, copyright 1980 by SIAM.]

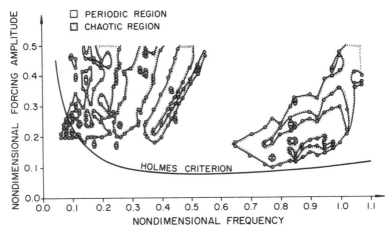

Figure 6-3   Chaos diagram for vibration of a mass in a double-well potential [Duffing's equation, Eq. (6-2.2)]. The smooth boundary represents the homoclinic orbit criterion (Section 6-3.).

periodic motions are predictable, whereas above this boundary one loses the ability to exactly predict which of the many periodic or chaotic modes the motion will be attracted. Above the theoretical criteria (based on homoclinic orbits), the motion is very sensitive to initial conditions, even when it is periodic (see Section 7.7).

Experimental Convection Loop: Lorenz Equations

Aside from the logistic equation, the Lorenz model for convection turbulence (see Chapters 1 and 4) is perhaps the most studied system of equations that admit chaotic solutions. Yet most mathematicians have focused on a few sets of parameters. These equations take the form (see also Sparrow, 1982)

$$\dot{x} = \sigma(y - x)$$
$$\dot{y} = rx - y - xz \qquad (6\text{-}2.3)$$
$$\dot{z} = xy - bz$$

An experimental realization of these equations can be obtained in a circular convection loop, also known as the *thermosiphon* (see Section 4.7) (Figure 6-4*a*). This experiment has received extensive study from a group at the University of Houston (Widmann et al., 1989; Gorman et al., 1986). A qualitative diagram of the various dy-

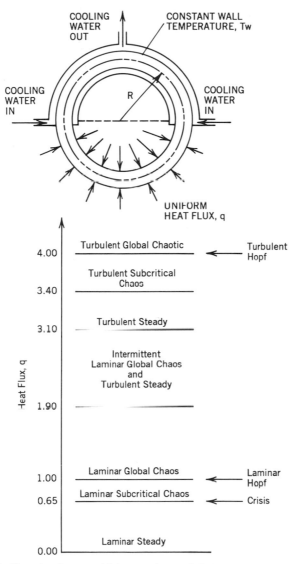

Figure 6-4 (a) Sketch of a toroidal container of fluid under gravity and thermal gradients, otherwise known as the *thermosiphon*. (b) Qualitative chaos diagram of the dynamic regimes for the thermosiphon. [From Widmann et al. (1989).]

namic regimes as a function of the applied heat flux is shown in Figure 6-4b as observed in their experiments. In their 1986 paper the Houston group showed that the Lorenz equation only gave good predictive results for the steady regime, and that an additional degree of freedom was needed to improve the agreement between theory and experiment (see also Yorke et al., 1985).

Forced Vibrations of Two-Coupled Pendulums

Figure 6-5 is a sketch of an experiment with two masses hung on a lightweight cable which is assumed to be inextensible. This problem is equivalent to two coupled spherical pendulums with a constraint. The effective number of degrees of freedom is three: one in-plane mode and two out-of-plane modes. The end points are excited with harmonic excitation. In mechanical engineering this represents a 3-D four-bar linkage, whereas in civil engineering it could represent a model for cable car dynamics. An experimental chart of four dynamic regimes shows both chaotic as well as quasiperiodic regions in the parameter space of excitation amplitude and frequency. Quasiperiodic motions are typical in multiple-degree-of-freedom systems of this kind. These experiments were performed in our laboratory by Professor F. Benedettini of the University of l'Aquila, Italy.

Forced Motions of a Rotating Dipole in Magnetic Fields: The Pendulum Equation

In this experiment, a permanent magnet rotor is excited by crossed steady and time harmonic magnetic fields (see Moon et al., 1987), as shown in Figure 4-6. The nondimensionalized equation of motion for the rotation angle θ resembles that for the pendulum in a gravitational potential:

$$\ddot{\theta} + \gamma\dot{\theta} + \sin\theta = f\cos\theta\cos\omega t \qquad (6\text{-}2.4)$$

The regions of chaotic rotation in the f–ω plane, for fixed damping, are shown in Figure 6-6. This was one of the first published examples where both experimental and numerical simulation data are compared with a theoretical criterion for chaos. The theory is based on the homoclinic orbit criterion and is discussed in Section 6.4. As in the case of the two-well potential, chaotic motions are to be found in the vicinity of the natural frequency for small oscillations ($\omega = 1.0$ in Figure 6-6). See Figure 5-13 for a sketch of the experiment.

Forced Oscillations of a Nonlinear *RLC* Circuit

There have been a number of experimental studies of chaotic oscillations in nonlinear circuits (e.g., see Chapter 4). One example is an *RLC* circuit with a diode. Shown in Figure 6-7 are the subharmonic and chaotic regimes in the driving voltage–frequency plane (Klinker

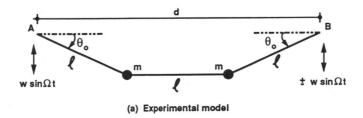

(a) Experimental model

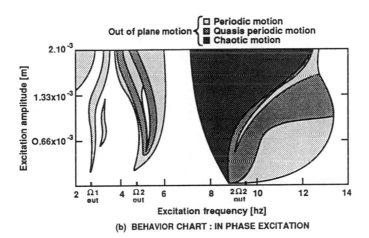

(b) BEHAVIOR CHART : IN PHASE EXCITATION

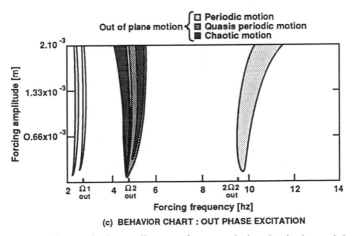

(c) BEHAVIOR CHART : OUT PHASE EXCITATION

Figure 6-5 Experimental chaos diagram for coupled spherical pendulum with a constraint. [From Benedettini and Moon (1992).] (*a*) Experimental model. (*b*) Behavior chart: in-phase excitation. (*c*) Behavior chart. Out-phase excitation.

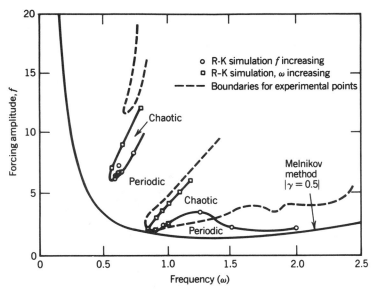

Figure 6-6 Experimental chaos diagram for forced motions of a rotor with nonlinear torque–angle property. Comparison with homoclinic orbit criterion calculated using the Melnikov method (Section 6.3). [From Moon et al. (1987) with permission of North-Holland Publishing Co., copyright 1987.]

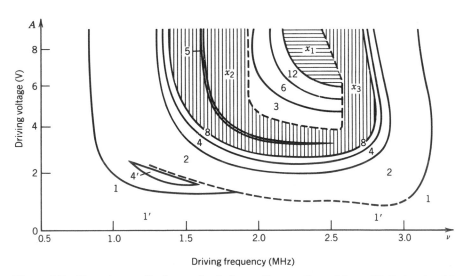

Figure 6-7 Experimentally determined chaos diagram for a driven *RLC* circuit with a varactor diode that acts as a nonlinear capacitor. The hatched regions are chaotic motions, and the numbers indicate the order of the subharmonic. Dashed lines indicate a hysteretic transition. [From Klinker et al. (1984) with permission of North-Holland Publishing Co., copyright 1984.]

et al., 1984). In this example, regions of period doubling are shown as precursors to the chaotic motions. However, in the midst of the hatched chaotic regime, a period-5 subharmonic was observed. Periodic islands in the center of chaotic domains are common observations in experiments on chaotic oscillations. [See a similar study by Bucko et al. (1984). See also Figure 4-32.]

Harmonically Driven Surface Waves in a Fluid Cylinder

As a final example, we present experimentally determined harmonic and chaotic regions of the amplitude–frequency parameter space for surface waves in a cylinder filled with water from a paper by Ciliberto and Gollub (1985). A 12.7-cm-diameter cylinder with 1-cm-deep water was harmonically vibrated in a speaker cone (Figure 6-8). The amplitude of the transverse vibration above the flat surface of the fluid can be written in terms of Bessel functions where the linear mode shapes are given by $U_{nm} = J_n(k_{nm}r)\sin(n\theta + d_{nm})$. Figure 6-8 shows the driving amplitude–frequency plane in a region where two modes can interact: $(n, m) = (4, 3)$ and $(7, 2)$. Below the lower boundary, the surface

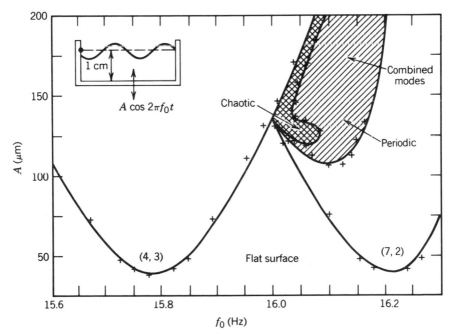

Figure 6-8 Experimental chaos diagram for surface waves in a cylinder filled with water. The diagram shows where two linear modes interact. [From Ciliberto and Gollub (1985).]

remains flat. A small region of chaotic regimes intersect. Presumably, other chaotic regimes exist where other modes (n, m) interact. (See also Figure 8-1.)

In summary, these examples show that, given periodic forcing input to a physical system, large regions of periodic or subharmonic motions do exist and presumably are predictable using classical methods of nonlinear analysis. However, these examples also show that *chaos is not a singular happening*; that is, it can exist for wide ranges in the parameters of the problem. Also, and perhaps most important, there are regions where both periodic and chaotic motions can exist and the precise motion that will result may be unpredictable.

6.3 THEORETICAL PREDICTIVE CRITERIA

The search for theoretical criteria to determine under what set of conditions a given dynamical system will become chaotic has tended to be ad hoc. The strategy thus far has been for theorists to find criteria for specific mathematical models and then use these models as analogs or paradigms to infer when more general or complex physical systems will become unpredictable. An example is the period-doubling bifurcation sequence discussed by May (1976) and Feigenbaum (1978) for the quadratic map (e.g., see Chapters 1 and 3). Although these results were generalized for a wider class of one-dimensional maps using a technique called *renormalization theory,* the period-doubling criterion is not always observed for higher-dimensional maps. In mechanical and electrical vibrations, a Poincaré section of the solution in phase space often leads to maps of two or more dimensions. Nonetheless, the period-doubling scenario is one possible route to chaos. In more complicated physical systems, an understanding of the May–Feigenbaum model can be very useful in determining when and why chaotic motions occur.

In this section, we briefly review a few of the principal theories of chaos and explore how they lead to criteria that may be used to predict or diagnose chaotic behavior in real systems. These theories include the following:

(a) Period doubling
(b) Homoclinic orbits and horseshoe maps
(c) Shil'nikov criterion

(d) Intermittency and transient chaos

(e) Overlap criteria for conservative chaos

(f) Ad hoc theories for multiple-well potential problems

Period-Doubling Criterion

This criterion is applicable to dynamical systems whose behavior can be described exactly or approximately by a first-order difference equation (see Chapter 3):

$$x_{n+1} = \lambda x_n (1 - x_n) \tag{6-3.1}$$

The dynamics of this equation were studied by May (1976), Feigenbaum (1978, 1980), and others. They discovered solutions whose period doubles as the parameter λ is varied (the period in this case is the number of integers p for x_{n+p} to return to the value x_n). One of the important properties of Eq. (6-3.1) that Feigenbaum discovered was that the sequence of critical parameters $\{\lambda_m\}$ at which the period of the orbit doubles satisfies the relation

$$\lim_{m \to \infty} \frac{\lambda_{m+1} - \lambda_m}{\lambda_m - \lambda_{m-1}} = \frac{1}{\delta}, \qquad \delta = 4.6692 \ldots \tag{6-3.2}$$

This important discovery gave experimenters a specific criterion to determine if a system was about to become chaotic by simply observing the prechaotic periodic behavior. It has been applied to physical systems involving fluid, electrical, and laser experiments. Although these problems are often modeled mathematically by continuous differential equations, the Poincaré map can reduce the dynamics to a set of difference equations. For many physical problems, the essential dynamics can be modeled further as a one-dimensional map (see e.g., Chapter 5).

$$x_{n+1} = f(x_n) \tag{6-3.3}$$

The importance of Feigenbaum's work is that he showed how period-doubling behavior was typical of one-dimensional maps that have a hump or zero tangent [i.e., the map is *noninvertible* or there exist two values of x_n which when put into $f(x_n)$ give the same value of x_{n+1}]. He also demonstrated that if the mapping function depends on some parameter Λ [i.e., $f(x_n; \Lambda)$], then the sequence of critical values of this parameter at which the orbit's period doubles $\{\Lambda_m\}$ satisfies the same relation (6-3.2) as that for the quadratic map. Thus the period-doubling phenomenon has been called *universal*, and δ has been called a universal constant (now known quite naturally as the *Feigenbaum number*).

The author must raise a flag of caution here. The term "universal" is used in the context of one-dimensional maps (6-3.3). There are many chaotic phenomena which are described by two- or higher-dimensional maps (e.g., see the buckled beam problem in Chapter 2). In these cases, period doubling may indeed be one route to chaos, but there are many other bifurcation sequences that result in chaos beside period doubling (see Holmes, 1984).

Renormalization and the Period-Doubling Criterion. There are two ideas that are important in understanding the period-doubling phenomenon. The first is the concept of *bifurcation* of solutions, and the second is the idea of *renormalization*. The concept of bifurcation was illustrated in Chapter 3. For example, in Figure 6-9 a steady periodic solution x_0 becomes unstable at a parameter value of λ, and the amplitude now oscillates between two values x^+ and x^-, completing a cycle in twice the time of the previous solution. Further changes in λ make x^+ and x^- unstable, and the solution branches to a new cycle with period 4.

A readable description of renormalization as it applies to period doubling may be found in Feigenbaum (1980). The technique recognizes the fact that a cascade of bifurcations exists (Figure 6-10) and that it might be possible to map each bifurcation into the previous one by a change in scale of the physical variable x and a transformation of

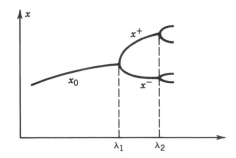

Figure 6-9   Diagram showing two branches of a bifurcation diagram near a period-doubling point.

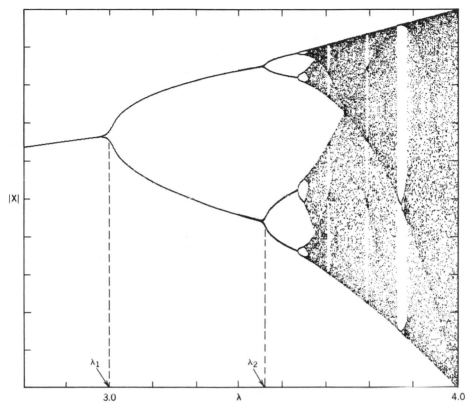

Figure 6-10 Bifurcation diagram for the quadratic map (3-6.2). Steady-state behavior as a function of the control parameter showing the period-doubling phenomenon.

the control parameter. To illustrate this technique, we outline an approximate scheme for the quadratic map (see also Lichtenberg and Lieberman, 1983).

One form of the quadratic map is given by

$$x_{n+1} = f(x_n) \qquad (6\text{-}3.4)$$

where $f(x) = \lambda x(1 - x)$. Period-1 cycles are just constant values of x given by fixed points of the mapping, that is, $x_n = f(x_n)$. Now a fixed point or equilibrium point can be stable or unstable. That is, iteration of x can move toward or away from the fixed point, x_0. The stability of the map depends on the slope of $f(x)$ at x_0; that is,

$$\left| \frac{df(x_0)}{dx} \right| > 1 \quad \text{implies instability} \qquad (6\text{-}3.5)$$

Because the slope $f' = \lambda(1 - 2x)$ depends on λ, x_0 becomes unstable at $\lambda_2 = \pm 1/|1 - 2x_0|$. Beyond this value, the stable periodic motion has period 2. The fixed points of the period-2 motion are given by

$$x_2 = f(f(x_2)) \quad \text{or} \quad x_2 = \lambda^2 x_2(1 - x_2)[1 - \lambda x_2(1 - x_2)] \quad (6\text{-}3.6)$$

The function $f(f(x))$ is shown in Figure 6-11.

Again there are stable and unstable solutions. Suppose the x_0 solution bifurcates and the solution alternates between x^+ and x^- as shown in Figure 6-9. We then have

$$x^+ = \lambda x^-(1 - x^-) \quad \text{and} \quad x^- = \lambda x^+(1 - x^+) \quad (6\text{-}3.7)$$

To determine the next critical value $\lambda = \lambda_2$ at which a period-4 orbit emerges, we change coordinates by writing

$$x_n = x^{\pm} + \eta_n \quad (6\text{-}3.8)$$

Putting Eq. (6-3.8) into (6-3.7), we get

$$\eta_{n+1} = \lambda \eta_n[(1 - 2x^+) - \eta_n]$$
$$\eta_{n+2} = \lambda \eta_{n+1}[(1 - 2x^-) - \eta_{n+1}] \quad (6\text{-}3.9)$$

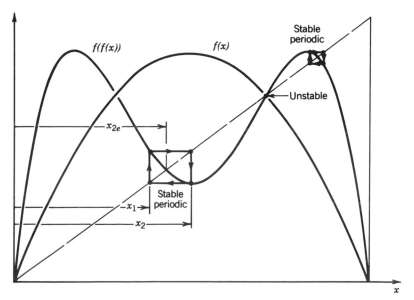

Figure 6-11   First and second iteration functions for the quadratic map (3-6.2).

We next solve for η_{n+2} in terms of η_n, keeping only terms to order η_n^2 (this is obviously an approximation), to obtain

$$\eta_{n+2} = \lambda^2 \eta_n [A - B\eta_n] \tag{6-3.10}$$

where A and B depend on x^+, x^-, and λ. Next, we rescale η and define a new parameter $\bar{\lambda}$ using

$$\bar{x} = \alpha \eta, \qquad \bar{\lambda} = \lambda^2 A, \qquad \alpha = B/A, \qquad \bar{x}_{n+2} = \bar{\lambda} \bar{x}_n (1 - \bar{x}_n)$$

This has the same form as our original equation [Eq. (6-3.1)]. Thus, when the solution bifurcates to period 4 at $\lambda = \lambda_2$, the critical value of $\bar{\lambda}$ equals λ_1. We therefore obtain an equation

$$\lambda_1 = \lambda_2^2 A(\lambda_2) \tag{6-3.11}$$

Starting from the point $x_0 = 0$, there is a bifurcation sequence for $\lambda < 0$. For this case Lichtenberg and Lieberman show that (6-3.11) is given by

$$\lambda_1 = -\lambda_2^2 + 2\lambda_2 + 4 \tag{6-3.12}$$

It can be shown that $\lambda_1 = -1$, so that $\lambda_2 = (1 - \sqrt{6}) = -1.4494$. If one is bold enough to propose that the recurrence relation (6-3.12) holds at higher-order bifurcations, then

$$\lambda_\kappa = -\lambda_{\kappa+1}^2 + 2\lambda_{\kappa+1} + 4 \tag{6-3.13}$$

At the critical value for chaos,

$$\lambda_\infty = -\lambda_\infty^2 + 2\lambda_\infty + 4$$
$$= (1 - \sqrt{17})/2 = -1.562 \tag{6-3.14}$$

One can also show that another bifurcation sequence occurs for $\lambda > 0$ (Figure 6-10) where the critical value is given by

$$\lambda = \hat{\lambda}_\infty = 2 - \lambda_\infty = 3.56 \tag{6-3.15}$$

The exact value is close to $\lambda_\infty = 3.56994$. Thus, the rescaling approximation scheme is not too bad.

This line of analysis also leads to the relation

$$\lambda_\kappa \simeq \lambda_\infty + \alpha\delta^{-\kappa} \tag{6-3.16}$$

which results in the scaling law (6-3.2). Thus, knowing that two successive bifurcation values can give one an estimate of the chaos criterion λ_∞, we obtain

$$\lambda_\infty \simeq \frac{1}{(\delta - 1)} [\delta\lambda_{\kappa+1} - \lambda_\kappa] \tag{6-3.17}$$

A final word before we leave this section: The fact that λ may exceed the critical value ($|\lambda| > |\lambda_\infty|$) does not imply that chaotic solutions will occur. They certainly are possible. But there are also many *periodic windows* in the range of parameters greater than the critical value in which periodic motions as well as chaotic solutions can occur.

We do not have space to do complete justice to the rich complexities in the dynamics of the quadratic map. It is certainly one of the major paradigms for understanding chaos, and the interested reader is encouraged to study this problem in the aforementioned references. (See also Appendix B for computer experiments.)

Homoclinic Orbits and Horseshoe Maps

One theoretical technique that has led to specific criteria for chaotic vibrations is a method based on the search for horseshoe maps and homoclinic orbits in mathematical models of dynamical systems. This strategy and a mathematical technique, called the *Melnikov method,* has led to Reynolds-type criteria for chaos relating the parameters in the system. In two cases, these criteria have been verified by numerical and physical experiments. Keeping with the tenor of this book, we do not derive or go into too much of the mathematical theory of this method, but we do try to convey the rationale behind it and guide the reader to the literature for a more detailed discussion of the method. We illustrate the Melnikov method with two applications: the vibrations of a buckled beam and the rotary dynamics of a magnetic dipole motor.

The homoclinic orbit criterion is a mathematical technique for obtaining a predictive relation between the nondimensional groups in the physical system. It gives one a necessary but not sufficient condition for chaos. It may also give a necessary and sufficient condition for predictability in a dynamical system (see Chapter 7, Section 7.7, "Fractal Basin Boundaries"). Stripped of its complex, somewhat arcane mathematical infrastructure, it is essentially a method to prove whether a model in the form of partial or ordinary differential equations has the properties of a horseshoe or a baker's-type map.

The horseshoe map view of chaos (see also Chapters 1, 3) looks at a collection of initial condition orbits in some ball in phase space. If a system has a horseshoe map behavior, this initial volume of phase space is mapped under the dynamics of the system onto a new shape in which the original ball is stretched and folded (Figure 6-12). After many iterations, this folding and stretching produces a fractal-like structure and the precise information as to which orbit originated where is lost. More and more precision is required to relate an initial condition to the state of the system at a later time. For a finite precision problem (as most numerical or laboratory experiments are), predictability is not possible.

Homoclinic Orbits. A good discussion of homoclinic orbits may be found in the books by Lichtenberg and Lieberman (1983), Guckenheimer and Holmes (1983), and Wiggins (1988). We have learned earlier that although many dynamics problems can be viewed as a continuous curve in some phase space (x versus $v = \dot{x}$) or solution space (x versus t), the mysteries of nonlinear dynamics and chaos are often deciphered by looking at a digital sampling of the motion such as a Poincaré map. We have also seen that although the Poincaré map is a sequence of points in some n-dimensional space, it can lie along certain continuous curves. These curves are called *manifolds*. A dis-

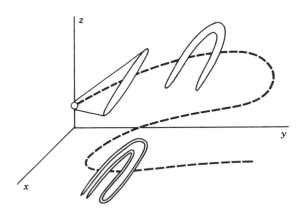

Figure 6-12 Evolution of an initial condition sphere.

cussion of homoclinic orbits refers to a sequence of points. This sequence of points is called an *orbit*.

In the dynamics of mappings, one can have critical points at which orbits move away from or toward. One example is a saddle point at which there are (a) two manifold curves on which orbits approach the point and (b) two curves on which the sequence of Poincaré points move away from the point, as illustrated in Figure 6-13 (see also Sections 3.2, 3.4). Such a point is similar to a *saddle point* in nonlinear differential equations.

To illustrate a homoclinic orbit, we consider the dynamics of the forced, damped pendulum. First, recall that for the unforced, damped pendulum, the unstable branches of the saddle point swirl around the

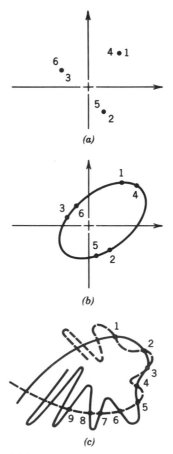

Figure 6-13   (a) Periodic orbit in a Poincaré map. (b) Quasiperiodic orbit. (c) Homoclinic orbit.

equilibrium point in a vortexlike motion in the θ–$\dot{\theta}$ phase plane as shown in Figure 6-14. Although it is not obvious, the Poincaré map synchronized with the forcing frequency also has a saddle point in the neighborhood of $\theta = \pm n\pi (n \text{ odd})$, as shown in Figure 6-15 for the case of the forced pendulum. For small forcing, the stable and unstable branches of the saddle do not touch each other. However, as the force is increased, these two manifolds intersect. It can be shown that *if they intersect once, they will intersect an infinite number of times*. [Another example of the intersection of stable and unstable manifolds is given in Section 3.4, Figure 3-10 for the Standard map, and Eq. (3-4.1).] The points of intersection of stable and unstable manifolds are called *homoclinic points*. A Poincaré point near one of these points will be mapped into all the rest of the intersection points. This is called a *homoclinic orbit* (Figure 6-13). Now why are these orbits important for chaos?

The intersection of the stable and unstable manifolds of the Poincaré map leads to a horseshoe-type map in the vicinity of each homoclinic point. As we saw in Chapter 1, horseshoe-type maps lead to unpredictability, and unpredictability or sensitivity to initial conditions is a hallmark of chaos.

To see why homoclinic orbits leads to horseshoe maps, we recall that for a dissipative system the areas get mapped into smaller areas. However, near the unstable manifold, the areas are also stretched. Because the total area must decrease, the area must also contact more than it stretches. Areas near the homoclinic points also get folded, as shown in Figure 6-16a.

A dynamic process can be thought of as a transformation of phase space; that is, a volume of points representing different possible initial

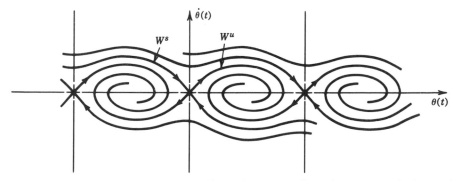

Figure 6-14 Stable and unstable manifolds for the motion of an unforced, damped pendulum.

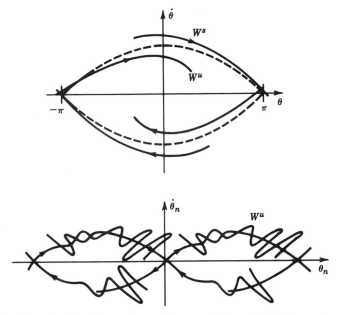

Figure 6-15 Sketch of stable and unstable manifolds of the Poincaré map for the harmonically forced, damped pendulum.

conditions is transformed into a distorted volume at a later time. Regular flow results when the transformed volume has a conventionally shaped volume. Chaotic flows result when the volume is stretched, contracted, and folded as in the baker's transformation or *horseshoe* map.

The Melnikov Method. The Melnikov function is used to measure the distance between unstable and stable manifolds when that distance is small [see Guckenheimer and Holmes (1983) or Wiggins (1988, 1990) for a mathematical discussion of the Melnikov method]. It has been applied to problems where the dissipation is small and where the equations for the manifolds of the zero dissipation problem are known. For example, suppose we consider the forced motion of a nonlinear oscillator where (q, p) are the generalized coordinate and momentum variables. We assume that both the damping and forcing are small and that we can write the equations of motion in the form

$$\dot{q} = \frac{\partial H}{\partial p} + \varepsilon g_1$$

$$\dot{p} = -\frac{\partial H}{\partial q} + \varepsilon g_2$$

(6-3.18)

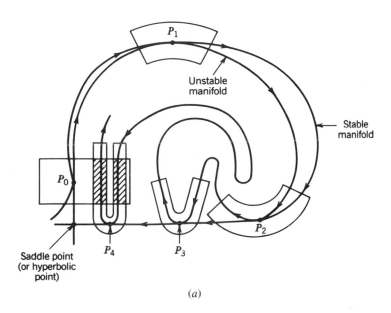

(a)

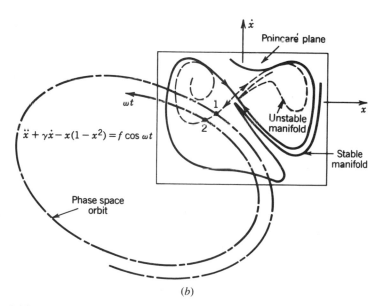

(b)

Figure 6-16 (a) The development of a folded horseshoe map for points in the neighborhood of a homoclinic orbit. (b) Saddle point of a Poincaré map and its associated stable and unstable manifolds before a homoclinic orbit develops.

where $\mathbf{g} = \mathbf{g}(p, q, t) = (g_1, g_2)$, ε is a small parameter, and $H(q, p)$ is the Hamiltonian for the undamped, unforced problems ($\varepsilon = 0$). We also assume that $\mathbf{g}(t)$ is periodic so that

$$\mathbf{g}(t + T) = \mathbf{g}(t) \tag{6-3.19}$$

and that the motion takes place in a three-dimensional phase space $(q, p, \omega t)$, where ωt is the phase of the periodic force and is modulo the period T.

In many nonlinear problems, a saddle point exists in the unperturbed Hamiltonian problem [$\varepsilon = 0$ in Eq. (6-3.18)], such as for the pendulum or the double-well potential Duffing's equation, Eq. (6-2.2). When $\varepsilon \neq 0$, one can take a Poincaré section of the three-dimensional torus flow synchronized with the phase ωt. It has been shown (see Guckenheimer and Holmes, 1983) that the Poincaré map also has a saddle point with stable and unstable manifolds, W^s and W^u, shown in Figure 6-16b.

The Melnikov function provides a measure of the separation between W^s and W^u as a function of the phase of the Poincaré map ωt. This function is given by the integral

$$M(t_0) = \int_{-\infty}^{\infty} \mathbf{g}^* \cdot \nabla \mathbf{H}(q^*, p^*) \, dt \tag{6-3.20}$$

where $\mathbf{g}^* = \mathbf{g}(q^*, p^*, t + t_0)$ and $q^*(t)$ and $p^*(t)$ are the solutions for the unperturbed homoclinic orbit originating at the saddle point of the Hamiltonian problem. The variable t_0 is a measure of the distance along the original unperturbed homoclinic trajectory in the phase plane. We consider two examples.

Magnetic Pendulum. A convenient experimental model of a pendulum may be found in the rotary dynamics of a magnetic dipole in crossed steady and time-periodic magnetic fields as shown in Figure 4-6 (see also Moon et al., 1987).

The equation of motion, when normalized, is given by

$$\ddot{\theta} + \gamma\dot{\theta} + \sin\theta = f_1 \cos\theta \cos\omega t + f_0 \tag{6-3.21}$$

The $\sin\theta$ term is produced by the steady magnetic field, and the f_1 term is produced by the dynamic field. We have also included linear damping and a constant torque f_0. Keeping with the assumptions of the theory, we assume that one can write $\gamma = \varepsilon\bar{\gamma}$, $f_0 = \varepsilon\bar{f}_0$, and $f_1 = \varepsilon\bar{f}_1$, where $0 \ll \varepsilon < 1$ and $\bar{\gamma}, \bar{f}_0$, and $\bar{f}_1$ are of order one.

The Hamiltonian for the undamped, unforced problem is given by

$$H = \tfrac{1}{2}v^2 + (1 - \cos\theta)$$

where $q \equiv \theta$ and $p \equiv v = \dot{\theta}$. The energy H is constant ($H = 2$) on the homoclinic orbit emanating from the saddle point ($\theta = v = 0$). The unperturbed homoclinic orbit is given by

$$\theta^* = 2\tan^{-1}(\sinh t)$$
$$v^* = 2\operatorname{sech} t \tag{6-3.22}$$

In Eq. (6-3.18), $g_1 = 0$ and $g_2 = f_0 + f_1\cos\theta\cos\omega t$. The resulting integral can be carried out exactly using contour integration [e.g., see Guckenheimer and Holmes (1983) for a similar example]. The result gives

$$M(t_0) = -8\bar{\gamma} + 2\pi\bar{f}_0 + 2\pi\bar{f}_1\omega^2\operatorname{sech}\left(\frac{\pi\omega}{2}\right)\cos\omega t_0 \tag{6-3.23}$$

The two perturbed manifolds will touch transversely when $M(t_0)$ has a simple zero, or when

$$f_1 > \left|\frac{4\gamma}{\pi} - f_0\right|\frac{\cosh(\pi\omega/2)}{\omega^2} \tag{6-3.24}$$

where we have canceled the ε factors. When $f_0 = 0$, the critical value of the forcing torque is given by

$$f_{1c} = \frac{4\gamma}{\pi\omega^2}\left(\frac{\cosh\pi\omega}{2}\right) \tag{6-3.25}$$

This function is plotted in Figure 6-6 along with experimental and numerical simulation data. The criterion (6-3.25) gives a remarkably good lower bound on the regions of chaos in the forcing amplitude–frequency plane.

Two-Well Potential Problem. Forced motion of a particle in a two-well potential has numerous applications such as postbuckling behavior of a buckled elastic beam (Moon and Holmes, 1979). Damped periodically forced oscillations can be described by a Duffing-type equation

$$\ddot{x} + \gamma\dot{x} - x + x_3 = f \cos \omega t \qquad (6\text{-}3.26)$$

The Hamiltonian for the unperturbed problem is

$$H(x, v) = \tfrac{1}{2}(v^2 - x^2 + \tfrac{1}{2}x^4)$$

For $H = 0$, there are two homoclinic orbits originating and terminating at the saddle point at the origin. The variables x^* and v^* take on values along the right half-plane curve given by

$$x^* = \sqrt{2} \text{ sech } t \quad \text{and} \quad v^* = -\sqrt{2} \text{ sech } t \tanh t$$

In this problem, $g_1 = 0$ and $g_2 = \bar{f}\cos \omega t - \bar{\gamma}v$, where $\gamma = \varepsilon\bar{\gamma}$ and $f = \varepsilon\bar{f}$ as in the previous example. The Melnikov function (6-3.20) then takes the form

$$M(t_0) = -\sqrt{2}f\int_{-\infty}^{\infty} \text{ sech } t \tanh t \cos \omega(t + t_0) \, dt$$

$$- 2\gamma \int_{-\infty}^{\infty} \text{ sech}^2 t \tanh^2 t \, dt$$

which can be integrated exactly using methods of contour integration. The solution was originally found by Holmes (1979), but an error crept into his paper. The correct analysis is in Guckenheimer and Holmes (1983):

$$M(t_0) = -\frac{4\gamma}{3} - \sqrt{2}f\pi\omega \text{ sech } \frac{\pi\omega}{2} \sin \omega t_0$$

For a simple zero we require

$$f > \frac{4\gamma}{3} \frac{\cosh(\pi\omega/2)}{\sqrt{2}\pi\omega} \qquad (6\text{-}3.27)$$

This lower bound on the chaotic region in (f, ω, γ) space has been verified in experiments by Moon (1980a) (see also Figures 6-2 and 6-3).

Melnikov Criterion for a One-Well Potential. In the study of the dynamics of ship capsize in a strong wind (see Chapter 4), Thompson

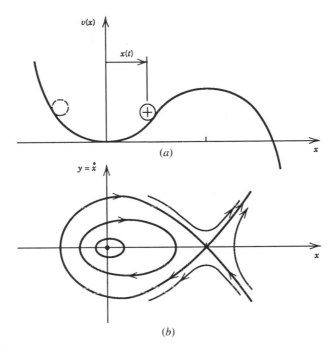

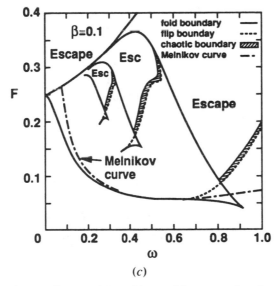

Figure 6-17 (*a*) One-well potential problem with escape barrier, (*b*) Phase plane portrait of unforced motion of a particle in a one-well potential, (*c*) Homoclinic orbit criterion (6-3.28b) for the one-well potential problem (6-3.28a) (from Thompson, 1989b).

and co-workers at University College, London, have used a one-well potential oscillator with a one-sided escape barrier:

$$\ddot{x} + \beta\dot{x} + x - x^2 = F \sin \omega t \qquad (6\text{-}3.28\text{a})$$

It is straightforward to show that the critical value of the force F as a function of frequency ω for homoclinic tangency is approximately given by (Thompson, 1989b; Thompson et al., 1990),

$$F = \frac{\beta \sinh \pi\omega}{5\pi\omega^2} \qquad (6\text{-}3.28\text{b})$$

This curve is plotted in Figure 6-17c. Also shown in this figure are the escape regions (capsize of the ship). These data from Thompson (1989b) also show narrow bands of chaotic regions just below the escape or capsize regime. The solid curves show values of F, ω where a stable and unstable solution coalesce, called a fold and also where a period doubling bifurcation occurs (shown as the flip boundary). These results demonstrate that while the Melnikov criteria sometimes provides a lower bound for chaotic or complex motions, the use of classical perturbation and bifurcation analysis may be useful to obtain more precise bounds on chaotic behavior [see Thompson and Stewart (1986) for a discussion of bifurcation theory and chaos].

Multiple Homoclinic Criteria: Three- and Four-Well Potentials. The homoclinic orbit criterion for a map is more easily applied if the underlying phase-space flow has homoclinic or heteroclinic orbits. The existence of such infinite time orbits are usually associated with the presence of saddle points—that is, equilibrium points with both stable (inflow) trajectories and unstable (outflow) trajectories. However, when two or more saddles are present, then there exist more than one mechanism for inflow and outflow trajectories in the Poincaré map to get tangled up. Thus, we can derive multiple criteria for chaos. The implications of such a phenomena are illustrated in two problems involving (a) a particle in a one-degree-of-freedom three-well potential and (b) a particle in a two-degree-of-freedom four-well potential. In each case, a small amount of periodic excitation is added along with a small amount of damping (see Li and Moon, 1990a,b). Experimentally, each case can be realized by placing three or four magnets below a steel cantilever beam as in Figure 4-1 (see Appendix B for a description of a two-well potential experiment). These problems represent multi-equilibrium systems when the excitation is absent.

Three-Well Potential. The three-well problem can be modeled by an equation of the form

$$\ddot{x} + \gamma\dot{x} + x(x^2 - x_0^2)(x^2 - 1) = f\cos\omega t \qquad (6\text{-}3.29)$$

which can be derived from an appropriate sixth-order polynomial potential function. When $f = 0$, this problem possesses three stable and two unstable (saddles) equilibrium positions. When the forcing is present ($f \neq 0$) and small, it can be demonstrated that the Poincaré map also has two saddle points.

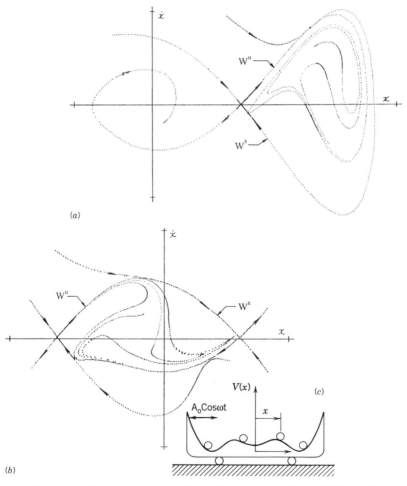

(a)

(b)

(c)

Figure 6-18 (a–c) The intersections of stable and unstable manifolds of the saddle points of a Poincaré map based on numerical integrations of the forced three-well potential oscillator (6-3.29). [From Li and Moon (1990a).]

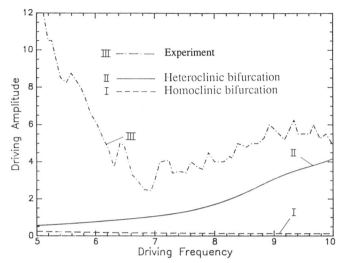

Figure 6-19   Two criteria for homoclinic orbits from the three-well potential oscillator (6-3.29). [From Li and Moon (1990a).]

There are two ways in which one can get homoclinic orbits in the map: Unstable manifolds from each saddle may intersect stable manifolds of the same saddle point, or an unstable manifold from one saddle may intersect the stable manifold from the other saddle point of the map. This is illustrated in Figure 6-18 for numerical integration of Eq. (6-3.29).

Criteria for chaos can then be found by procedures similar to that for the two-well potential problem, but with the aid of numerical computation (see Li and Moon, 1990a,b). An example is shown in Figure 6-19 as a function of driving amplitude and frequency (f, ω). interpretation of this diagram is that below both criteria, the problem is predictable—that is, insensitive to small changes in initial conditions. Between the two criteria, there are regions in the initial condition space which are respectively sensitive and insensitive to small changes. Above both criteria the problem is strongly sensitive to initial conditions. These results were confirmed by experimental observations as well as by numerical studies of basins of attractions (e.g., see color plate CP-7,8).

Four-Well Potential.   The four-well potential problem with harmonic forcing and damping is similar to a particle on a roulette table, as the level curves of the potential function show (Figure 6-20). The dynamics of this problem takes place in a five-dimensional phase space

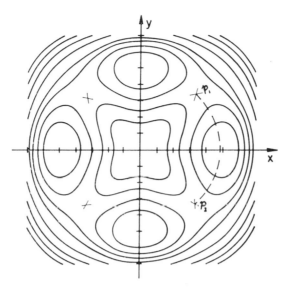

Figure 6-20 Level curves of a two-degree-of-freedom oscillator (6-3.30) in a four-well potential for the oscillator shown in Figure 4-1. [From Li (1987).]

in contrast to the three-dimensional phase spaces of the two- and three-well potential problems presented above. The Poincaré map triggered on the forcing phase lives in a four-dimensional phase space which defies the imagination. In this space the symmetric four-well problem has five saddles and many opportunities for entanglement of stable and unstable manifolds. To date we have no systematic way to determine all the possible ways of generating homoclinic orbits in this problem. Instead we guess at two obvious mechanisms based by analogy with the two-well and pendulum examples treated above. [The advanced reader should consult Wiggins (1988) for a treatment of homoclinic orbits in higher-dimensional phase spaces.]

The structure of the mathematical model for this problem is of the form (see Li and Moon, 1990b and Li, 1987)

$$\ddot{x} + \gamma_1 \dot{x} + \frac{\delta V(x, y)}{\partial x} = f_1 \cos(\Omega t + \varphi_0)$$

$$\ddot{y} + \gamma_2 \dot{y} + \frac{\delta V(x, y)}{\partial y} = f_2 \cos \Omega t$$

(6-3.30)

The two criteria are derived from a guess at two restricted classes of motions: radial motion through two of the wells and circumferential motion through four of the potential wells. Motion restricted to radial

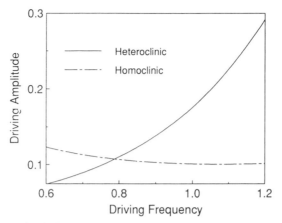

Figure 6-21   Two criteria for homoclinic orbits from the four-well potential oscillator (6-3.30). [From Li and Moon (1990b).]

motion is exactly similar to the two-well problem with one degree of freedom studied by Holmes (1979). Therefore, one obtains a criterion for homoclinic tangles in the Poincaré map similar to (6-3.27).

Experimental observation of a cantilevered steel rod with four magnets below the end of the rod shows that sometimes the rod will exhibit circumferential motions through each of the wells. However, the precise orbit is not circular. But numerical calculations show that a nearly circular orbit is possible. Thus, we artificially restrict the radial motion ($\dot{r} = 0, r = \sqrt{x^2 + y^2}$). The problem is similar to a rotor or pendulum with four potential wells, and a criterion similar to (6-3.25) may be possible.

Using these analogies, two criteria were derived numerically as shown in Figure 6-21. Numerical and experimental observations seem to indicate that chaos results when the parameters (f_1, Ω) exceed both criteria. Basin boundary studies (see color plate on this book's jacket) show increasing complexity as each criterion is crossed.

These two studies show that even in simple problems in Newtonian particle mechanics, extremely complex dynamic phenomena are possible; they also show that our knowledge to date is still extremely primitive, especially as regards phase spaces of dimension four or higher (see also Kittel et al., 1990).

Shil'nikov Chaos

The above discussion shows how a homoclinic orbit in the Poincaré map can lead to a horseshoe map structure and eventually chaos. But

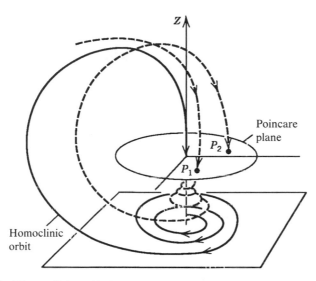

Figure 6-22   Homoclinic orbit in three-dimensional phase space generated by a saddle-focus fixed point.

what about homoclinic orbits in a flow described by a set of differential equations? For a 2-D phase plane, homoclinic or heteroclinic orbits cannot lead to chaos without periodic forcing. But what about homoclinic orbits in 3-D? In 1965, two years after the publication of Lorenz's paper on nonperiodic solutions for fluid convecture problems, L. P. Shil'nikov of the Soviet Union proposed a theorem which suggests that the existence of a homoclinic orbit in a 3-D flow would imply the existence of nonperiodic trajectories. In this work he chose a system of three first-order nonlinear differential equations with a fixed point which is characterized by a saddle focus (see Chapter 1; also see Guckenheimer and Holmes, 1983). A saddle focus has eigenvalues $(\rho \pm i\omega, \lambda)$, as shown in Figure 6-22. Near the fixed point, trajectories spiral out or in on some 2-D surface characterized by $\rho \pm i\omega$, or they approach or depart the fixed point in an exponential manner with time exponent λ along a direction transverse to the spiral surface. If the trajectory that spirals out (or in) eventually joins the trajectory coming into (or out of) the fixed point, then one has a homoclinic orbit. (Note however, that from any point on this orbit it takes an infinite time forward or backward to reach the fixed point.)

Shil'nikov proposed a criteria for the existence of these nonperiodic orbits that are generated by the homoclinic orbit:

$$|\rho/\lambda| < 1 \qquad\qquad (6\text{-}3.31)$$

Several experimental studies have been published which purport to have measured chaotic dynamics originating from a Shil'nikov homoclinic orbit. Argoul et al. (1987a) studied a continuous chemical flow reactor for the Belousov–Zhabotinski reaction, and Bassett and Hudson (1988) performed an experiment on the electrodissolution of a rotating copper disk in a $H_2SO_4/NaCl$ solution. In both papers the experimental results were compared to a third-order model of the form

$$\dot{x} = y$$

$$\dot{y} = z \tag{6-3.32}$$

$$\dot{z} = -\eta z - \nu y - \mu x - k_1 x^2 - k_2 y^2 - k_3 xy - k_4 xz - k_5 x^2 z$$

For example, in Argoul et al. (1987a) the following parameters are chosen: $k_1 = -1$, $k_2 = 1.425$, $k_3 = 0$, $k_4 = -0.2$, $k_5 = 0.01$. This system has two equilibrium points, one at the origin and the other at $(x, y, z) = (-\mu/k, 0, 0)$. For $\eta = 1.3$ and $\mu \geq 1.3$, a spiral-type strange attractor can be found as shown in Figure 6-23 which is qualitatively similar to that obtained from the experiments using a reconstructed phase space with variables $C(t)$, $C(t + T)$, $C(t + 2T)$, where C

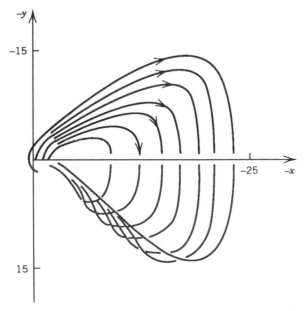

Figure 6-23 Shil'nikov-type strange attractor based on numerical integration of a model for a chemical flow reactor (6-3.32). $(\eta, \nu, \mu) = (1, 1.3, 1.38)$ [From Argoul et al. (1987a).]

represents the concentration of *Ce* in the continuously stirred tank reactor. The Poincaré map is obtained by intersecting the attractor with a plane in the 3-D space and shows a linear structure which suggests the use of a 1-D return map. In this numerical simulation of (6-3.32) the 1-D map derived from the Poincaré section shows an intersecting *multibranched map* (Figure 6-24) which is also obtained from the experiments. Each branch is labeled with an integer and represents the number of turns the trajectory orbits around the saddle focus in between two successive Poincaré map times. Argoul et al. (1987) also proposed a scaling law for the distance between two successive branches:

$$\lim_{n\to\infty} \frac{X^{(n+1)} - X^{(n)}}{X^{(n)} - X^{(n-1)}} = \exp(-2\pi\rho/\omega) = \delta \sim 0.8 \qquad (6\text{-}3.33)$$

The dynamics can thus be described in terms of an infinite set of symbols (each representing the number of turns around the saddle focus between mapping times).

There have also been claims and counterclaims about Shil'nikov chaos in laser dynamics (e.g., see Arecchi et al., 1987 and Swetits and Buoncristiani, 1988).

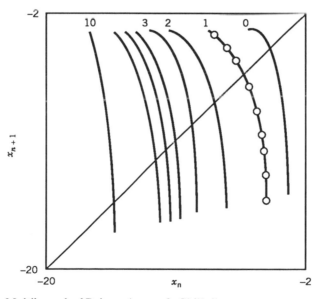

Figure 6-24   Multibranched Poincaré map of a Shil'nikov-type strange attractor based on numerical integration of a model for a chemical flow reactor (6-3.32). [From Argoul et al. (1987).]

Intermittent and Transient Chaos

Thus far we have discussed what one might call "steady-state" chaotic vibration. Two other forms of unpredictable, irregular motions are intermittency and transient chaos. In the former, bursts of chaotic or noisy motion occur between periods of regular motion (see Figure 6-25). Such behavior was even observed by Reynolds in pipe flow preturbulence experiments in 1883 (see Sreenivasan, 1986). Transient chaos is also observed in some systems as a precursor to steady-state chaos. For certain initial conditions, the system may behave in a randomlike way, with the trajectory moving in phase space as if it were on a strange attractor. However, after some time, the motion settles onto a regular attractor such as a periodic vibration. Scaling properties of nonlinear motion can sometimes be used to determine experimentally a critical parameter for these two types of chaotic motion. In the case of intermittency, where the dynamic system is close to a periodic motion but experiences short bursts of chaotic transients, an explanation of this behavior has been posited by Manneville and Pomeau (1980) in terms of one-dimensional maps or difference equations.

From numerical experiments on maps, the mean time duration of the periodic motion between chaotic bursts $\langle \tau \rangle$ has been found to be

$$\langle \tau \rangle \sim \frac{1}{|\lambda - \lambda_c|^{1/2}} \tag{6-3.34}$$

where λ is a control parameter (e.g., fluid velocity, forcing amplitude, or voltage) and λ_c is the value at which a chaotic motion occurs. As $\lambda - \lambda_c$ increases, the chaotic time interval increases and the periodic interval decreases. Thus, one might call this *creeping chaos*.

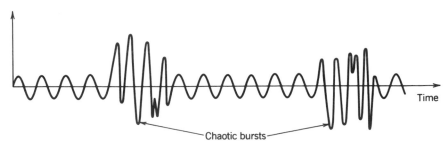

Figure 6-25   Sketch of intermittent chaotic motion.

To measure λ_c experimentally, one must measure two average times $\langle \tau \rangle_1$ and $\langle \tau \rangle_2$ at corresponding values of the control parameter, that is, λ_1 and λ_2. This should determine the proportionality constant in Eq. (6-3.34) as well as λ_c. Having obtained a candidate value for λ_c, however, one should then measure other values of $(\langle \tau \rangle, \lambda)$ to validate the scaling relation (6-3.34).

The case of transient chaos has been studied by Grebogi et al. (1983a,b, 1985b) of the University of Maryland in a series of papers describing numerical experiments on two-dimensional maps. In one study (1983), they investigated a two-dimensional extension of the one-dimensional quadratic difference equation called the *Henon map* (see also Section 1.3):

$$x_{n+1} = 1 - \alpha x_n^2 + y_n$$

$$y_{n+1} = -J x_n$$

where J is the determinant of the Jacobian matrix which controls the amount of area contraction of the map. In the Maryland group's research on transient chaos, the case of $J = -0.3$ with the parameter a varied was investigated. For example, for $\alpha > \alpha_0 = 1.062371838$, a period-6 orbit gave birth to a six-piece strange attractor that exists in the region

$$\alpha_0 < \alpha < \alpha_c = 1.080744879$$

For $\alpha > \alpha_c$, the orbit under the iteration of the Henon map was found to wander around the ghost of the strange attractor in the x–y plane, sometimes for over 10^3 iterations, before settling onto a period-4 motion.

They also discovered that the average time for the transient chaos $\langle \tau \rangle$ followed a scaling law:

$$\langle \tau \rangle \sim (\alpha - \alpha_c)^{-1/2} \tag{6-3.35}$$

The average was found by choosing 10^2 initial conditions for each choice of α. The initial conditions were chosen in the original basin of attraction of the defunct strange attractor. These transients can be very long. For example, in the case of the Henon map, Grebogi and co-workers found $\langle \tau \rangle \approx 10^4$ for $\alpha - \alpha_c, = 5 \times 10^{-7}$ and $\langle \tau \rangle \approx 10^3$ for $\alpha - \alpha_c = 10^{-5}$.

Chirikov's Overlap Criterion for Conservative Chaos

The study of chaotic motions in conservative systems (no damping) predates the current interest in chaotic dissipative systems. Because the practical application of conservative dynamical systems is limited to areas such as planetary mechanics, plasma physics, and accelerator physics, engineers have not followed this field as closely as other advances in nonlinear dynamics. In this section, we focus on the bouncing ball chaos described in Chapter 4 (Figure 4-11). However, the resulting difference equations are relevant to the behavior of coupled nonlinear oscillators (e.g., see Lichtenberg and Lieberman, 1983) as well as to the behavior of electrons in electromagnetic fields. The equations for the impact of a mass, under gravity, on a vibrating table are given by (4-2.19); with a change of variables, these become (see also Section 3.4)

$$v_{n+1} = v_n + K \sin \varphi_n$$
$$\varphi_{n+1} = \varphi_n + v_{n+1}$$

(6-3.36)

where v_n is the velocity before impact and φ_n is the time of impact normalized by the frequency of the table (i.e., $\varphi = \omega t \bmod 2\pi$). K is proportional to the amplitude of the vibrating table in Figure 4-11. These equations differ from those in (4-2.19) by the assumption that there is no energy loss on impact. This implies that regions of initial conditions in the phase space (v, φ) preserve their area under multiple iteration of the map (6-3.36).

Orbits in the (v, φ) plane for different initial conditions are shown in Figure 6-26 for two different values of K.

Consider the case of $K = 0.6$. The dots at $v = 0, 2\pi$ correspond to period-1 orbits; that is,

$$v_1 = v_1 + K \sin \varphi_1$$
$$\varphi_1 = \varphi_1 + v_1$$

whose solution is given by $\varphi_1 = 0, \pi, v_1 = 0$ (both mod 2π). The solution near $\varphi = \pi$ is stable for $|2 - K| < 2$. The solution near $\varphi = 0, 2\pi$, however, can be shown to be unstable for $|2 + K| < 2$ and can represent saddle points of the map.

Near $v = \pi$ one can see a period-2 orbit given by the solution to

$$v_2 = v_1 + K \sin \varphi_1, \qquad \varphi_2 = \varphi_1 + v_2$$
$$v_1 = v_2 + K \sin \varphi_2, \qquad \varphi_1 = \varphi_2 + v_1$$

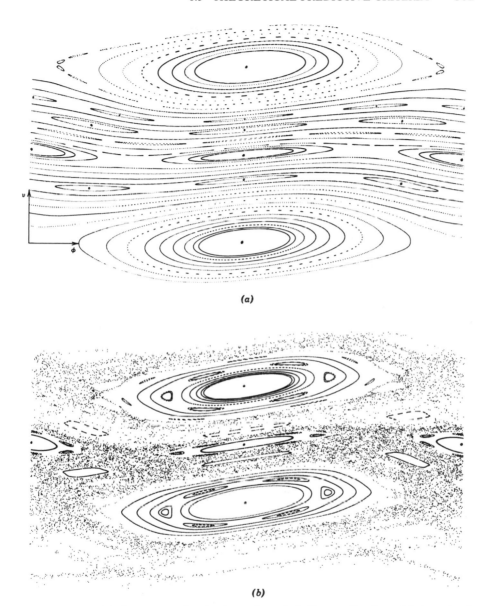

(a)

(b)

Figure 6-26   (a) Poincaré map for elastic motion of a ball on a vibrating table (standard map) for the parameter $\gamma = 0.6$ in Eq. (6-3.36) showing periodic and quasiperiodic orbits. (b) The case of $\gamma = 1.2$ showing the appearance of stochastic orbits.

Again one can show that there are both stable and unstable period-2 points. One can also show that the stable points exist as long as $K < 2$.

The rest of the continuous-looking orbits in Figure 6-26 represent quasiperiodic solutions where the ball impact frequency is incommensurate with the driving period. Finally, a third type of motion is present in Figure 6-26b ($K = 1.2$). Here we see a diffuse set of dots near where saddle points and the saddle separatrices used to exist. This diffuse set of points represents *conservative chaos*. For $K < 1$, it is localized around the saddle points. However, for $K \approx 1$, this wandering orbit becomes global in nature. (See also Figures 1-13, 3-35.)

The reader should note that in Figure 6-26 ($K = 0.6$) one can obtain all types of motion by simply choosing different initial conditions (because there is no damping, there are no attractors).

A criterion for global chaos in this system was proposed by the Soviet physicist Chirikov (1979). He observed that as K is increased, the vertical distance between the separatrices associated with both period-1 and period-2 motion decreased. If chaos did not intervene, these separatrices would overlap (Figure 6-27)—thus the name *overlap criterion*.

If one performs a small-K analysis of the standard map (6-3.36) near one of these periodic resonances, the size of each separatrix region is found to be

$$\Delta_1 = 4K^{1/2}$$
$$\Delta_2 = K$$

$$(6\text{-}3.37)$$

Each analysis ignores the effect of the other resonance. The condition

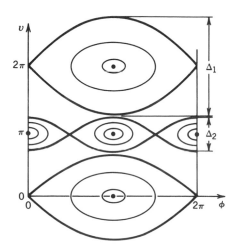

Figure 6-27   Sketch of period-1 and period-2 orbits and concomitant quasiperiodic orbits for the standard map used in the deivation of Chirikov's criterion.

for overlap is that $\Delta_1 + \Delta_2 = 2\pi$, or

$$4K_c^{1/2} + K_c = 2\pi \qquad (6\text{-}3.38)$$

The solution to this equation is $K_c = 1.46$. This value overestimates the critical value of $K = K_c$ for global chaos which is found numerically to be around $K_c \approx 1.0$. The reader is referred to Lichtenberg and Lieberman (1983) for further details concerning the overlap criterion.

The more practical-minded reader might ask: *What happens when we have a small amount of damping present?* For that case, some of the multiperiod subharmonics become attractors and the ellipses surrounding these attractors become spirals that limit the periodic motions. *What of the conservative chaos?* Initial conditions in regions where there was conservative chaos become long chaotic transients which wander around phase space before settling into a periodic motion. *And what about real chaotic motions?* When damping is present, one needs a much larger force, $K > 6$, for which a fractal-like strange attractor appears (see Figure 4-11). Thus, the overlap criterion discussed above is only useful for strictly conservative, Hamiltonian systems.

Criteria for a Multiple-Well Potential

In this section, we describe an ad hoc criterion for chaotic oscillations in problems with multiple potential energy wells. Such problems include the buckled beam (Chapter 2) and a magnetic dipole motor with multiple poles. In solid-state physics, interstitial atoms in a regular lattice can have more than one equilibrium position. Often the forces that create such problems can be derived from a potential. Let $\{q_i\}$ be a set of generalized coordinates and $V(q_i)$ be the potential associated with the conservative part of the force such that $-\partial V/\partial q_j$ is the generalized force associated with the q_j degree of freedom. For one degree of freedom, a special case might have the following equation of motion:

$$\ddot{q} + \gamma \dot{q} + \frac{\partial V}{\partial q} = f \cos \omega t \qquad (6\text{-}3.39)$$

where linear damping and periodic forcing have been added. $V(q_i)$ has as many local minima as stable equilibrium positions, as shown in Figure 6-28. For small periodic forcing, the system oscillates periodi-

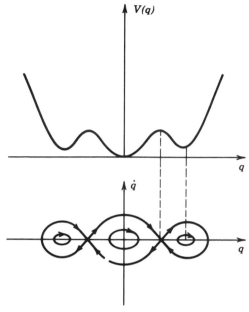

Figure 6-28 Multiple-well potential energy function and associated phase plane.

cally in one potential well. But for larger forcing, the motion "spills over" into other wells and chaos often results. This criterion then seeks to determine *what value of the forcing amplitude will cause the periodic motion in one well to jump into another well.*

To illustrate the method, consider the particle in a two-well symmetric potential (i.e., the buckled beam problem of Chapter 2):

$$\ddot{q} + \gamma\dot{q} - \tfrac{1}{2}q(1 - q^2) = f \cos \omega t \qquad (6\text{-}3.40)$$

Because we are seeking a criterion that governs the transition from periodic to chaotic motion, we use standard perturbation theory to find a relation between the amplitude of forced motion $\langle q^2 \rangle$ (where $\langle \, \rangle$ indicates a time average) and the parameters $\gamma, f,$ and ω. We then try to find a *critical* value of $\langle q^2 \rangle = A_c$ independent of the forcing amplitude; that is,

$$\langle q^2 \rangle = g(\gamma, \omega, f) = A_c(\omega) \qquad (6\text{-}3.41)$$

The left-handed equality in Eq. (6-3.41) is found using classical perturbation theory, whereas the right-hand equality is based on a heuristic postulate. To carry out this program for the two-well potential, we must write Eq. (6-3.40) in coordinates centered about one of the equilibrium positions:

$$\eta = q - 1$$

To obtain a perturbation parameter, we write $\eta = \mu X$, so that the equation of motion takes the form

$$\ddot{X} + \gamma \dot{X} + X(1 + \tfrac{3}{2}\mu X + \tfrac{1}{2}\mu^2 X^2) = \frac{f}{\mu}\cos(\omega t + \phi_0) \qquad (6\text{-}3.42)$$

The phase angle ϕ_0 is adjusted so that the first-order motion is proportional to $\cos \omega t$. The resulting periodic motion for small f is assumed to take the form

$$X = C_0 \cos \omega t + \mu(C_1 + C_2 \cos \omega t) + \mu^2 X_1(t) \qquad (6\text{-}3.43)$$

Using either Duffing's method or Lindstedt's perturbation method (e.g., see Stoker, 1950), the resulting amplitude force relation can be found to be

$$(\mu C_0)^2\{[(1 - \omega^2) - \tfrac{3}{2}(\mu C_0)^2]^2 + \gamma^2\omega^2\} = f^2 \qquad (6\text{-}3.44)$$

Based on numerical experiments, we postulate the existence of a *critical velocity*. We propose that chaos is imminent when the maximum velocity of the motion is near the maximum velocity on the sparatrix for the phase plane of the undamped, unforced oscillator. In terms of the original variables, this criterion becomes (see Figure 6-29)

$$\mu C_0 = \frac{\alpha}{2\omega} \qquad (6\text{-}3.45)$$

where α is close to unity. Substituting Eq. (6-3.45) into Eq. (6-3.44), we obtain a lower bound on the criterion for chaotic oscillations:

$$f_c = \frac{\alpha}{2\omega}\left\{\left[(1 - \omega^2) - \frac{3\alpha^2}{8\omega^2}\right]^2 + \gamma^2\omega^2\right\}^{1/2} \qquad (6\text{-}3.46)$$

This expression has been checked against experiments by the author (Moon, 1980a), and a factor of $\alpha \approx 0.86$ seemed to give excellent

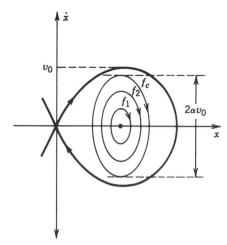

Figure 6-29 Overlap criteria for a multiple-well problem using semiclassical analytic methods.

agreement with experimental chaos boundaries as shown in an earlier figure (Figure 6-2). For low damping, this criterion gives a much better bound than does the homoclinic orbit criterion using the Melnikov function.

As illustrated in Figure 6-29, this criterion is similar to the Chirikov overlap criterion—namely, that chaos results when a regular motion becomes too large.

The method outlined in this section has also been used on a three-well potential problem, (6-3.29), and has been tested successfully in experiments on a vibrating beam with three equilibria by Li (1984).

Criteria Derived from Classical Perturbation Analysis. The novitiate to the field of nonlinear dynamics may be misled by the current interest in chaos to conclude that the field lay dormant in the prechaos era. However, a large literature exists describing (a) mathematical perturbation methods for calculating primary and subharmonic resonances and (b) the stability characteristics of solutions to nonlinear systems (e.g., see Nayfeh and Mook, 1979). Thus, it is no surprise that studies are beginning to emerge that attempt to use the more classical analyses in the effort to find criteria for chaotic motion. For example, Nayfeh and Khdeir (1986) use perturbation techniques to predict the occurrence of period-doubling or period-tripling bifurcations as precursors to chaotic oscillations of ships in regular sea waves (see also Chapter 4, Eq. (4-2.17)).

In another study, Szemplinska-Stupnicka and Bajkowski (1986) have studied the Duffing oscillator of Ueda [Eq. (4-6.1)]. They found subharmonic solutions using perturbation techniques and link the on-set of chaos to the loss of stability of the subharmonics using classical stability analysis. They use analog computer experiments to check their results. They conclude that for the Duffing–Ueda attractor [Eq. (4-6.1)], the chaotic motion is a transition zone between the subharmonic and resonant harmonic solutions. See Szemplinska-Stupnicka (1992).

Although the author believes that the fundamental nature of chaotic motion is more closely related to such mathematical paradigms as horseshoe maps, fractals, and homoclinic orbits, the use of semiclassical methods of perturbation analysis may provide more practical analytic chaos criteria for certain classes of nonlinear systems.

6.4 LYAPUNOV EXPONENTS

Thus far we have discussed mainly predictive criteria for chaos. Here we describe a tool for *diagnosing* whether or not a system is chaotic. Chaos in deterministic systems implies a sensitive dependence on initial conditions. This means that if two trajectories start close to one another in phase space, they will move exponentially away from each other for small times on the average. Thus, if d_0 is a measure of the initial distance between the two starting points, at a small but later time the distance is

$$d(t) = d_0 2^{\lambda t} \tag{6-4.1}$$

If the system is described by difference equations or a map, we have

$$d_n = d_0 2^{\Lambda n} \tag{6-4.2}$$

[The choice of base 2 in Eqs. (6-4.1) and (6-4.2) is convenient but arbitrary.] The symbols Λ and λ are called *Lyapunov exponents.*[1]

An excellent review of Lyapunov exponents and their use in experiments to diagnose chaotic motion is given by Wolf et al. (1985). This review also contains two useful computer programs for calculating Lyapunov exponents. Another review is Abarbanel et al. (1991).

The divergence of chaotic orbits can only be locally exponential, because if the system is bounded, as most physical experiments are,

---

[1] Lyapunov was a Russian mathematician (1857–1918) who introduced this idea around the turn of the century.

$d(t)$ cannot go to infinity. Thus, to define a measure of this divergence of orbits, we must average the exponential growth at many points along a trajectory, as shown in Figure 6-30. One begins with a reference trajectory [called a *fiduciary* by Wolf et al. (1985)] and a point on a nearby trajectory and measures $d(t)/d_0$. When $d(t)$ becomes too large (i.e., the growth departs from exponential behavior), one looks for a new "nearby" trajectory and defines a new $d_0(t)$. One can define the Lyapunov exponent by the expression

$$\lambda = \frac{1}{t_N - t_0} \sum_{k=1}^{N} \log_2 \frac{d(t_\kappa)}{d_0(t_{\kappa-1})} \qquad (6\text{-}4.3)$$

Then the criterion for chaos becomes

$$\lambda > 0 \quad \text{(chaotic)} \qquad (6\text{-}4.4)$$

$$\lambda \leq 0 \quad \text{(regular motion)}$$

The reader by now has surmised that this operation can only be done with the aid of a computer whether the data are from a numerical simulation or from a physical experiment.

Only in a few pedagogical examples can one calculate λ explicitly. To examine one such case, consider the extension of the concept of Lyapunov exponents to a one-dimensional map (see Chapter 3),

$$x_{n+1} = f(x_n) \qquad (6\text{-}4.5)$$

Following Chapter 3, we define the Lyapunov or characteristic expo-

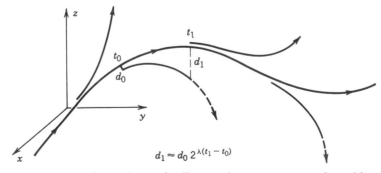

Figure 6-30 Sketch of the change in distance between two nearby orbits used to define the largest Lyapunov exponent.

nent as

$$\Lambda = \lim_{\substack{n \to \infty \\ k \to \infty}} \frac{1}{N} \sum^{N} \log_2 \left| \frac{df(x_n)}{dx} \right| \qquad (6\text{-}4.6)$$

An illustrative example given in Chapter 3 is the Bernoulli map (3-7.3)

$$x_{n+1} = 2x_n (\text{mod } 1) \qquad (6\text{-}4.7)$$

s shown in Figure 3-24. Except for the switching value at $x = \frac{1}{2}$, $|f'| = 2$. Applying the definition (6-4.6), we find $\Lambda = 1$. Thus, on the average, the distance between nearby points grows as

$$d_n = d_0 2^n \qquad (6\text{-}4.8)$$

The units of Λ are one bit per iteration. One interpretation of Λ is that one bit of information about the initial state is lost every time the map is iterated (see Section 3.7). So if we start out with m significant decimal places of information, we lose one for each iteration; that is, we lose one bit of information. *After m iterations we have lost knowledge of the initial state of the system.*

Earlier in this chapter, we learned that the solution for the logistic or quadratic map becomes chaotic when the control parameter α is greater than 3.57:

$$x_{n+1} = ax_n(1 + x_n) \qquad (6\text{-}4.9)$$

This can be verified by calculating the Lyapunov exponent as a function of a as shown in Figure 3-25. Beyond $a = 3.57$, the exponent becomes nonpositive in the periodic windows $3.57 < a < 4$. When $a = 4$, it has been shown that $\lambda = \ln 2$ (e.g., see Schuster, 1984).

Another example of a map for which one can calculate the Lyapunov exponent is the *tent map;* (3-7.4). As in the Bernoulli map (6-4.7), $|f'(x)|$ is a constant and the Lyapunov exponent is found to be (Lichtenberg and Lieberman, 1983, pp. 416–417)

$$\lambda = \log 2r$$

where $2r$ is the slope in (3-7.4). When $2r > 1$, $\lambda > 0$ and the motion is chaotic, but when $2r < 1$, $\lambda < 0$ and the orbits are regular; in fact, all points in $0 < x < 1$ are attracted to $x = 0$.

Numerical Calculation of the Largest Lyapunov Exponent

For every dynamical process, be it a continuous time history or discrete time evolution, there is a spectrum of Lyapunov or characteristic numbers that tells how lengths, areas, and volumes change in phase space. The idea of a spectrum of such numbers is discussed in the following section. However, inasfar as a criterion for chaos is concerned, one need only calculate the largest exponent, which tells whether nearby trajectories diverge ($\lambda > 0$) or converge ($\lambda < 0$) on the average. As yet there is no analog instrument that will measure the Lyapunov exponent, although if this measure of chaotic motion continues to prove useful, some clever person will probably invent one. At the present time, however, calculations of Lyapunov exponents must be done by digital computer, preferably a midsized laboratory computer.

There are two general methods: One is for data generated by a known set of differential or difference equations (flows and maps), and the other is to be used for experimental time series data. The Wolf et al. (1985) paper discusses both methods, but our experience to date reveals that more research on finding a reliable algorithm for experimental data is needed (see also Abarbanel et al. 1991). We will review briefly techniques for a set of differential equations of the form

$$\dot{\mathbf{x}} = \mathbf{f}(\mathbf{x}; \mathbf{c}) \tag{6-4.10}$$

where $\mathbf{x}$ is a set of n state variables and $\mathbf{c}$ is a set of n parameters. More complete discussion of these techniques may be found in Shimada and Nagashima (1979), the works of Benettin et al. (see the 1980 reference for a complete list), and Ueda (1979).

The main idea in calculating using (6-4.3) is to be able to determine the length ratio $d(t_k)/d(t_{k-1})$. One method is to numerically integrate the above set of equations to obtain a reference solution $\mathbf{x}^*(t; \mathbf{x}_0)$, where $\mathbf{x}_0$ is the initial condition. Then at each time step t_k integrate the equation again, using as an initial condition some nearby point $\mathbf{x}^*(t_k) + \boldsymbol{\eta}$. However, a more direct method is to use the equation to find the variation of trajectories in the neighborhood of the reference trajectory $\mathbf{x}^*(t)$. That is, at each time step t_k we solve the variational equations

$$\dot{\boldsymbol{\eta}} = \mathbf{A} \cdot \boldsymbol{\eta} \tag{6-4.11}$$

where $\mathbf{A}$ is the matrix of partial derivatives $\nabla \mathbf{f}(\mathbf{x}^*(t_k))$. We note that, in general, the elements of $\mathbf{A}$ depend on time. However, if $\mathbf{A}$ were constant, the solution of $\eta(t)$ between $t_k < t < t_{k+1}$ would depend on the initial condition. If this initial condition is chosen at random, then it is likely to have a component that lies in the direction of the largest positive eigenvalue of $\mathbf{A}$. It is the change in length in this direction that the largest Lyapunov exponent measures.

Thus, the numerical scheme goes as follows. Integrate (6-4.10) to find $\mathbf{x}^*(t)$. Allow a certain time to pass before calculating $d(t)$ in order to get rid of transients. After all, we are assuming we are on a stable attractor. After the transients are judged to be small, begin to integrate (6-4.11) to find $\eta(t)$. One can choose $|\eta(0)| = 1$, but choose the initial direction to be arbitrary. Then numerically integrate $\dot{\eta} = \mathbf{A}(x^*(t)) \cdot \eta$, taking into account the change in $\mathbf{A}$ through $x^*(t)$. [In practice one can integrate both (6-4.10) and (6-4.11) simultaneously. [After a given time interval $t_{k+1} - t_k = \tau$, take

$$\frac{d(t_{k+1})}{d(t_k)} = \frac{|\eta(t; t_k)|}{|\eta(0; t_k)|} \tag{6-4.12}$$

To start the next time step in (6-4.3), use the direction of $\eta(\tau; t_k)$ for the new initial condition, that is,

$$\eta(0; t_{k+1}) = \frac{\eta(\tau; t_k)}{|\eta(\tau; t_k)|} \tag{6-4.13}$$

where we have normalized the initial distance to unity.

An example of this calculation is shown in Figure 6-31, where we have numerically integrated the Duffing equation [Eq. (6-2.1)] in the chaotic state as a function of the elapsed time. The equations used were

$$\dot{x} = y$$
$$\dot{y} = -ky - x^3 + B \cos z \tag{6-4.14}$$
$$\dot{z} = 1$$

The resulting matrix becomes

$$\mathbf{A} = \begin{bmatrix} 0 & 1 & 0 \\ -3x^2 & -k & -B \sin z \\ 0 & 0 & 0 \end{bmatrix} \tag{6-4.15}$$

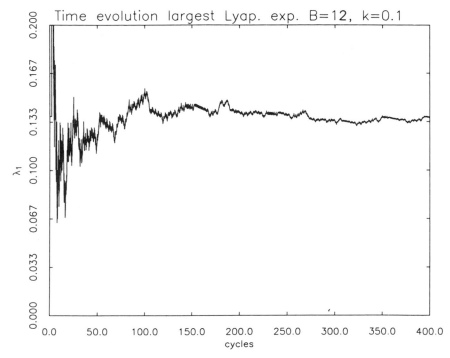

Figure 6-31   Calculation of the largest Lyapunov exponent for chaotic motion of the Duffing attractor (6-2.1) as a function of the total time record.

Because this really is a periodically driven oscillator, changes of lengths in the phase space direction $z = t$ are zero, as manifested by the row of zeroes in the matrix **A**. Thus, to find the largest Lyapunov exponent in this problem, one can work in the *projection* of the phase space (x, y, z) onto the phase plane (x, y), using the inner bracketed matrix in (6-4.15).

For the data in Figure 6-31, the time step for numerical integration was $\Delta t = 0.01$ and the number of time steps to integrate $\eta(t)$ was chosen to be 10, or $\tau = 0.1$. The inner matrix in **A**, (6-4.15), was updated at every Runge–Kutta time step Δt.

It is clear from Figure 6-29 that λ is a statistical property of the motion; that is, one must average the changes in lengths over a long time in order to get reliable values. Also, one has to be careful in choosing the Runge–Kutta step size Δt as well as the Lyapunov exponent step size τ.

A comparison of Lyapunov exponents for different parameters in the Duffing equation is shown in Table 6-1. This algorithm for calculat-

TABLE 6-1   Comparison of Calculated Lyapunov Exponent for Duffing's Equation $\ddot{x} + k\dot{x} + x^3 = B \cos t$

k	B	λ_1 (this book)[a]	Ueda (1979)	
			λ_1	λ_2
0.1	9.9	0.012	0.065	−0.166
0.1	10	0.094	0.102	−0.202
0.1	11	0.114	0.114	−0.214
0.1	12	0.143	0.149	−0.249
0.1	13	0.167	0.182	−0.282
0.1	13.3	0.174	0.183	−0.284

[a] Runge–Kutta integration time step, $\Delta t = 0.01$; Lyapunov restart time $= 10\Delta t$; total time $= 400$ cycles $= 800\pi$.

ing Lyapunov exponents has proved very useful in constructing empirical chaos criteria or chaos diagrams. If one has access to a really fast computer such as the so-called supercomputers, then one can calculate λ as a function of the parameters in the problem [**c** in Eq. (6-4.10)]. For example, one can choose $\mathbf{c} = (k, B)$ in the Duffing problem and find λ for 100×100 values of k and B. If $\lambda > 0$, then one prints out a symbol; otherwise, if $\lambda \sim 0$ or $\lambda < 0$, one leaves a blank. Such numerically determined chaos diagrams are useful to search for possible regions of parameter space where chaotic motion may exist (see Figure 6-3). Given the vagaries of numerical calculation, however, one should not rely solely on this technique to certify a region as chaotic. Other tests such as spectral analysis, Poincaré maps, or fractal dimension should also be used to confirm suspected regions of chaotic motion.

Lyapunov Exponents and Distribution Functions. The calculation of the Lyapunov exponent (6-4.3) may be thought of as an average over time or iterates of the mapping (6-4.5). If one has a probability density function that tells the probability that certain trajectories will be in a given region of phase space, then it is possible to replace this time average by a spatial average in phase space. This idea has been explored by several researchers (Everson, 1986; Hsu, 1987). The idea is illustrated for a two-dimensional map following Everson. The case for a one-dimensional map was discussed in Section 3.6, Eq. (3-7.12).

We recall that when the system is chaotic, at least one Lyapunov exponent will be greater than zero. Start with the distance between two neighboring trajectories $\mathbf{x}_n$ and $\mathbf{y}_n$. This distance is given by $d_n =$

$|\mathbf{x}_n - \mathbf{y}_n|$ and the Lyapunov exponent is given by

$$\Lambda = \lim_{N \to \infty} \frac{1}{N} \sum_{}^{N} \log \frac{d_{n+1}}{d_n} \tag{6-4.16}$$

If an invariant probability distribution function $\rho(\mathbf{x})$ is assumed, then Λ can be calculated by

$$\Lambda = \iint \log \frac{d_{n+1}}{d_n} \rho(u, v) \, du \, dv \tag{6-4.17}$$

where a two-dimensional phase space is assumed with $\mathbf{x} = (u, v)$.

The invariant density function is assumed to satisfy the normalization condition

$$\iint \rho(u, v) \, du \, dv = 1$$

where the integral is taken over all of phase space.

Everson (1986) applied this idea to a map related to the bouncing ball problem (4-2.19) and the standard map (6-3.32),

$$\theta_{n+1} = \theta_n + BV_n, \qquad \text{mod } 2\pi$$
$$V_{n+1} = \varepsilon V_n + (1 + \varepsilon)(1 + \sin \theta_{n+1}) \tag{6-4.18}$$

This is similar to the problem examined by Holmes (1982), where $0 < \varepsilon < 1$ represents dissipation and BV_n represents the velocity of the ball as it leaves the platform at the nth bounce (see Figure 4-11a).

Everson (1986) used two observations to apply (6-4.17) to (6-4.18) to calculate the largest Lyapunov exponent. First, he notes that from numerical experiments the invariant distribution function appears to be independent of the phase θ, so that in polar coordinates (V, θ)

$$\int_0^\infty \rho \, dV = \frac{1}{2\pi} \tag{6-4.19}$$

Second, he was able to obtain an approximate expression for the expression d_{n+1}/d_n; that is, for $B \gg 1$,

$$\frac{d_{n+1}}{d_n} \to |B(1 + \varepsilon) \cos \theta| \tag{6-4.20}$$

which is independent of the velocity. Using (6-4.19) and (6-4.20), he was able to calculate

$$\Lambda = \log \frac{B(1 + \varepsilon)}{2} \tag{6-4.21}$$

which agrees quite well with numerical calculations.

In another application of this technique, Hsu (1987) used (6-4.17) but found the probability density function numerically using a technique called *cell mapping* [e.g., see Hsu (1981, 1987), Kreuzer (1985), and Tongue (1987)]. Further study of the determination of invariant probability distribution functions in the future may allow more general application of this method of determining Lyapunov exponents.

Lyapunov Spectrum

Thus far we have talked only of the stretching of distance between orbits in a chaotic process. However, in three or more dimensions we know that regions of phase space may contract as well as stretch under a dynamic process. In particular, for dissipative systems, a small volume of initial conditions gets mapped into a smaller volume at a later time. This is illustrated in Figure 6-32, where a small sphere of initial conditions of radius δ is mapped at a later time into an ellipsoid with principal axes $(\mu_1^n\delta, \mu_2^n\delta, \mu_3^n\delta)$. Thus, for every dynamical system there is a spectrum of Lyapunov exponents or numbers $\{\lambda_i\}$, $\lambda_i = \log \mu_i$.

Computationally, this spectrum can be calculated from a time history of a motion in phase space by finding out how lengths, areas, volumes, and hypervolumes change under a dynamic process. Wolf

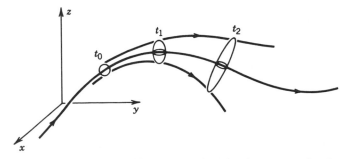

Figure 6-32 Sketch showing the divergence of orbits from a small sphere of initial conditions for a chaotic motion.

et al. (1985) used this idea to develop a computation algorithm to calculate the $\{\lambda_i\}$. If the λ_i are ordered such that $\lambda_1 > \lambda_2 \cdots > \lambda_n$, then they show that lengths vary as $d(t) \approx d_0 2^{\lambda_1 t}$, areas (formed from one point on the reference trajectory, and two nearby points) vary as $A(t) \approx A_0 2^{(\lambda_1 + \lambda_2)t}$, and small volumes vary as $V(t) \approx V_0 2^{(\lambda_1 + \lambda_2 + \lambda_3)t}$, and so on.

Farmer et al. (1983) provided an analytic definition for the complete Lyapunov spectrum along with one example for which one can calculate the $\{\lambda_i\}$ exactly. In the remainder of this chapter we give a sketch of the calculation of Lyapunov exponents for a two-dimensional map. Many of the details are omitted, and the interested reader is referred to the original Farmer et al. paper. To begin, we consider a general N-dimensional map

$$\mathbf{x}_{n+1} = \mathbf{F}(\mathbf{x}_n) \tag{6-4.22}$$

where $\mathbf{x}_n$ is a vector in an N-dimensional phase space. Then the change in shape of some small hypersphere will depend on the derivatives of the functions $\mathbf{F}(\mathbf{x}_n)$ with respect to the different components of $\mathbf{x}_n$. The relevant matrix is called a *Jacobian matrix*. For example, if

$$F = (f(x, y, z), g(x, y, z), h(x, y, z))$$

then

$$J = \begin{bmatrix} \dfrac{\partial f}{\partial x} & \dfrac{\partial f}{\partial y} & \dfrac{\partial f}{\partial z} \\[2mm] \dfrac{\partial g}{\partial x} & \dfrac{\partial g}{\partial y} & \dfrac{\partial g}{\partial z} \\[2mm] \dfrac{\partial h}{\partial x} & \dfrac{\partial h}{\partial y} & \dfrac{\partial h}{\partial z} \end{bmatrix} = [\nabla \mathbf{F}] \tag{6-4.23}$$

After n iterations of the map, the local shape of the initial hypersphere depends on

$$[J_n] = [\nabla \mathbf{F}(\mathbf{x}_n)][\nabla \mathbf{F}(x_{n-1})] \cdots [\nabla \mathbf{F}(x_1)] \tag{6-4.24}$$

In general, one can find the eigenvalues of J_n which one orders according to $j_1(n) \geq j_2(n) \geq \cdots \geq j_N(n)$, where the $j_K(n)$ are the absolute values of the eigenvalues. The Lyapunov exponents are then defined by

$$\lambda_i = \lim_{n \to \infty} \frac{1}{n} \log_2 j_i(n) \tag{6-4.25}$$

Farmer et al. illustrated the use of this definition for a two-dimensional map called a *baker's transformation* (Figure 6-33), named for its analogy to rolling and cutting pie dough. It is similar to the horseshoe map described in Chapter 1. The equations for this map are

$$x_{n+1} = \begin{cases} \lambda_a x_n, & y < \tfrac{1}{2} \\ \tfrac{1}{2} + \lambda_b x_n, & y > \tfrac{1}{2} \end{cases}$$

$$y_{n+1} = \begin{cases} 2y_n, & y < \tfrac{1}{2} \\ 2(y - \tfrac{1}{2}), & y > \tfrac{1}{2} \end{cases} \tag{6-4.26}$$

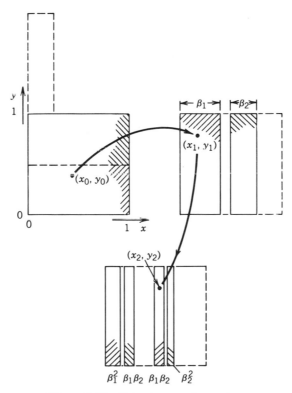

Figure 6-33 Baker's transformation.

This map is a generalization of the Bernoulli map in the previous section (6-4.8). In this case, the Jacobian matrix becomes

$$J = \begin{bmatrix} S_1 & 0 \\ 0 & 2 \end{bmatrix} \tag{6-4.27}$$

where $S_1 = \lambda_a$ for $y < \frac{1}{2}$ and $S_1 = \lambda_b$ for $y > \frac{1}{2}$.

For iterations of the map, the magnitudes of the eigenvalues become

$$j_1(n) = 2^n, \qquad j_2(n) = \lambda_a^k \lambda_b^l, \qquad k + l = n$$

where one assumes that there are k iterations in the left half-plane and l iterations in the right half-plane. Applying the definition (6-4.25),

$$\lambda_1 = \lim_{n \to \infty} \frac{1}{n} \log_2 2^n$$

$$\lambda_1 = \lim_{n \to \infty} \left\{ \frac{k}{n} \log_2 \lambda_a + \frac{l}{n} \log_2 \lambda_b \right\}$$

TABLE 6-2   Lyapunov Exponents for Dynamical Models

System	Parameter Values	Lyapunov Spectrum (bits/s)	Lyapunov Dimension (see Chapter 7)
Henon			
		$\lambda_1 = 0.603$	
$X_{n+1} = 1 - aX_n^2 + Y_n$	$\begin{cases} a = 1.4 \\ b = 0.3 \end{cases}$	$\lambda_2 = -2.34$	1.26
$Y_{n+1} = bX_n$		(bits/iteration)	
Rössler chaos			
$\dot{X} = -(Y + Z)$	$a = 0.15$	$\lambda_1 = 0.13$	
$\dot{Y} = X + aY$	$b = 0.20$	$\lambda_2 = 0.00$	2.01
$\dot{Z} = b + Z(X - c)$	$c = 10.0$	$\lambda_3 = -14.1$	
Lorenz			
$\dot{X} = \sigma(Y - X)$	$\sigma = 16.0$	$\lambda_1 = 2.16$	
$\dot{Y} = X(R - Z) - Y$	$R = 45.92$	$\lambda_2 = 0.00$	2.07
$\dot{Z} = XY - bZ$	$b = 4.0$	$\lambda_3 = -32.4$	
Rössler hyperchaos			
$\dot{X} = -(Y + Z)$	$a = 0.25$	$\lambda_1 = 0.16$	
$\dot{Y} = X + aY + W$	$b = 3.0$	$\lambda_2 = 0.03$	3.005
$\dot{Z} = b + XZ$	$c = 0.05$	$\lambda_3 = 0.00$	
$\dot{W} = cW - dZ$	$d = 0.5$	$\lambda_4 = -39.0$	

Source: Wolf et al. (1985).

Here we invoke an assumption that after many iterations an orbit spends as much time in the left half-plane as in the right half-plane, or

$$\frac{k}{n} = \frac{1}{2}, \qquad \frac{l}{n} = \frac{1}{2}$$

so that

$$\lambda_1 = 1, \qquad \lambda_2 = \tfrac{1}{2}\log_2\lambda_a\lambda_b < 0 \qquad (6\text{-}4.28)$$

Knowing these two Lyapunov exponents, one can then calculate a fractal dimension for this map. The relation between Lyapunov exponents and fractal dimensions has been examined by Farmer et al. (1983) and is discussed briefly in Chapter 7.

The spectra of Lyapunov exponents for several dynamics flows and maps are shown in Table 6-2, taken from Wolf et al. (1985).

Lyapunov Exponents for a Continuous Time Dynamics

When the dynamics are governed by a set of N ordinary differential equations $\dot{x} = f(x)$, the spectrum of Lyapunov exponents is related to the integration of the linearly perturbed dynamics, $\eta(t)$, about a solution $x(t)$, as in (6-4.11). The matrix of partial derivatives $A = \nabla f$ is time-dependent, because it depends in general on the reference solution $x^*(t)$. The solution at time $t = \tau$, $\eta(\tau)$ can then be written formally in the form

$$\eta(\tau) = \Phi(\eta^*(t)) \cdot \eta(0) \qquad (6\text{-}4.29)$$

where Φ is an $N \times N$ matrix and where $\eta(0)$ is the initial perturbation. A formal definition of the Lyapunov exponent λ_i follows from the construction of a positive and symmetric matrix

$$[\Phi^T\Phi]^{1/2\tau} \qquad (6\text{-}4.30)$$

whose eigenvalues are denoted by μ_i. Then the Lyapunov exponents are defined by the limit

$$\lambda_i = \lim_{\tau \to \infty} \log \mu_i \qquad (6\text{-}4.31)$$

(e.g., see Geist et al., 1990 and Abarbanel et al., 1991). This formal definition does not provide an obvious and practical way to determine the complete set of λ_i from numerical or experimental data. The reader is referred to the above references for the latest techniques and is also referred to Wolf et al. (1985) and Parker and Chua (1989). The latter book gives algorithms for calculating many of the measures of chaotic dynamics.

Typically, one has a set of discrete time sampled data $x_k = x(t = k\tau)$, where $x(t)$ is a measured state variable and τ is the sampling time. One numerical technique for a set of data $\{\cdots x_{i-1}, x_i, x_{i+1}, \cdots\}$ is to construct a set of vectors in an embedding space of N dimensions; $\mathbf{x} = (x_j, x_{j+1}, ..., x_{j+N-1})$. N is chosen at least as large as the dimension of the space of the chaotic attractor. Two methods are then used to calculate Lyapunov exponents. One is based on calculating the change in a small hypervolume as discussed above as the dynamics evolves. For example, an $M \leq N$-dimensional volume will change on the average according to $V(t) = V_0 \exp[(\lambda_1 + \lambda_2 + \cdots + \lambda_M)t]$. This method has been adopted by Wolf et al. (1985). Another method is based on estimating the local Jacobian matrices $\nabla \mathbf{f}$ (e.g., see Abarbanel et al., 1991 or Geist et al., 1990).

In most of these methods, several factors can raise questions as to the accuracy of the exponents. These factors include:

- Ill-conditioned matrices Φ
- Small number of data points (e.g., $< 10^4$)
- Accuracy of data
- Peculiar geometry of the attractor
- Spurious exponents when N is too high

To date, reliable algorithms for experimental calculation of all the λ_i are wanting. The researcher should always use measured Lyapunov exponents with some suspicion, especially where the dimension of the attractor is six or higher.

Hyperchaos

In the introductory chapter of this book, we defined chaotic dynamics as a sensitivity of the time history of a system to initial conditions. This sensitivity is exemplified by the horseshoe map (Chapter 1) in which a small ball or cube of initial conditions in phase space is stretched and folded back on itself. This stretching is measured by a positive Lyapunov exponent. Actually, for an n-dimensional phase space there are n Lyapunov exponents each measuring the relative stretching and contraction of the various axes of the ball of initial conditions. In many systems this stretching occurs along one direction and results in one positive Lyapunov exponent. However, in some systems, two or more directions in the phase space suffer stretching under the dynamic process. The occurrence of two or more positive Lyapunov exponents is called *hyperchaos*. One example that has been studied numerically is the example of two coupled Van der Pol oscillators (Kapitaniak and Steeb, 1991):

$$\ddot{x} - a(1 - x^2)\ddot{x} + x^3 = b(\sin \omega t + y)$$
$$\ddot{y} - a(1 - y^2)\dot{y} + y^3 = b(\sin \omega t + x)$$

(6-4.32)

Hyperchaos was studied for the values $a = 0.2$, $\omega = 4.0$ and for a variety of control parameters, for example, $b = 6.5, 7.0, 8.0$. This system must be described in a five-dimensional phase space. For $b = 7.0$, Kapitaniak and Steels numerically calculated a set of Lyapunov exponents $(0.69, 0.23, 0, -0.66, -0.94)$.

This system also has multiple solutions dependent on the initial conditions for the case $b = 7.0$; the initial conditions that led to hyperchaos were $x(0) = 1.0$, $\dot{x}(0) = \dot{y}(0) = y(0) = 0$. In the case of $b = 7.0$, the above authors also show that there are four separate solutions; time T periodic, $3T$ periodic, chaotic, and hyperchaotic solutions. Each has its own basin of attraction whose boundaries may be fractal (see Chapter 7 for a discussion of basin boundaries).

This example shows the complexity that arises when the dimension of the phase space becomes greater than three. Some mathematicians (e.g., Rössler, 1979) believe that there are new dynamical phenomena to be discovered in four or more dimensions. This author believes that in spite of the flood of books and papers on nonlinear and chaotic dynamics, we are still at the beginning of an era of new knowledge in this field. For novitiates to the field of chaos, there is still much to be discovered. (See Table 6-2 for another example.)

PROBLEMS

6-1 *Period Doubling*. Use a small computer to enumerate the critical values of λ in the logistic equation (6-3.1) and show that the sequences of values of λ_n and a_n approach the universal numbers (6-3.2) and (3-6.5).

6-2 Suppose an experiment on an electrical circuit exhibits a period-doubling bifurcation with two successive critical control voltages of 10.8 and 11.0 volts. Assume these bifurcation points are close to the limit point of the period-doubling sequence. Then (i) find the next value in the sequence and (ii) estimate the value of the limit point λ_∞.

6-3 Consider the logistic map (6-3.1). Find the period-2 map [i.e., $x_{n+2} = F(x_n)$]. [*Hint:* see Eq. (6-3.6)]. Show either analytically or numerically that this map has two maxima. Show either analytically [using Eq. (6-3.5)] or numerically that the magnitude of the slopes at the fixed points becomes greater than unity when $\lambda = \lambda_2$.

6-4 Another first-order iterated map that exhibits period doubling is the sine map

$$x_{n+1} = \lambda \sin \pi x_n, \qquad 0 \le x_n < 1$$

With a small computer, show that this map exhibits a period-doubling sequence and verify (6-3.2). Also show (numerically) that the period-2 map, $x_{n+2} = \lambda \sin \pi[\lambda \sin \pi x_n]$, has a double hump similar to the quadratic or logistic map (6-3.1).

6-5 *Circle Map and Intermittency.* The circle map was shown to be the high damping limit of a rotary impact oscillator [see Section 3.5, (3-5.7)]:

$$\theta_{n+1} = \theta_n + \Omega - \frac{\kappa}{2\pi} \sin 2\pi\theta_n, \qquad \theta_n = (\text{mod } 1)$$

Here the angle of rotation is normalized by 2π. As an exercise, show that this map becomes multivalued when $\kappa = 1$. This map has also been used as a model for intermittency. In this model, intermittency can occur when the above function $\theta_{n+1} = F(\theta_n)$ is close to the identity map $\theta_{n+1} = \theta_n$. First assume a subinterval $a \le \theta \le b$ where $F(\theta) > \theta$ and $|F(\theta) - \theta| < \varepsilon$, where ε is a small value. Estimate the number of iterations the orbit would take to go through this region. Second, with a small computer, find a value of $\kappa > 1$ which exhibits this intermittent trapping phenomena (see also Bergé et al., 1985).

6-6 *Homoclinic Orbits.* The following cubic map was used by

Holmes (1979) to model the dynamic snap-through buckling of a thin rod under compressive load:

$$x_{i+1} = y_i$$

$$y_{i+1} = -bx_i + dy_i - y_i^3, \qquad d \geq 0, \quad b > 0$$

This is also a model for a particle in a double-well potential. Examine the fixed point $(x, y) = (0, 0)$ and show that it is a saddle point when $d > 1 + b$. For $b = 0.2$, use d as a parameter. Use a small computer to show that the stable and unstable manifolds do not touch for $d - 2, 2.5$, but become tangled for $d > 2.6$; for example, try $d = 2.72, 2.77$. [*Hint:* derive a linear map near $(x, y) = (0, 0)$ and find the slopes of the stable and unstable manifold. Choose many initial conditions near the origin along these directions to iterate the nonlinear map to generate the two manifolds.]

6-7 *Melnikov Criteria.* The equation for the capsize of ships in seas with high winds and periodic beam waves has been shown to be modeled by a one-well potential oscillator with a saddle (6-3.28a). Rewriting this equation in the form (6-3.18)

$$\dot{x} - y = 0$$

$$\dot{y} - x + x^2 = \varepsilon[-\beta'y + F' \sin \omega t]$$

assume that ε is small and derive the criteria (6-3.28b) using the Melnikov function (6-3.20). As a hint, note that the unperturbed Hamiltonian is given by

$$H = \tfrac{1}{2}\dot{y}^2 - \frac{x^2}{2} + \frac{x^3}{3}$$

and that the unperturbed homoclinic orbit (i.e., the orbit that goes through the saddle when $\varepsilon = 0$) is given in parametric form

$$x_h(t) = 1 - \frac{3}{1 + \cosh t}$$

$$y_h(t) = \frac{3 \sinh t}{(1 + \cosh t)^2}$$

Following Thompson (1989b), the Melnikov function can be written as two integrals on the line $(-\infty < t < \infty)$ I_1, I_2. The value of I_1 is 2/15, while the value of the second I_2 is solved using the calculus of complex variables and the method of residues. Plot the critical value of forcing amplitude versus frequency and show that the lowest values occur at a value of $\omega < 1$.

6-8 *Multiple-Well Criterion.* Consider the one-well potential problem with an escape saddle as the limit of an unsymmetric two-well potential problem. Derive a criterion for chaos from classical nonlinear perturbation theory (e.g., Nayfeh and Mook, 1979 or Hagedorn, 1988 similar to Eq. (6-3.46). Equate the largest velocity on the period-1 orbit to some large fraction (0.8–0.9) of the highest velocity on the unperturbed homoclinic orbit through the saddle (see previous problem). Compare this criterion with the Melnikov criterion (6-3.28b). Alternative problem: For those readers not familiar with perturbation theory, numerically integrate the one-well equation (6-3.28a) (e.g., using a Runge–Kutta algorithm) for larger and larger values of F near $\omega = 1$ and compare the value of F when the orbit is close to the separatrix, and compare their value with the Melnikov value (6-3.28b).

6-9 *Lyapunov Exponents.* Given $\Lambda = 1.2$ for a certain first-order map. Then how many iterations does it take for the average distance between two initially close orbits to grow by 60 db?

6-10 *Lyapunov Exponents.* Suppose a physical system admits a piecewise linear first-order map in three nonoverlapping subintervals $a_1 < x < a_2$; $b_1 < x < b_2$; $c_1 < x < c_2$. Also assume that the probability density function for a chaotic attractor generated by this map has piecewise constant values in these three subintervals and is zero elsewhere. Use (6-4.17) to estimate the Lyapunov exponent.

7

FRACTALS AND
DYNAMICAL SYSTEMS

*Do you see O my brothers and sisters? It is not chaos or death—
it is form, union, plan—it is eternal life—it is Happiness.*

Walt Whitman
Leaves of Grass

7.1 INTRODUCTION

Both "chaotic" and "strange attractor" have been used to describe
the nonperiodic, randomlike motions that are the focus of this book.
Whereas "chaotic" is meant to convey a loss of information or loss of
predictability, the term "strange" is meant to describe the unfamiliar
geometric structure on which the motion moves in phase space. In
Chapter 6, we described a quantitative measure of the chaotic aspect
of these motions using Lyapunov exponents. In this chapter, we will
describe a quantitative measure of the strangeness of the attractor.
This measure is called the *fractal dimension*. To do this, we will have
to describe the concept of fractal as it pertains to our applications. In
addition to the application of fractal ideas to the description of the
attractor itself, it has been discovered that other geometric objects in
the study of chaos, such as the boundary between chaotic and periodic
motions in initial condition or parameter space, may also have fractal properties. Thus, we will also include a section on *fractal basin
boundaries*.

At the beginning of this book, we noted that the revolution in nonlinear dynamics has been sparked by the introduction of new geometric, analytic, and topological ideas which have given experimentalists (including numerical analysts) new tools to analyze dynamical processes. This in some ways parallels the earlier Newtonian revolution which introduced the calculus into dynamics. (Of course, Newton contributed much more by proposing new physical laws along with new mathematics.) Thus, in some sense, we are entering the second phase of the Newtonian revolution in dynamics, and new geometric concepts like fractals must be mastered if one is to use the results of the new dynamics in practical problems.

Perhaps the most singular characteristic of chaotic vibrations in dissipative systems is the Poincaré map. These pictures provide a cross section of the attractor on which the motion rides in phase space and when the motion is chaotic, a mazelike, multisheeted structure appears. We have learned that this threadlike collection of points seems to have further structure when examined on a finer scale. To characterize such Poincaré patterns, we have used the term *fractal*. In this chapter we will try to make the mathematical meaning of fractal more precise. However, this treatment is not rigorous. Instead, what follows is one engineer's attempt to understand fractal structures and how to apply them to chaotic dynamics.

In the following section, we will begin with a few simple examples of fractal curves and sets, namely, *Koch curves* and *Cantor sets*. We will also introduce a quantitative measure of fractal qualities: the fractal dimension. Then we will illustrate these concepts in several applications in nonlinear and chaotic vibrations.

The author presumes that the reader has no prior knowledge of set theory or topology beyond engineering mathematics at the baccalaureate level.

For the reader who wants to study more about fractals, there are now several excellent texts to use. Two books which have already become classics are the treatise by Mandelbrot (1982) and the beautiful colorful tour of fractal sets by Peitgen and Richter (1986). However, those who desire a more mathematical treatment may find a very readable book in Falconer (1990) or Barnsley (1988). The latter book provides both mathematical and computational tools for the reader who wishes to play with fractals on the computer. A very readable introductory text on fractals in that by Peitgen et al. (1992). Finally, there is a treatment oriented toward applications of fractals in the physical sciences by Feder (1988).

Koch Curve

This example is chosen from the book by Mandelbrot (1977) and was originally described by von Koch in 1904. One begins with a geometric construction that starts with a straight line segment of length 1. After dividing the line into three segments, one replaces the middle segment by two lines of length 1/3 as shown in Figure 7-1. Thus, we are left with four sides, each of length 1/3, so that the total length of the new boundary is 4/3. To get a fractal curve, one repeats this process for each of the new four segments and so on. At each step, the length is increased by 4/3 so that the total length approaches infinity. After many steps, one can see that the curve looks fuzzy. In fact, in the limit one has a continuous curve that is nowhere differentiable. In some sense, this new curve is trying to cover an area as would a young child scribbling with crayons. Thus, we have the apparent paradox of a continuous curve that has some properties of an area. It is not surprising that one can define a dimension of this fractal curve which results in a value between 1 and 2.

Cantor Set

The Cantor set is attributed to George Cantor (1845–1918), who discovered it in 1883. It is a very important concept in modern nonlinear dynamics. If the Koch curve can be considered a process of adding finer and finer length structure to an initial line segment, then the Cantor set is the complement operation of removing smaller and smaller segments from a set of points initially on a line.

The construction begins as in the previous example with a line segment of length 1 which is subdivided into three sections as in Figure 7-2. However, instead of adding two more segments as in the Koch curve, one removes the middle segment of points so that the total number of segments is increased to two, and the total length is reduced

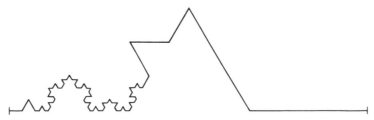

Figure 7-1 Partial construction of a fractal Koch curve.

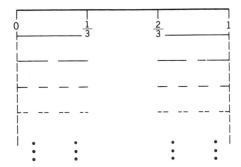

Figure 7-2   *Top to bottom:* Sequential steps in the construction of a Cantor set.

to 2/3. This process is continued for the remaining line segments and so on. At each stage one throws away the middle segments of points, creating twice as many line segments but reducing the total length by 2/3. In the limit the total length approaches 0, although as we shall see below, the fractal dimension of this set of points is between 0 and 1.

The Devil's Staircase

The discontinuous fractal Cantor set can be used to generate a continuous fractal function by integrating an appropriate distribution function defined on the set. For example, we imagine a distribution of mass on the interval $0 \leq x \leq 1$ with total mass equal to 1 in some units. Then if we redistribute the mass on the remaining Cantor intervals, at each step of the limiting process the mass density increases on the decreasing Cantor intervals such that the total mass is 1. At the nth step, the number of intervals is 2^n each of length $(1/3)^n$ so that the density is $(3/2)^n$. Integrating the mass density along x, we obtain the mass as a function of x:

$$M_n(x) = \int_0^x \rho_n(x)\, dx$$

where $\rho_n = (3/2)^n$ on the Cantor intervals and $\rho_n = 0$ otherwise. The limit of this process as $n \to \infty$ is a function called the *devil's staircase* which has an infinite number steps. One intermediate function $M_n(x)$ is shown in Figure 7-3.

In the limit, $M(x) = \lim_{n \to \infty} M_n(x)$. The expression $dM(x)/dx$ is an infinite set of delta functions.

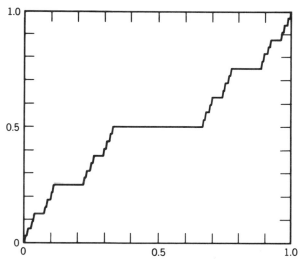

Figure 7-3 Devil's staircase function.

Fractal Dimension

Thus far we have two examples of fractal sets but do not have any test to determine if a set of points is fractal. To classify the Poincaré map of some nonlinear system, we need some quantitative measure of the fractal nature of the attractor.

There are many measures of the dimension of a set of points. We will describe a very intuitive or geometric definition called the *capacity* or *box-counting* dimension. Other definitions, which incorporate deeper mathematical subtleties, may be found in Mandelbrot[1] (1977), Farmer et al. (1983), or Feder (1988) as well as in the next section. We begin with the measurement of the dimension of points along a line or distributed on some area.

First consider a *uniform* distribution of N_0 points along some line or one-dimensional manifold in a three-dimensional space, as shown in Figure 7-4. We then ask how we can *cover* this set of points with small cubes with sides of length ε. (One can also use spheres of radius ε.) To be more specific, we calculate the minimum number of such cubes $N(\varepsilon)$ to cover the set ($N(\varepsilon) < N_0$). When N_0 is large and ε small enough, the number of cubes to cover a line will scale as

$$N(\varepsilon) \approx \frac{1}{\varepsilon}$$

[1] B. Mandelbrot is a mathematician with IBM Corp., Yorktown Heights, New York.

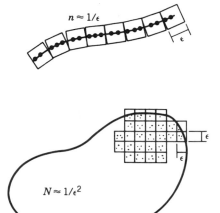

Figure 7-4   Covering procedure for linear and planar distributions of points.

Similarly, if we distribute points uniformly on some two-dimensional surface in three-dimensional space, one will find that the minimum number of cubes to cover the set will scale in the following way:

$$N(\varepsilon) \approx \frac{1}{\varepsilon^2}$$

If the reader is convinced that this is intuitive, then it is natural to define the dimension by the following scaling law:

$$N(\varepsilon) \approx \frac{1}{\varepsilon^d} \tag{7-1.1}$$

Taking the logarithm of both sides of Eq. (7-1.1) and adding a subscript to denote *capacity dimension,* we have

$$d_c = \lim_{\varepsilon \to 0} \frac{\log N(\varepsilon)}{\log(1/\varepsilon)} \tag{7-1.2}$$

Implicit in this definition is the requirement that the number of points in the set be large or $N_0 \to \infty$.

A set of points is said to be fractal if its dimension is non-integer—hence the term *fractal dimension.*

In the two examples of the Koch curve or Cantor set, the fractal dimension can be calculated exactly. For example, consider the nth iteration of the generation of the Koch curve where we let the size of the cubes be equal to the length of a straight line segment. At the nth step in the construction, the number of segments is

$$N_n = 4^n$$

where the size ε is given by

$$\varepsilon_n = \left(\frac{1}{3}\right)^n$$

Replacing the limit $\varepsilon \to 0$ with $n \to \infty$ in Eq. (7-1.2), one can easily see that for the *Koch curve*

$$d_c = \frac{\log 4}{\log 3} = 1.26185 \ldots \tag{7-1.3}$$

Similarly, one can show that for the *Cantor set*

$$d_c = \frac{\log 2}{\log 3} = 0.63092 \ldots \tag{7-1.4}$$

One way to interpret the fractal dimension of the Koch curve is that the distribution of points cover more than a line but less than an area. Another fractal-producing process that begins with an area distribution of points is shown in the exercise in Figure 7-5 called the *Sierpinski triangle* (Named after the Polish mathematician Waclaw Sierpinski, 1882–1969). At each step one removes a triangular area, creating three new triangles, but the scale is half the size of the original. One can show that this process leads to a fractal dimension of $d_c = \log 3 / \log 2$.

The connection between dynamics and fractals may not be evident so far, but in each of the three examples above, one has an iterative process. The relationship between fractals and iterative maps is made more explicit with the following two examples.

The *horseshoe map* has been discussed earlier in Chapters 1, 3 and 6 and is shown graphically in Figure 7-6. It is perhaps the simplest example of an iterative dynamical process in the plane that leads to a loss of information and fractal properties.

The calculation of the capacity fractal dimension is similar to that for the Cantor set except that the vertical direction leads to a contribu-

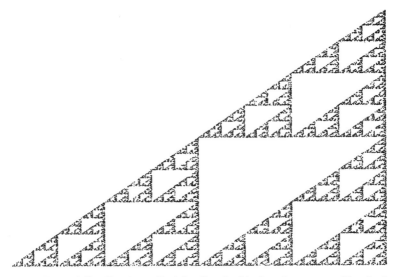

Figure 7-5 Fractal distribution called the *Serpinski triangle* generated by the iterated map (7-3.5), (7-3.6).

tion of "one" to the dimension. Using the definition (7-1.2), one can show that

$$d_c = \frac{\log 2}{\log |\varepsilon|} + 1 \qquad (7\text{-}1.5)$$

where ε is the contraction parameter and $0 < \varepsilon < 1/2$. [See also Bergé et al. (1985) for a discussion of this example.]

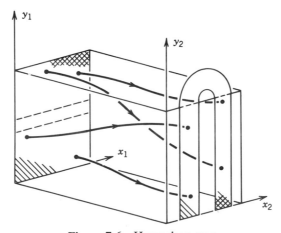

Figure 7-6 Horseshoe map

Another example for which one can calculate the fractal properties is the *baker's transformation* two-dimensional map (Figure 6-33). This example may be found in Farmer et al. (1983) and is similar to the horseshoe map. Its name derives from the idea of a baker rolling, stretching, and cutting pastry dough as shown in Figure 6-33. In this example, one can write out the specific difference equation or mapping relating a piece of dough at position (x_n, y_n) to its new position in one iteration:

$$x_{n+1} = \begin{cases} \lambda_a x_n & \text{if } y_n < \alpha \\ \frac{1}{2} + \lambda_b x_n & \text{if } y_n > \alpha \end{cases} \tag{7-1.6}$$

$$y_{n+1} = \begin{cases} y_n/\alpha & \text{if } y_n < \alpha \\ \dfrac{1}{1-\alpha}(y_n - \alpha) & \text{if } y_n > \alpha \end{cases}$$

where $0 \le x_n \le 1$ and $0 \le y_n \le 1$.

The article by Farmer et al. (1983) is very readable, so we will not present the details but will quote the results. The problem is used by Farmer et al. to show the difference between different definitions of fractal dimension. They define the following function:

$$H(\alpha) = \alpha \log\frac{1}{\alpha} + (1 - \alpha)\log\frac{1}{1-\alpha} \tag{7-1.7}$$

Using the definition of capacity, they find that

$$d_c = 1 + \hat{d}_c \tag{7-1.8}$$

where $\hat{d}_c$ satisfies a transcendental equation

$$1 = \lambda_a^{\hat{d}_c} + \lambda_b^{\hat{d}_c} \tag{7-1.9}$$

When $\lambda_a = \lambda_b = \lambda$,

$$d_c = 1 + \frac{\log 2}{\log |\lambda|} \tag{7-1.10}$$

which is independent of α and identical to that for the horseshoe map (7-1.5).

Other examples of iterated maps which produce fractal distribution of points are found in Barnsley (1988). Barnsley showed how simple maps can produce natural-looking fractal objects such as trees, ferns, and clouds.

It is probably safe to say that artists have intuitively understood the nature of fractal properties of nature, especially the impressionists in the way they used dots of color to achieve different effects of filling Euclidean space. In a more recent example, an advertisement in a popular magazine featured a Japanese artist whose design for a kimono material shows these fractal properties quite clearly (Figure 1-28).

7.2 MEASURES OF FRACTAL DIMENSION

There are two criticisms of the use of capacity as a measure of fractal dimension of strange attractors—one theoretical and the other computational. First, capacity dimension is a geometric measure; that is, it does not account for the frequency with which the orbit might visit the covering cube or ball. Second, the process of counting a covering set of hypercubes in phase space is very time-consuming computationally. In this section we will discuss three alternative definitions of fractal dimension which will address the shortcomings of the capacity or box-counting dimensions. However, it should be pointed out that for many strange attractors these different dimensions give roughly the same value.

Pointwise Dimension

Let us consider a long time trajectory in phase space as shown in Figure 7-7. First, we time sample the motion so that we have a large number of points per orbit. Second, we place a sphere or cube of radius or length r at some point on the orbit and count the number of points within the sphere $N(r)$. The probability of finding a point in this sphere is then found by dividing by the total number of points in the orbit N_0; that is,

$$P(r) = \frac{N(r)}{N_0} \qquad (7\text{-}2.1)$$

For a one-dimensional orbit, such as a closed periodic orbit, $P(r)$ will be linear in r as $r \rightarrow 0$, $N_0 \rightarrow \infty$; $P(r) \approx br$. If the orbit were quasiperiodic, that is it moves on a two-dimensional toroidal surface in a three-dimensional phase space, then the probability of finding a point on the orbit in a small cube or sphere of radius r would be $P(r) \approx br^2$. This leads one to define a dimension of an orbit at a point x_i (here x_i is a

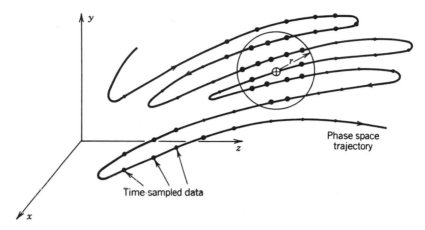

Figure 7-7 Long-time trajectory of motion in phase space showing the time-sampled data points and the counting sphere.

vector in phase space) by measuring the relative percentage of time that the orbit spends in the small sphere; that is

$$d_P = \lim_{r \to 0} \frac{\log P(r; \mathbf{x}_i)}{\log r} \qquad (7\text{-}2.2)$$

For some attractors, this definition will be independent of the point $\mathbf{x}_i$. But for many, d_P will depend on $\mathbf{x}_i$ and an averaged pointwise dimension is best used. Also, for some sets of points such as a Cantor set, there will be gaps in the distribution of points so that $P(r)$ is not a smooth function of r as $r \to 0$, as can be seen in the Devil's staircase in Figure 7-3.

   To obtain an averaged pointwise dimension, one randomly chooses a set of points $M < N_0$ and calculates $d_p(\mathbf{x}_i)$ at each point. The averaged pointwise dimension is given by

$$\hat{d}_P = \frac{1}{M} \sum_{i=1}^{M} d_P(\mathbf{x}_i) \qquad (7\text{-}2.3)$$

   As an alternative, one can average the probabilities $P(r; \mathbf{x}_i)$. Choose a random *subset* of M points distributed around the attractor, where $M \ll N_0$. We then conjecture that

$$\lim_{r \to 0} \frac{1}{M} \sum_{i=1}^{M} P(r; \mathbf{x}_i) = ar^{d_P}$$

or

$$d_P = \lim_{r \to 0} \frac{\log(1/M)\Sigma P(r)}{\log r}$$

In practice, if $N_0 \approx 10^3 – 10^4$ points, then $M \approx 10^2 – 10^3$.

Correlation Dimension

This measure of fractal dimension has been successfully used by many experimentalists [e.g., see Malraison et al. (1983), Swinney (1985), Ciliberto and Gollub (1985), and Moon and Li (1985a)] and in some ways is related to the pointwise dimension. An extensive study of this definition of dimension has been given by Grassberger and Proccacia (1983).

As in the definition of pointwise dimension, one discretizes the orbit to a set of N points $\{x_i\}$ in the phase space. (One can also create a pseudo-phase-space; see Chapter 5 and next section.) One then calculates the distances between pairs of points, say $s_{ij} = |x_i - x_j|$, using either the conventional Euclidean measure of distance (square root of the sum of the squares of components) or some equivalent measure such as the sum of absolute values of vector components. A correlation function is then defined as

$$C(r) = \lim_{N \to \infty} \frac{1}{N^2} \left(\begin{array}{l} \text{number of pairs } (i, j) \\ \text{with distance } s_{ij} < r \end{array} \right) \qquad (7\text{-}2.4)$$

For many attractors this function has been found to exhibit a power law dependence on r as $r \to 0$; that is,

$$\lim_{r \to 0} C(r) = ar^d$$

so that one may define a fractal or correlation dimension using the slope of the $\ln C$ versus $\ln r$ curve:

$$d_G = \lim_{r \to 0} \frac{\log C(r)}{\log r} \qquad (7\text{-}2.5)$$

It has been shown that $C(r)$ may be calculated more effectively by constructing a sphere or cube at each point x_i in phase space and counting the number of points in each sphere; that is,

$$C(r) = \lim_{r \to 0} \frac{1}{N^2} \sum_i^N \sum_j^N H(r - |\mathbf{x}_i - \mathbf{x}_j|) \qquad (7\text{-}2.6)$$

where $H(s) = 1$ if $s > 0$ and $H(s) = 0$ if $s < 0$. This differs from the pointwise dimension in that the sum here is performed about *every* point.

Information Dimension

Many investigators have suggested another definition of fractal dimension that is similar to the capacity (7-1.2) but tries to account for the frequency with which the trajectory visits each covering cube. As in the definition of capacity, one covers the set of points, whose dimension one wishes to measure, by a set of N cubes of size ε. This set of points is again a uniform discretization of the continuous trajectory. (It is assumed that a long enough trajectory is chosen to effectively cover the attractor whose dimension one wants to measure. For example, if the motion is quasiperiodic, the trajectory has to run long enough to "visit" all regions on the toroidal surface of the attractor.)

To calculate the information dimension, one counts the number of points N_i in each of the N cells and determines the probability of finding a point in that cell P_i, where

$$P_i \equiv \frac{N_i}{N_0}, \qquad \sum_i^N P_i = 1 \qquad (7\text{-}2.7)$$

where N_0 is the total number of points in the set. Note that $N_0 \neq N$. The *information entropy* is defined by the expression

$$I(\varepsilon) = -\sum^N P_i \log P_i \qquad (7\text{-}2.8)$$

[When the log function is with respect to base 2, $I(\varepsilon)$ has the units of bits.] For small ε it is found that I behaves as

$$I \approx d_I \log(1/\varepsilon)$$

so that for small ε we may define a dimension

$$d_I = \lim_{\varepsilon \to 0} \frac{I(\varepsilon)}{\log(1/\varepsilon)} = \lim \frac{\Sigma P_i \log P_i}{\log \varepsilon} \qquad (7\text{-}2.9)$$

To see that this definition is related to the capacity, we note that if the probabilities P_i were equal for all cells, that is,

$$P_i = \frac{N_i}{N_0} = \frac{1}{N} \qquad (7\text{-}2.10)$$

then

$$I = P_i \log P_i = -N\frac{1}{N}\log\frac{1}{N} = \log N$$

so that $d_I = d_c$. In general, it can be shown that (see Farmer et al., 1983)

$$d_I \le d_c \qquad (7\text{-}2.11)$$

Further discussion of the information dimension may be found in Farmer et al. (1983), Grassberger and Proccacia (1983), and Shaw (1984).

The information entropy is a measure of the *unpredictability* in a system. That is, for a uniform probability in each cell, $P_i = 1/N$, I is at a maximum. If all the points are located in one cell (maximum predictability), $I = 0$ as can be seen by the calculation

For $P_i = 1/N$, $I = \log N$

For $P_1 = 1, P_i = 0, i \ne 1$, $I = 1 \cdot \log 1 = 0$

Definition (7-2.8) and the use of the symbol $I(\varepsilon)$ are confusing in the literature. Shaw (1981) used the symbol H to denote entropy and I to denote the negative entropy $(-H)$ or *information*. Thus, for Shaw, a more predictable system (i.e., sharper P_i distribution) has *higher* information.

Relationship Between Fractal Dimension and Lyapunov Exponents

Thus far we have defined the following fractal dimensions:

d_c the capacity (7-1.2)
d_P pointwise dimension (7-2.2)
d_G correlation dimension (7-2.5)
d_I information dimension (7-2.9)

Grassberger and Proccacia (1983) have shown that the information dimension and the correlation dimension are lower bounds on the capacity definition; that is,

$$d_G \le d_I \le d_c \tag{7-2.12}$$

For many of the standard strange attractors, however, all three were very close (see Table 7-1).

In summary, one can say that the capacity dimension takes no account of the distribution of points between covering cells, whereas the information entropy dimension measures the probability of finding a point in a cell. Finally, the correlation dimension accounts for the probability of finding two points in the same cell (e.g., see Grassberger and Proccacia, 1984).

A further relationship betweeen fractal dimension, information entropy, and Lyapunov exponents was made by Kaplan and Yorke (1978). We recall from Chapter 6 that the Lyapunov exponents measure the rate at which trajectories *on* the attractor diverge from one another and trajectories *off* the attractor converge toward the attractor (e.g., see Figure 6-32). Thus, a small sphere of initial conditions centered at some point on the attractor in phase space is imagined to deform in time under the dynamical process into an ellipse. For example, for a chaotic two-dimensional map,

$$\mathbf{x}_{n+1} = f(\mathbf{x}_n) \tag{7-2.13}$$

a circle of initial conditions (with radius ε) deforms into an ellipse after M iterations of the map. The major and minor radii are given by

TABLE 7-1 Fractal Dimension of Selected Dynamical Systems

Name of Systems	Dimension	Type	Source of Data
Henon map (1-3.8)	1.26	Capacity	Grassberger and
($\alpha = 1.4$, $b = 0.3$)	1.21 $\pm$ 0.01	correlation	Proccacia (1983)
Logistic map (1-3.6)	0.538	Capacity	Grassberger and
($\lambda = 3.5699456$)	0.500 $\pm$ 0.005	correlation	Proccacia (1983)
Lorenz equations	2.06 $\pm$ 0.01	Capacity	Grassberger and
(1-3.9)	2.05 $\pm$ 0.01	correlation	Proccacia (1983)
Two-well potential	2.14 ($\gamma = 0.15$)	Correlation	Moon and Li (1985a)
[Eq. (6-3.7)], $f = 0.16$,			
$\omega = 0.8333$)	2.61 ($\gamma = 0.06$)		
Chua's circuit	2.82	Lyapunov	Matsumoto et al.
			(1985)

$L_1^M \varepsilon$ and $L_2^M \varepsilon$. When L_1 and L_2 are averaged over the whole attractor, they are referred to as *Lyapunov numbers*, and $\lambda_i = \log L_i$ are called the *Lyapunov exponents*.

Kaplan and Yorke (1978) (see also Farmer et al., 1983)[2] have suggested that one can calculate a dimension for a fractal attractor based on the Lyapunov exponents. For a two-dimensional map this dimension becomes

$$d_L = 1 + \frac{\log L_1}{\log (1/L_2)} = 1 - \frac{\lambda_1}{\lambda_2} \qquad (7\text{-}2.14)$$

For higher-dimensional maps in an N-dimensional phase space, the relation is more complicated. First we order the Lyapunov numbers; that is,

$$L_1 > L_2 > \ldots > L_k > \ldots > L_N \qquad (7\text{-}2.15)$$

Then find L_k such that the product is

$$L_1 L_2 \cdots L_k \geq 1$$

The Lyapunov dimension is defined to be

$$d_L = k + \frac{\log(L_1 L_2 \cdots L_k)}{\log(1/L_{k+1})} \qquad (7\text{-}2.16)$$

Kaplan and Yorke (1978) suggested that this is a lower bound on the capacity dimension; that is,

$$d_L \leq d_c \qquad (7\text{-}2.17)$$

As an example, consider a three-dimensional set of points generated by a Poincaré map of a fourth-order set of first-order differential equations with dissipation. If the attractor is strange, we assume

$$L_1 > 1, \qquad L_2 = 1, \qquad L_3 < 1$$

For example, one principal axis of the ellipsoid of initial conditions grows, one stays the same length, and one axis contracts. Also, be-

---

[2] Note that Farmer et al. (1983) used λ to denote the Lyapunov *number*, not the Lyapunov *exponent*.

cause the system is dissipative, the volume of the ellipsoid must be less than that of the original sphere of initial conditions so that $L_1 L_2 L_3 < 1$. This leads us to use $k = 2$ in Eq. (7-2.16) and

$$d_L = 2 + \frac{\log L_1}{\log(1/L_3)} = 2 + \frac{\lambda_1}{|\lambda_3|} \qquad (7\text{-}2.18)$$

The usefulness of this formula for experimental data is unclear at this time because it is not easy to obtain a measurement of the contraction Lyapunov number L_3 (e.g., see Wolf et al., 1985).

A comparison of the different definitions of fractal dimension for the baker's transformation (7-1.6) has been given by Farmer et al. (1983). This example is one of the few dynamical systems for which one can analytically calculate the properties of the chaotic dynamics. They show that the Lyapanov dimension (7-2.20) is equal to the information dimension (7-2.9) and is given by

$$d_I = d_L = 1 + \frac{H(\alpha)}{\alpha \log(1/\lambda_a) + \beta \log(1/\lambda_b)} \qquad (7\text{-}2.19)$$

where $\beta = 1 - \alpha$. When $\lambda_a = \lambda_b$ one can show that

$$d_I = d_L = 1 + \frac{H(\alpha)}{\log(1/\lambda)} \qquad (7\text{-}2.20)$$

Furthermore, if $\alpha = \frac{1}{2}$, then $H(\alpha) = \log 2$ and

$$d_I = d_1 = d_c$$

In some ways, α and λ_a/λ_b represent inhomogenity factors in the map. When $\alpha = \frac{1}{2}$ and $\lambda_a/\lambda_b = 1$, the map is like the horseshoe or Cantor maps and all these definitions of dimension d_I, d_L, d_c become equal. The implications are that different definitions of fractal dimension are likely to yield different results when the dynamical process leads to a "nonuniform" Poincaré map.

7.3 FRACTAL-GENERATING MAPS

The title of this text, *Chaotic and Fractal Dynamics*, may be provocative to some dynamicists. The meaning of this term has two interpretations. First, a modern understanding of chaotic dynamics requires some knowledge of fractal mathematics. Second, whereas fractals can be studied independently of dynamics, the creation of fractal sets is closely linked with iterative processes as illustrated above for the baker's map. And these iterative processes which lead to the unpredictability inherent in fractal mathematics are close analogs of the dynamic processes in physics that also lead to fractal structures.

Iterated Linear Maps

It is now accepted that many geometric objects in the natural world have fractal-like shapes and surfaces such as coastlines, clouds, mountain ranges, certain trees, and leaves. In a recent book, Barnsley (1988) showed how one can recreate these shapes using iterated linear maps and made a very nice connection between the static fractal objects and the dynamical equations that generate them. In this section, we try to outline a few of these ideas in the hope that it may inspire the reader to delve deeper into these techniques. One potential application of these dynamical methods of generating fractals is the concept of data compression. Thus, if one wants to send a good picture of a fractal-like object (e.g., a landscape) instead of using a high-resolution image scanner (TV camera) with upwards of 10^6 pixels of data, Barnsley and his associates propose to send the mathematical equations (with perhaps only 10^2 bytes of information) which can dynamically generate an approximation of the landscape after transmission.

To get an idea of this technique, we have to recall some of the properties of linear maps. These maps take the form

$$A' = TA \qquad (7\text{-}3.1)$$

where for 2-D planar maps A represents a point in the initial area and A' represents the new point under the matrix operation T:

$$T = \begin{bmatrix} a & c \\ b & d \end{bmatrix} \qquad (7\text{-}3.2)$$

As discussed in Chapter 3, a linear map can contract or expand, rotate, shear, or reflect an area collection of points. Of course, the iteration of one linear map cannot create a fractal object or a chaotic orbit; however, a *sequence* of different linear maps can. One example is the Cantor set discussed above. The step-by-step process of contracting the current set of points along the line and replicating it twice can be written as two linear maps ω_1, ω_2; that is

$$A' = WA$$

$$W = \bigcup_{i=1}^{2} \omega_i$$

$$\omega_1 = \tfrac{1}{3}x$$

$$\omega_2 = \tfrac{1}{3}x + \tfrac{2}{3} \qquad (7\text{-}3.3)$$

The notation $\bigcup \omega_i$ means that first ω_i is applied to the set of points A, and then ω_2 is applied to A and the new set of points A' is the union of the two sets $\omega_1 A$, $\omega_2 A$. In this case there is no overlap (see Figure 7-8). [See Barnsley (1988) for a discussion of overlapping linear maps.] Thus, the dynamical process that generates the Cantor set can be written as a map that acts on a *set* of points:

$$A_{n+1} = WA_n \qquad (7\text{-}3.4)$$

This differs from Chapter 3, where the map acts on a position vector $\mathbf{x}$ thereby generating a single orbit $\{\mathbf{x}_n; \ n = 1, 2, \ldots, \infty\}$. The map (7-3.4) generates a *dense bundle* of orbits.

Under suitable assumptions, repeated application of the mapping $A_{n+1} = WA_n$ leads to an attractor. This means that starting from different sets A_1, A_1' one ends up with the same set A. This property is illustrated in Figure 7-10 for the Sierpinski triangle. After many iterations, each point in the initial set A_1 undergoes an orbit. However,

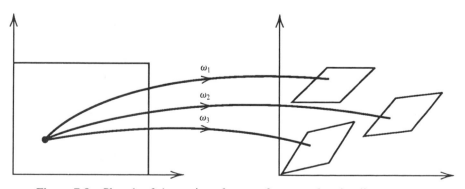

Figure 7-8 Sketch of the action of a set of nonoverlapping linear maps.

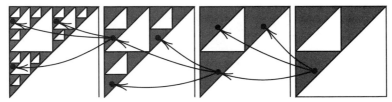

Figure 7-9   Serpinski triangle generated by application of a set of three linear maps (7-3.5).

any attempt to trace this orbit back through the order of transformations $[\omega_i, \omega_j, \omega_k, \ldots]$ is very complex (see Figure 7-9). It has been shown (Barnsley, 1988) that such an orbit looks random. This is similar to the result for the horseshoe map.

Another example of the generation of a fractal set in the plane is given by the following map: (Barnsley, 1988)

$$\begin{Bmatrix} x' \\ y' \end{Bmatrix} = \begin{bmatrix} A & B \\ C & D \end{bmatrix} \begin{Bmatrix} x \\ y \end{Bmatrix} + \begin{Bmatrix} E \\ F \end{Bmatrix} \qquad (7\text{-}3.5)$$

	A	B	C	D	E	F
ω_1:	0.5	0	0	0.5	1	1
ω_2:	0.5	0	0	0.5	50	1
ω_3:	0.5	0	0	0.5	50	50

Iteration of this set of linear transformations generates the fractal called the *Sierpinski gasket,* which is somewhat like a planar sponge (Figures 7-5, 7-9, 7-10). This method of generating a fractal can be computationally very time-consuming, because at every iteration cycle all three linear maps must act on all the points which define A. Another method, however, makes use of the chaotic nature of the orbits in this iteration process.

The generation of the Cantor set using the sum of two linear transformations is not unlike the dynamical process of the horseshoe map described in Chapters 1 and 3. Here the contraction operation of the horseshoe is represented by the $\frac{1}{3}x$ terms in ω_1, ω_2 [Eq. (7-3.3)], whereas the bending operation is represented by adding the second map ω_2, which replicates the set $\omega_1 A$ and shifts it by $\frac{2}{3}$. It can be shown (e.g., Guckenheimer and Holmes, 1983 or Barnsley, 1988) that horseshoe-type maps contain an infinite set of chaotic orbits which jump from one half of the domain ($0 \le x \le \frac{1}{2}$) to the other half ($\frac{1}{2} < x \le 1$) as if it were equivalent to a random coin toss operation.

Using this property, Barnsley then constructed an algorithm to generate fractal sets based on a single orbit $\mathbf{x}_n = (x_n, y_n)$. If the generating functions contain k linear maps $\{w_i; i = 1, 2, ..., k\}$, then the orbit is given by

$$\mathbf{x}_{n+1} = \omega_I \mathbf{x}_n \qquad (7\text{-}3.6)$$

where the particular linear map at each iteration step is chosen at random, that is

$$I = 1 + \text{Integer}[k * \text{Random Number } [0, 1] - 10^{-4}]$$

One can also bias some of the ω_i more than others by using a set of probabilities $\{p_i\}$ where $\Sigma p_i = 1$ so that each ω_i is given a different probability weight. An example is shown in Figure 7-5, which shows a sequence of images as the iteration progresses. Further discussion of these fascinating ideas is beyond the scope of this book, and the

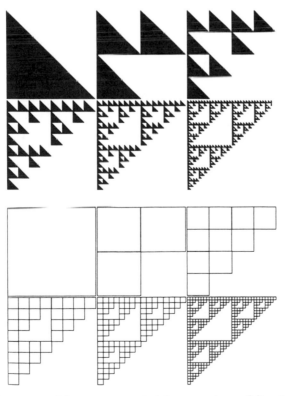

Figure 7-10 Sequence of images generated by a system of iterated linear maps (7-3.5) starting from two different initial sets of points (From Barnsley, 1988).

reader is encouraged to look at the many color images in the Barnsley text.

It is interesting to note that fractal objects can be created with both deterministic and random dynamic processes. It is this author's belief that the random processes are substitutes for unknown deterministic dynamics as is the case with the two iterated map algorithms. What is amazing, in either the deterministic chaotic models or the random models, is the global fractal structure that results. These mysteries between determinism, chaos, randomness, and fractals will keep both dynamicists and philosophers busy into the next century.

Analytic Maps on the Complex Plane

Many readers have perhaps seen the beautiful multicolor fractal pictures associated with the name of Mandelbrot (1982) (e.g., see Peitgen and Richter, 1986). These pictures are associated with a two-dimensional map involving the complex variable $z = x + iy$,

$$z_{n+1} = z_n^2 + c \tag{7-3.7}$$

where $c = a + ib$ is complex. In terms of real variables, this map becomes

$$
\begin{aligned}
x_{n+1} &= x_n^2 - y_n^2 + a \\
y_{n+1} &= 2x_n y_n + b
\end{aligned}
\tag{7-3.8}
$$

This map looks similar to other 2-D maps studied in this book where

$$
\begin{aligned}
x_{n+1} &= f(x_n, y_n) \\
y_{n+1} &= g(x_n, y_n)
\end{aligned}
\tag{7-3.9}
$$

However, in the case of the complex map (7-3.8), $F = f + ig$ is an analytic function of z. This means that a derivative $dF(z)/dz$ exists and that the functions $f(x, y)$ and $g(x, y)$ satisfy

$$\frac{\partial f}{\partial x} = \frac{\partial g}{\partial y}$$

$$\frac{\partial f}{\partial y} = -\frac{\partial g}{\partial x} \tag{7-3.10}$$

or

$$\nabla^2 f = \nabla^2 g = 0$$

In general, the 2-D maps studied earlier in the book do not satisfy these conditions. Thus, the quadratic complex map (7-3.7) and more general complex maps $z_{n+1} = F(z_n)$ are very special maps and have been found to have incredible complex dynamics and geometric properties (e.g., see Devaney, 1989).

Because this is an introductory book, we will briefly describe two geometric properties of complex maps: Julia sets and the Mandelbrot set.

Julia Sets. As with other maps, one can define fixed points and periodic or cycle points by the relations

$$z = F(z) \quad \text{and} \quad z = F^p(z) \tag{7-3.11}$$

where the superscript p indicates the application or composition of the map p times. Also, one can study the stability of each of these fixed points by looking at the derivative

$$\lambda = \frac{d}{dz} F^p(z) \tag{7-3.12}$$

It can be shown that the fixed point is attracting or repelling depending on whether $|\lambda| < 1$ or $|\lambda| > 1$ (see Devaney, 1989).

The Julia set of a complex map $F(z)$, sometimes denoted by $J(F)$, is the set of all the repelling fixed or periodic points. In the case of the map $F(z) = z^2$, one can show that $J(F)$ is a circle about the origin. That a dynamical system can have a continuous ring of unstable fixed points is not unusual. For example, a particle in a cylindrically symmetric potential $U = ar^2(1 - \frac{1}{2}r^2)$ has a circle of unstable saddles on $r = 1$.

The interesting property about these complex maps, however, is that by adding a constant to $F(z) = z^2 + c$, the Julia set becomes wrinkled or fractal. For example, if $|c| < \frac{1}{4}$, the $J(F)$ is still a closed curve but contains no smooth arcs (Devaney, 1989). This is illustrated in Figure 7-11. For larger values of $|c|$ the Julia set becomes even more interesting as illustrated by the case $c = -1$ in Figure 7-12. Here we see a fractal necklace with infinitely many loops. One can also show that once on the Julia set, further iterations of the map keep one on the

Figure 7-11   Julia set for $|c| < \frac{1}{4}$ in $z_{n+1} = z_n^2 + c$.

set [i.e., $J(F)$ is invariant]. It has also been shown that the dynamics on this set are chaotic; that is, there is sensitivity to initial conditions.

Because the real and imaginary parts of the mapping function $F(z)$ satisfy Laplace's equation [Eq. (7-3.10)], attempts have been made to interpret these Julia sets in terms of electric charge potentials. However, it is difficult to find a physical analog to the dynamic equations $z_{n+1} = F(z_n)$, when z is complex. However, the chaotic dynamics of repeller potentials, as in particle scattering problems, have received attention in recent years.

Although the study of chaos and complex maps appears to be a modern subject, the mathematics of repelling sets in complex maps has its origins in the work of mathematicians Julia and Fatou around the close of the 19th century.

Mandelbrot Sets. Whereas the Julia set is described in the plane of the state variables (x, y) of the complex map $z_{n+1} = F(z_n)$, the Mandelbrot set is described in the parameter space (a, b) of the control variable $c = a + ib$ in the complex quadradic map $F(z) = z^2 + c$. In constructing the Mandelbrot set, one fixes the initial conditions $z_0 = 0$ or $(x, y) = (0, 0)$ and looks for complex parameter values for which the iterates of the map do *not* go to infinity. This set is shown as the dark pattern in Figure 7-13 and Color Plate 3. Each color outside the set represents a given number of iterations for the vector z to go beyond a certain prefixed radius. What is remarkable about this set is the fractal nature of the boundary, which contains smaller versions of

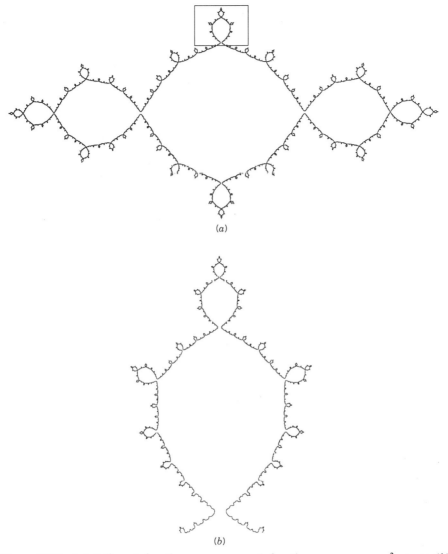

Figure 7-12 (*a*) Julia set for the case $c = -1$ for the map $z_{n+1} = z_n^2 + c$; (*b*) Enlargement of points in the box in (*a*).

the Mandelbrot set as one looks at the surface with a larger and larger computer microscope.

When c is real, then initial conditions on the real z axis, $y = 0$, yield a one-dimensional map $x_{n+1} = x_n^2 + a$. One can show that this map is equivalent to the logistic map $x_{n+1} = \lambda x_n(1 - x_n)$ and that period-doubling bifurcations occur as one moves along the real axis in the Mandelbrot set.

Again, although the physical relevance of these complex maps is not transparent at this time, they have served as a dramatic visual

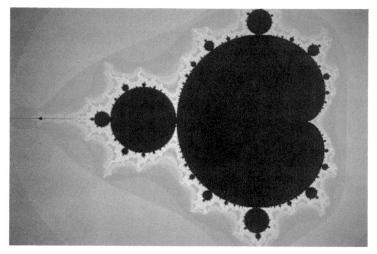

Figure 7-13 Mandelbrot set for the map $z_{n+1} = z_n^2 + c$. (Courtesy of J. Hubbard, Cornell University.)

paradigm about the intimate connection between dynamical systems and fractals and how incredible patterns of complexity can occur from simple mathematical models.

7.4 FRACTAL DIMENSION OF STRANGE ATTRACTORS

There are two principal applications of fractal mathematics to nonlinear dynamics: characterization of strange attractors and measurement of fractal boundaries in initial condition and parameter space. In this section, we discuss the use of the fractal dimension in both numerical and experimental measurements of motions associated with strange attractors.

As yet, there are no instruments, electronic or otherwise, which will produce an output proportional to the fractal dimension, although electro-optical methods may achieve this end in the future (see Section 7.5). To date, in both numerical and experimental measurements, the fractal dimension and Lyapunov exponents are found by discretizing the signals at uniform time intervals and the data are processed with a computer. There are three basic methods:

(a) Time discretization of phase-space variables
(b) Calculation of fractal dimension of Poincaré maps
(c) Construction of pseudo-phase-space using single variable measurements (sometimes called the *embedding space method*)

In both the first and third methods, the variables are measured and stored at uniform time intervals $\{\mathbf{x}(t_0 + n\tau)\}$, where n is a set of integers. The time interval τ is chosen to be a fraction of the principal forcing period or characteristic orbit time. If the Poincaré map in (b) is based on a time signal, then the τ is just the period of the time-based Poincaré map. However, if the Poincaré map is based on other phase-space variables, then the data are collected at variable times depending on the specific type of Poincaré map (see Chapter 5).

There are three principal definitions of fractal dimension used today: averaged pointwise dimension, correlation dimension, and Lyapunov dimension. In most of the experience with actual calculation of fractal dimension, 20,000 or more points are used, though several papers claim to have reliable algorithms based on as little as 1000 points (e.g., see Abraham et al., 1986). Direct algorithms for calculating fractal dimension based on N_0 points generally take N_0^2 operations so that superminicomputers or mainframe computers are often used. However, clever use of basic machine operations can reduce the number of operations to order $N_0 \ln N_0$ and significantly speed up calculation (e.g., see Grassberger and Proccacia, 1983).

Discretization of Phase-Space Variables

Suppose we know or suspect a chaotic system to have an attractor in three-dimensional phase space based on the physical variables $\{x(t), y(t), z(t)\}$. For example, in the case of the forced motion of a beam or particle in a two-well potential (see Chapter 2), $x = $ position, $v = \dot{x}$ is the velocity, and $z = \omega t$ is the phase of the periodic driving force. In this method, time samples of $(x(t), y(t), z(t))$ are obtained at a rate that is smaller than the driving force period. To each time interval there corresponds a point $\mathbf{x}_n = (x(n\tau), y(n\tau), z(n\tau))$ in phase space.

To calculate an averaged pointwise dimension, one chooses a number of random points $\mathbf{x}_n$. About each point one calculates the distances from $\mathbf{x}_n$ to the nearest points surrounding $\mathbf{x}_n$. (Note that these points are not the nearest in time, but in distance.) One does not need to use a Euclidean measure of distance. For example, the sum of absolute values of the components of $(\mathbf{x}_n - \mathbf{x}_m)$ could be used; that is,

$$s_{nm} = |x(n\tau) - x(m\tau)| + |y(n\tau) - y(m\tau)| + |z(n\tau) - z(m\tau)| \quad (7\text{-}4.1)$$

Then the number of points within a ball, cube, or other geometric shape of order ε is counted and a probability measure is found as a function of ε:

$$P_n(\varepsilon) = \frac{1}{N_0} \sum_{m=1}^{N_0} H(\varepsilon - s_{nm}) \tag{7-4.2}$$

where N_0 is the total number of sampled points and H is the Heaviside step function; $H(r) = 1$ if $r > 0$; $H(r) = 0$ if $r < 0$. The averaged pointwise dimension, following Eq. (7-2.3), is then

$$d_n = \lim_{\varepsilon \to 0} \frac{\log P_n(\varepsilon)}{\log \varepsilon}$$

$$\tag{7-4.3}$$

$$d \equiv \frac{1}{M} \sum_{n=1}^{M} d_n$$

where the limit defining d_n exists. For some attractors, the function P_n versus ε is not a power law but has steps or abrupt changes in slope. Then one can calculate a modified average pointwise dimension by first averaging P_n. For example, let

$$\hat{C}(\varepsilon) = \frac{1}{M} \sum_{n=1}^{M} P_n(\varepsilon)$$

$$\tag{7-4.4}$$

$$d = \lim_{\varepsilon \to 0} \frac{\log \hat{C}(\varepsilon)}{\log \varepsilon}$$

This is similar to the correlation dimension discussed in the previous section.

The example of the two-well potential (6-2.2) is shown in Figure 7-14a,b using the correlation dimension. This dimension is computed from numerically generated data using the equation $\dot{x} = y$, $\dot{y} = -\delta y - \frac{1}{2}x(1 - x^2) + f \cos z$, $\dot{z} = \omega$ for values of δ, f, ω in the chaotic regime. Figure 7-14a shows the logarithm of the correlation function, whereas Figure 7-14b shows the local slope versus the logarithm of the size of the test volume. The slope for the intermediate values of ε is around 2.5. This is consistent with the fact that the attractor lives in a three-dimensional space (x, y, z).

In practice, $N_0 \approx 3 \times 10^3 - 10^4$ points and $M \approx .20 N_0$. One should experiment with the choice of M by starting with a small value and increasing it until d reaches some limit.

The choice of ε also requires some judgment. The upper limit of ε is much smaller than the maximum size of the attractor yet large enough to capture the large-scale structure in the vicinity of the point

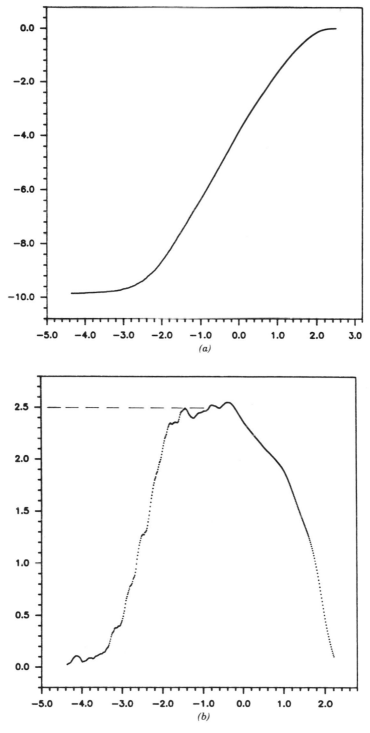

Figure 7-14 (*a*) Log *C* versus log *ε* for chaotic motion in a two-well potential (4-2.2). Data obtained from numerical integration. (*b*) Local slope of (*a*) showing fractal dimension in linear region of (*a*) of around 2.5.

$\mathbf{x}_n$. The smallest value of ε must be such that the associated sphere or cube contains at least one sample point.

Another constraint on the minimum size of ε is the "real noise" or uncertainty in the measurements of the state variables (x, y, z). In an actual experiment, there is a sphere of uncertainty surrounding each measured point in phase space. When ε becomes smaller than the radius of this sphere, the theory of fractal dimension discussed above comes into question because for smaller ε one cannot expect a self-similar structure.

Fractal Dimension of Poincaré Maps

In systems driven by a periodic excitation, as in the Duffing–Ueda strange attractor (4-6.1) or the two-well potential strange attractor (4-2.2), time or the phase $\phi = \omega t$ becomes a natural phase-space variable. In most cases, this time variable will lie in the attractor subspace and time can be considered as one of the contributions to the dimension of the attractor. In the case of a periodically forced, nonlinear, second-order oscillator, the Poincaré map based on periodic time samples produces a distribution of points in the plane. To calculate the fractal dimension of the complete attractor, it is sometimes convenient to calculate the fractal dimension of the Poincaré map $0 < D < 2$. If D is independent of the phase of the Poincaré map (remember $0 \leq \omega t \leq 2\pi$), then the dimension of the complete attractor is just

$$d = 1 + D \tag{7-4.5}$$

As an example, we present numerical and experimental data for the two-well potential or Duffing–Holmes strange attractor (Chapter 2):

$$\ddot{x} + \gamma \dot{x} - \tfrac{1}{2}x(1 - x^2) = f \cos \omega t \tag{7-4.6}$$

In this example, we are interested in two questions:

1. Does the fractal dimension of the strange attractor vary with the phase of the Poincaré map?
2. How does the fractal dimension vary with the damping γ?

The fractal dimension was calculated for a set of Poincaré maps and are listed in Table 7-2. This table shows an almost constant value around the attractor. Thus, the assumption $d = 1 + D$ in Eq. (7-4.5) appears to be a good one.

TABLE 7-2   Dimension of Experimental Poincaré Map Versus Phase for Vibration of a Buckled Beam[a]

ϕ	$D(1, 4)$[b]	$D(1, 7)$[c]
0	1.741	1.628
45	1.751	1.627
90	1.742	1.638
135	1.748	1.637
180	1.730	1.637

[a] Nondimensional damping, $\gamma = 0.013$; forcing frequency, 8.5 Hz; natural frequency about buckled state, 9.3 Hz; from Moon and Li (1985a).
[b] Based on four smallest log r points in log C versus log r.
[c] Based on seven smallest log r points in log C versus log r.

A numerically generated Poincaré map for the case of a particle in a two-well potential under periodic excitation is shown in Figure 7-15. The correlation function (Figure 7-16a) $C(\varepsilon)$ versus ε is shown plotted in a log–log scale and shows a linear dependence as assumed in the theory.

The data in Figure 7-15 was the same as that used in Figure 7-14. From Figure 7-16b, $D \approx 1.5$ or $d = 2.5$, which agrees with that calculated directly from the attractor in the phase space $(x, \dot{x}, \omega t)$ as in Figure 7-14.

The effect of damping on the fractal dimension of the two-well potential strange attractor was determined from Runge–Kutta numerical simulation. This dependence is shown in Figure 7-17. The data show that low damping yields an attractor that fills phase space $(D = 2, d = 3)$ as would a Hamiltonian (zero damping) system. As damping is increased, however, the Poincaré map looks one-dimensional and the attractor has a dimension close to $d = 2$, as in the case of the Lorenz equations.

The fractal dimension of a chaotic circuit (diode, inductor, and resistor in series driven with an oscillator) has been measured by Linsay (1985) using a Poincaré map. He measured the current at a sampling time equal to the period of the oscillator and constructed a three-dimensional pseudo-phase-space using $(I(t), I(t + \tau), I(t + 2\tau))$ (see next section). He obtained a fractal dimension of the Poincaré map of $D = 1.58$ and infers a dimension of the attractor of 2.58.

Dimension Calculation from Single Time Series Measurement

The methods discussed above assume that (a) one knows the dimension of the phase space wherein the attractor lies and (b) one has the

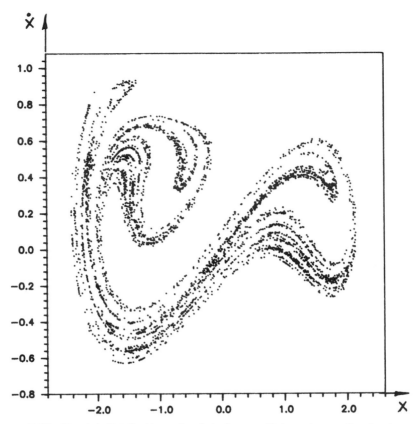

Figure 7-15 Fractal distribution of points from a Poincaré map for the two-well potential problem (4-2.2), using the same data as in Figure 7-14.

ability to measure all the state variables. However, in many experiments, the time history of only one state variable may be available or possible. Also in continuous systems involving fluid or solid continua, the number of degrees of freedom or minimum number of significant modes contributing to the chaotic dynamics may not be known a-priori. In fact, one of the important applications of fractal mathematics is to allow one to determine the smallest number of first-order differential equations that may capture the qualitative features of the dynamics of continuous systems. This has already had some success in thermofluid problems such as Rayleigh–Benard convection (see Malraison et al., 1983).

In early theories of turbulence (e.g., Landau, 1944), it was thought that chaotic flow was the result of the interaction of a very large or infinite set of modes or degrees of freedom in the fluid. At the present time, it is believed that the chaos associated with the transition to

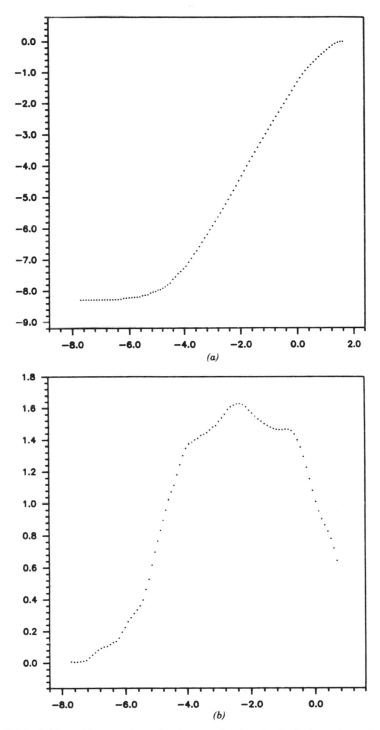

Figure 7-16 (a) Log C versus log ε for the set of points in the Poincaré map in Figure 7-15. (b) Local slope of (a) showing a fractal dimension in the linear region of (a) of around 1.5.

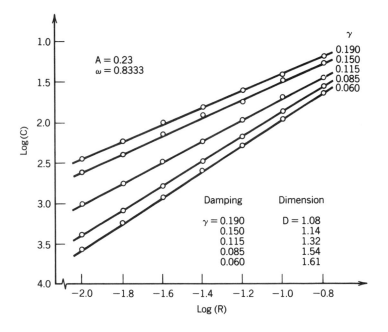

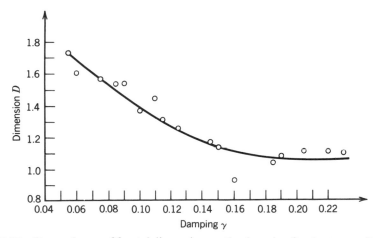

Figure 7-17   Dependence of fractal dimension on the damping for the two-well potential oscillator (4-2.2).

some forms of turbulence can be modeled by a finite set of ordinary differential equations (see e.g., Aubry et al. (1988)).

Thus, suppose that the number of first-order equations required to simulate the dynamics of a dissipative system is N. Then the fractal dimension of the attractor would be $d < N$. Then if we were to determine d by some means, we would then determine the minimum N.

Not knowing N, we cannot know how many physical variables $x(t)$, $y(t)$, $z(t)$, ...) to measure. Instead we construct a pseduo-phase-space using time-delayed measurements of one physical variable, say $(x(t), x(t + \tau), x(t + 2\tau), ...)$ (see Chapter 5 and also see Packard et al., 1980). For example, three-dimensional pseudo-phase-space vectors are calculated using three successive components of the digitized $x(t)$ (Figure 7-18), that is,

$$\mathbf{x}_n = \{x(t_0 + n\tau), x(t_0 + (n + 1)\tau), x(t_0 + (n + 2)\tau)\} \quad (7\text{-}4.7)$$

With these position vectors, one can use the correlation function (7-2.5) or averaged probability function (7-2.3) to calculate a fractal dimension.

To determine the minimum N, one constructs higher-dimensional pseudo-phase-spaces based on the time-sampled $x(t)$ measurements until the value of the fractal dimension reaches an asymptote, say, $d = M + \mu$, where $0 < \mu < 1$. Then the minimum phase-space dimension for this chaotic attractor is $N = M + 1$.

In reconstructing a dynamical attractor from the time history measurements of a single variable, the question arises of how many dimensions are required in the embedding space in order to capture all the topological features of the original attractor. A mathematician named Takens has proved several theorems about this question. If the original phase-space attractor lives in an N-dimensional space, then in general one must reconstruct an embedding space (our pseudo-phase-space) of $2N + 1$ dimensions.

To illustrate these ideas we have applied the embedding space method to find the dimension of the two-well potential (or buckled

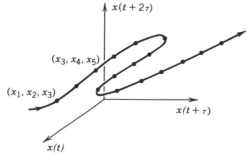

Figure 7-18 Sketch of an orbit in a three-dimensional pseudo-phase-space constructed from a single time series measurement.

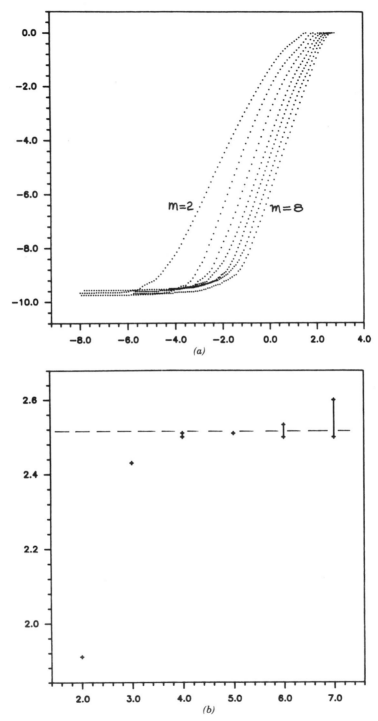

Figure 7-19 (*a*) Log C versus log ε for the two-well potential problem for different dimension embedding spaces. The time history data are identical to those in Figures 7-14 and 7-16. (*b*) Fractal dimensions of attractor versus the dimension of the embedding space.

beam) attractor (4-2.2). Earlier we saw that this attractor lives in a three-dimensional phase space $(x, \dot{x}, \omega t)$ and has a fractal dimension of $d = 2.5$ (Figure 7-14). Using the same data we also saw that we could calculate d from the Poincaré map (Figures 7-15 and 7-16). Using the same numerical data from a Runge–Kutta integration, we reconstructed the motion in a pseudo-phase-space using digitized values of $x(t)$ and embedding space dimensions of $m = 2$–8. The graphs in Figure 7-19a,b show the correlation function as well as the calculated dimension of the attractor in each embedding space.

One can see in Figure 7-19a,b that the dimension reaches an asymptote of $d = 2.5$ after $M \sim 4$–5, which is in agreement with Taken's theorem.

An example of calculating the fractal dimension from experimental data is shown in Figure 7-20 for the periodic excitation of a long, thin cantilevered beam with rectangular cross-section (Cusumano and Moon, 1990). In this problem, the resonant vibrations near the natural frequencies can couple into the torsional out-of-plane modes. The result is a dynamic snapping back and forth from one torsion-bending motion to another in a chaotic way. The data were obtained from both strain gage measurements on the beam and optical measurements of the tip displacement. In these calculations a random set of 20,000–25,000 points were selected from 100,000 time series points.

As another example using experimental data, we describe the work of a group at the French research laboratory at Saclay (e.g., see Malraison et al., 1983 and Bergé et al., 1984). They measured the

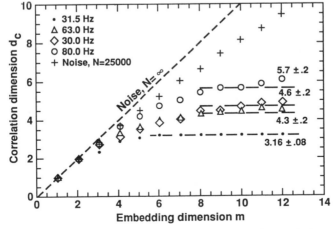

Figure 7-20 Calculation of fractal dimension from experimental data for periodic vibration of a thin cantilever beam. [From Cusumano and Moon (1990).]

fractal dimension of a convective fluid cell under a thermal gradient (Rayleigh–Benard convection, see Chapter 4). They calculated the fractal dimension using an averaged pointwise dimension (7-2.3) for different sizes of pseudo-phase-spaces. The fractal dimension saturated at a value of $d = 2.8$ when the embedding dimension of the phase space reached 5 or greater. They used 15,000 points and averaged $P_n(\varepsilon)$ over 100 random points. However, they also found regimes of chaotic flow where no clear slope of log $C(\varepsilon)$ versus log ε existed.

Similar results for the flow between two cylinders (Taylor–Couette flow) has been reported by a group from the Soviet Union (L'vov et al., 1981). They claim to measure the information dimension. Figure 7-21 shows the value of the slope of log $C(\varepsilon)$ versus log ε as a function of ε. This is characteristic of these measurements. The slope values at small ε reflect instrumentation noise, whereas the values at large ε are those for which the size of the covering sphere or hypercube reaches the scale of the attractor.

Using such techniques, one can determine how the fractal dimension changes as some control parameter in the experiment is varied. For example, in the case of Taylor–Couette flow (see Figure 4-42), Swinney and co-workers have measured the change in d as a function of Reynolds number (see Swinney, 1985).

In another fluid experiment, Ciliberto and Gollub (1985) have studied chaotic excitation of surface waves in a fluid. The surface wave chaos was excited by a 16-Hz vertical amplitude frequency; 2048 points were sampled with a sampling time of 1.5 s or around 300 orbits. Using the embedding space technique, they measured both the correlation dimension ($d_c = 2.20 \pm 0.04$) and the information dimension ($d_I = 2.22 \pm 0.04$), both of which reached asymptotic values when the embedding space dimension was 4 or greater.

Holzfuss and Mayer-Kress (1986) have examined the probable errors in estimating dimensions from a time series data set. The three methods studied involved the correlation dimension, averaged point wise dimension, and the averaged radius method of Termonia and Alexandrowicz (1983). They tested each on a set of 20,000 points from a quasiperiodic motion on a 5-torus, which consists of a time history with five incommensurate frequencies. Using the pseudo-phase-space method for embedding dimensions of 2–20, they found that the averaged pointwise dimension had the smallest standard deviation of the three. The average was taken over 20% of the reference points, and curves that did not show scaling behavior over a significant portion of the range of r were rejected.

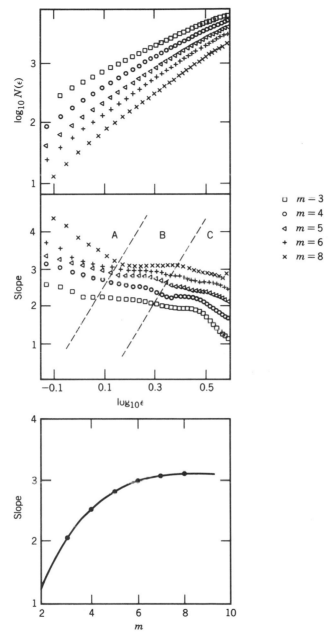

Figure 7-21 Calculation of fractal dimension for chaotic flow of fluid between two rotating cylinders: Taylor–Couette flow (see Chapter 4). [From L'vov et al. (1981) with permission of Elsevier Science Publishers, copyright 1981.]

7.5 MULTIFRACTALS

Fractals Within Fractals

As we have seen above, the fractal dimension measures the way in which a distribution of points fills a geometric space on the average. But, what if the distribution is highly inhomogeneous? Can a set of points have a distribution of fractal dimensions? A set of points with multiple fractal dimensions is not only possible, but is also common in a number of experimental as well as simple mathematical maps. [See Feder (1988) or Falconer (1990) for a more complete discussion of multifractals.] One remarkable example is the observation of multi-fractal properties of a Poincaré map from a quasiperiodic route to chaos in a Rayleigh–Benard thermal convection experiment reported by Jensen et al. (1985). Equally intriguing about this example is the ability of the circle map (Chapter 3) to quantitatively predict the correct distribution of fractal dimensions.

In a discussion of multifractal properties of a distribution of points, one must distinguish between the amplitude or *measure* of the distribution and the geometric set or so-called *support* of the distribution. To illustrate these ideas we describe two examples of simple dynamical systems that generate multifractals and show another in Figure 7-22.

Example I.

Imagine a uniform mass distribution along a line [0, 1]. The iteration rule that creates a multifractal set of points is as follows: Divide the line into two segments, say $[0, \frac{1}{2})$ and $(\frac{1}{2}, 1]$, and redistribute the mass so that P is distributed uniformly on the right segment and (1 − P) is distributed uniformly on the left segment. As this process is iterated, the original uniform mass distribution becomes highly inhomogeneous. The dimension of the support, which remains a continuous line, is unity, yet the distribution clearly has a fractal nature to it. For example, by picking some small interval of this distribution and rescaling the abscissa and ordinate scales, one can recover the overall distribution; that is, the distribution obeys the following scaling relation:

$$f(x) = \lambda^B f(bx) \tag{7-5.1}$$

Example II.

Take the same example as in Example I, but instead of redistributing the mass over the whole line, distribute the mass elements P and

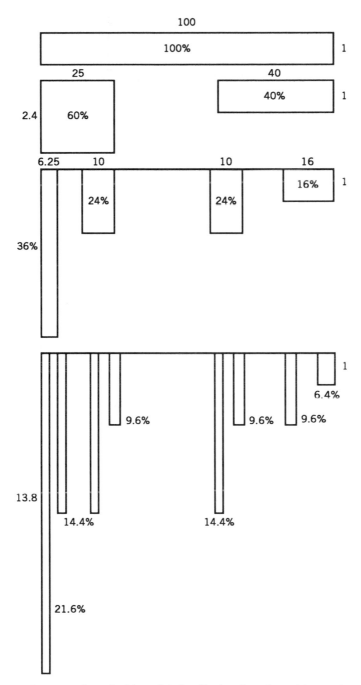

Figure 7-22 Construction of a binomial distribution function with two length scales (see also Feder (1988), Chapter 6).

$(1 - P)$ over the left third and right third, respectively, as in the Cantor set. After iterating this rule, both the distribution and the geometric support on which it lives will look fractal. In fact, the fractal dimension of the support is $D_0 = \log 2/\log 3$. But, clearly D_0 does not describe the measure or distribution of mass itself.

A third example with two length scales is shown in Figure 7-22.

The above examples are called "binomial multiplicative processes." Suppose we denote the fraction of the mass on the ith line segment as μ_i. Then for Example I with equal length segments $\delta = (\frac{1}{2})^n$, one can show that μ_i has the following form after n iterations:

$$\mu_i = P^k(1 - P)^{(n-k)}, \qquad k = 0, 1, 2, \dots , n$$

Now define $k = \xi n$ $(0 \le \xi \le 1)$ so that

$$\mu_i(\xi) = [P^\xi(1 - P)^{(1-\xi)}]^n \qquad (7\text{-}5.2)$$

To examine the multifractal properties, look at the set of $\mu_i(\xi)$ for $\xi =$ constant, and count the number of cells or segments with the same ξ value or the same measure $\mu_i(\xi) \equiv \mu_\xi$. Then for n very large, the number of cubes of length $\delta = (\frac{1}{2})^n$ to cover the set of line segments of *equal measure* $\mu\xi$ is assumed to scale as

$$N(\xi) \sim \delta^{-d(\xi)} \qquad (7\text{-}5.3)$$

One can show that $d = f(\xi)$ for equal line segments is given by

$$f(\xi) = -[\xi \ln \xi + (1 - \xi) \ln(1 - \xi)]/\ln 2$$

(see Figure 7-23 for Example I; also see Feder, 1988).

It is remarkable that a simple binomial process leads to a continuous distribution of fractal dimensions. Note that the maximum value of $f(\xi)$ equals the fractal dimension of the entire support, which remains the original line element, $[0, 1]$.

This description, however, is not suitable for experiments, so a change of variables is performed using

$$\mu_\xi = \delta^\alpha \quad \text{or} \quad \alpha = \log \mu_\xi/\log \delta \qquad (7\text{-}5.4)$$

where δ is the length of the line segment after the nth iteration.

To determine $f(\alpha)$ experimentally, one uses the moments of the distribution $\mu_i(\alpha)$. Remember that it is assumed that there is a probabil-

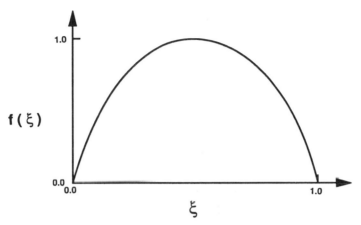

Figure 7-23 Distribution of fractal dimensions $d(\xi)$ [Eq. (7-5.3)] for a binomial distribution function in Example I. (From Feder, 1988.)

ity or mass measure which is approximated by partitioning the domain of support into cells of size δ. For a given δ, one can calculate a probability moment function

$$\chi(q, \delta) = \sum_i \mu_i(\delta)^q \tag{7-5.5}$$

(e.g., see Halsey et al., 1986). These moments generate a set of dimensions by assuming a scaling relation

$$\chi(q, \delta) \sim \delta^{-\tau(q)} \tag{7-5.6}$$

or

$$\tau(q) = -\lim_{\delta \to 0} \frac{\log \chi(q, \delta)}{\log \delta} \tag{7-5.7}$$

The motivation for taking moments and introducing another variable q is not obvious at a first reading, but suffice it to say that taking moments of the distribution gives one more information. Thus, using (7-5.5) and (7-5.6), one finds a spectrum of dimensions $\tau(q)$. If one treats q as a continuous variable, it has been shown (e.g., see Feder, 1988) that one can derive the $f(\alpha)$ curve from the implicit equations

$$\alpha(q) = -\frac{d\tau}{dq}$$

$$f(\alpha) = q\alpha(q) + \tau(q) \tag{7-5.8}$$

Some authors introduce another symbol, D_q, which is related to $\tau(q)$:

$$D_q = \frac{\tau(q)}{q - 1} \tag{7-5.9}$$

Then it has been shown that D_0 is the fractal dimension of the support (our box-counting dimension), D_1 is the information dimension, and D_2 is the correlation dimension (see Hentschel and Proccacia, 1983).

Multifractals, Quasiperiodicity and the Circle Map. To illustrate the application of multifractals to dynamics consider the application to the circle map

$$\theta_{n+1} = \theta_n + \Omega - \frac{K}{2\pi} \sin(2\pi\theta_n) \tag{7-5.10}$$

When the winding number Ω is chosen as the so-called "golden mean" $\Omega = (\sqrt{5} - 1)/2$, then the critical value for chaos is $K = 1$. Then iteration of this map shows an inhomogeneous distribution of points around the circle whose probability measure (i.e., the mass density of

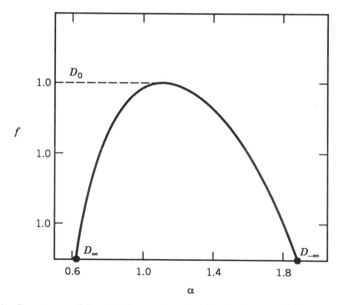

Figure 7-24 Spectrum of fractal dimensions for the circle map (7-5.10) at the critical value $K = 1$ and golden mean winding number. (From Halsey et al., 1986.)

points) has multi-fractal dimensions. Application of the above formuli, in fact, yields a spectrum of dimensions shown in Figure 7-24. One notes that the maximum value of $f(x)$ is $D_0 = 1$, which is the dimension of the support, i.e., the circle.

Experimental Multifractals in Dynamics

One of the criteria for selection of topics in this book was the applicability of new mathematical methods in dynamics to experiments. The example of a periodically forced Rayleigh–Benard convection problem is a beautiful example of the application of multifractals to fluid dynamics. In this work, first published by Jensen et al. (1985), a parallelepiped of mercury $0.7 \times 0.7 \times 1.4$ cm^3 was subjected to a thermal gradient at the same time periodically forced by pumping a small amount of current through the fluid in the presence of a magnetic field. The idea was to generate a quasiperiodic motion based on a natural convection oscillation (0.23 Hz), and the driven oscillation was chosen so that the ratio of frequencies was at the "golden mean."

The measured state variable was obtained from a thermal probe placed on the bottom plate of the cell. A time series was generated by taking a Poincaré map synchronous with the forcing period. The resulting set of data $\{... T_{n-1}, T_n, T_{n+1} ...\}$ was plotted as a return map T_n versus T_{n+1} (shown in Figure 7-25) which shows a characteristic of the cross section of a torus. What is not characteristic, however, is the inhomogeneous distribution of the density around the section of the torus. To avoid spurious bunching effects due to a projection of

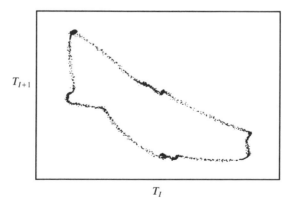

T_{I+1}

T_I

Figure 7-25 Experimental return map for periodically forced Rayleigh–Benard thermal convection showing the toroidal nature of the attractor. [From Jensen et al. (1985).]

the map onto the plane, actual calculations were carried out in a three-dimensional space (T_n, T_{n+1}, T_{n+2}).

To analyze the multifractal nature of this distribution, a set of cubes of size δ was used to cover the attractor and the probability of being in each cube was measured.

The results of calculation of the spectrum of fractal dimensions $f(\alpha)$ are shown in the same manner (see Figure 7-26) as for the *circle map*. What is truly remarkable are the identical spectrum curves for both the circle map and the Rayleigh–Benard convection experimental data. It suggests that there is something "universal" in the way dynamical systems make the transition from quasiperiodic motion to chaotic motion.

Another application of multifractal or interwoven sets of different fractal dimensions has been published by Meneveau and Sreenivasan (1987), who applied the theory to fully developed turbulence behind the wake of a circular cylinder (see also Sreenivasan, 1991).

The same fractal mathematics was also applied to a nonlinear electronic solid-state device using a crystal of p-type Ge by Gwinn and Westervelt (1987). The crystal was biased with a dc voltage to operate in a region of negative resistance where it achieved a stable limit cycle. A second sinusoidal signal was applied to the circuit so that the ratio

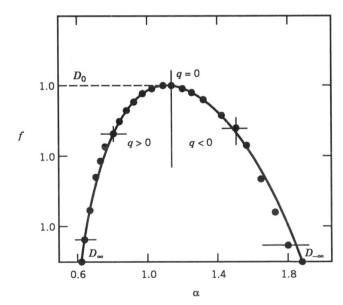

Figure 7-26   Spectrum of fractal dimensions $f(\alpha)$ for the periodically forced thermal convection experiment of Figure 7-25. [From Jensen et al. (1985).]

of the limit cycle frequency to the driven frequency was the "golden mean." Excellent agreement with the $f(\alpha)$ spectrum of the circle map was also obtained.

7.6 OPTICAL MEASUREMENT OF FRACTAL DIMENSION

All the methods for calculating the fractal dimension of strange attractors discussed above require the use of a powerful micro- or minicomputer. From an experimental point of view, however, it is natural to ask whether the fractal properties in dynamical systems can be directly measured using *analog devices* in the same way that other dynamical properties such as velocity or acceleration are measured. For general, multiple-degree-of-freedom systems, the answer is not known; but for simple nonlinear problems, the fractal dimension of a two-dimensional Poincaré map can be measured using optical techniques (Lee and Moon, 1986). This method is based on an optical interpretation of the correlation function (7-2.4). The use of the scattering of waves to measure fractal dimension of material fractals in three dimensions is described by several authors (e.g., see Schaefer and Keefer, 1984a,b, 1986).

A diagram illustrating this optical method for planar fractals embodied in Poincaré maps is shown in Figure 7-27. We recall that the correlation function involves counting the number of points in a cube

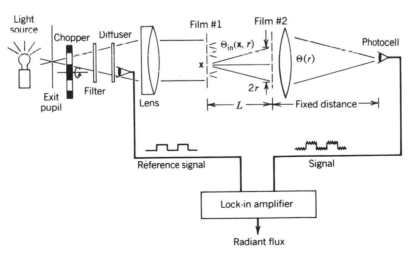

Figure 7-27 Experimental setup of the optical method for measuring fractal dimension. [From Lee and Moon (1986) with permission of Elsevier Science Publishers, copyright 1986.]

or ball surrounding each point in the fractal set of points. The optical method uses a parallel processing feature to perform all the sums at once. Light coming from one film creates a disk of light on another film. If each film is an identical copy of the Poincaré map of the strange attractor, the total light emanating from the second film is proportional to the correlation function. By changing the distance between the two films in Figure 7-27, the radius of the small circles changes and one can obtain the correlation sum as a function of the radius r. A plot of log $C(r)$ versus log r then yields the fractal dimension of the Poincaré map D.

If the map is a time-triggered Poincaré map, the dimension of the attractor is $1 + D$.

An Optical Parallel Processor for the Correlation Function

A sketch of the experimental setup is shown in Figure 7-27, displaying the optical path of light in this method. The method makes use of two properties of classical optics. First, if light is passed through a small aperture of diameter D in the region of Fraunhofer diffraction (if λ is the wavelength, $D \gg \lambda$), then light will cast a circle of radius r, with uniform intensity, on a plane located at a distance L from the aperture. This radius is given by $r = 1.22L\lambda/D$. In our method, the aperture orginates from a small dot on the negative of a planar Poincaré map and the small circle of light falls on an identical copy of this negative located at a distance L (Figure 7-27). Second, for incoherent light, the amount of light that emanates from the second negative is proportional to the number of small dots or circles within the circle of illumination. The total amount of light passing through both films is thus proportional to the correlation function $C(r)$. To calculate or vary r, we simply measure and vary L, the distance between the two negatives.

To make these ideas more concrete, let $\Phi(\mathbf{x}, r)$ be the radiant flux behind film #2 due to the flux $\Phi_{in}(\mathbf{x})$ entering the circular aperture at $\mathbf{x}$ on film #1:

$$\Phi(\mathbf{x}, r) = n(x, r)\mathrm{A}\frac{\Phi_{in}(\mathbf{x})}{\pi r^2} \tag{7-6.1}$$

where $n(\mathbf{x}, r) = \Sigma_j H(r - |\mathbf{x} - \mathbf{x}_j|)$ is the number of apertures located within the circle of light illuminated by the flux in the aperture at $\mathbf{x}$, and A is the area of the aperture of a point on film #1. One can see that Φ depends on both n and r explicitly. However, we would like a measure of n alone. Using the linear relation between r and L, we define an adjusted radiant flux $\Phi^* = (r/r_0)^2\Phi$, where r is the radius of the illuminated area when $L = L_0$ (L_0 is a convenient reference distance). Summing over all points in film #1, we obtain

$$\sum_{k=1}^{N} \Phi^*(x, r) = \left(\frac{r}{r_0}\right)^2 \sum \Phi(\mathbf{x}, r) = \frac{A}{\pi r_0^2} \sum \Phi_{in}(\mathbf{x})n(r) \qquad (7\text{-}6.2)$$

When the incident light intensity is uniform over film #1, we find

$$\left(\frac{L}{L_0}\right)^2 \sum_{k=1}^{N} \Phi_0(\mathbf{x}_k, r) \approx \sum_{k=1}^{N} n(r) \approx C(r) \qquad (7\text{-}6.3)$$

The maps can be obtained either from a numerical solution of a third-order system of equations or from experimental data. The light passing through film #2 was focused onto a photocell for the light flux measurement. A light filter (orange-amber color filter) was used at the

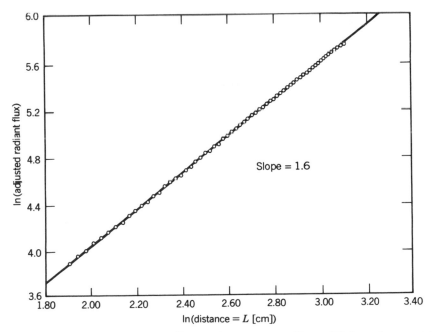

Figure 7-28 Radiant flux versus distance between two films of Poincaré maps on a log–log scale for data from the vibration of a buckled beam. [From Lee and Moon (1986) with permission of Elsevier Science Publishers, copyright 1986.]

light source to optimize the photocell response around 6328 Å. The dot size on the negatives was less than 0.2 mm; thus $D/\lambda \simeq 300$, which satisfies the Fraunhofer diffraction criterion.

The output voltage from the photocell contained a lot of noise. To extract the signal from the noise, a mechanical light chopper and a lock-in amplifier were used in the signal processing. The chopper was operated at approximately 100 Hz to avoid power-line noise.

The radiant flux behind film #2 was measured at the photocell as a function of the distance between films, and the adjusted radiant flux (7-4.2) versus L was plotted on a log–log scale as shown in Figure 7-28. Theoretically, the slope of this curve should give the fractal dimension (7-2.5).

The data were obtained from a Runge–Kutta simulation of the forced two-well potential equation (7-4.6). The 4000 points were generated by taking a Poincaré map synchronous with the driving frequency. The adjusted radiant flux output was measured at approximately 200 values of L. However, only the linear section of log C versus log L is plotted in Figure 7-28.

A comparison of the optically measured fractal dimension with those calculated from the numerical data of Moon and Li (1985a) is shown in Table 7-3 for several values of the damping. The results, as one can see, are remarkably good.

TABLE 7-3   Optically Measured Fractal Dimension for Computer-Simulated and Experimental Poincaré Maps

Numerical Poincaré Map [Eq. (7-4.6)]		
Damping	Calculated[a]	Measured
0.075	1.565[b]	1.558
0.105	1.393	1.417
0.135	1.202	1.162

Experimental Poincaré Map			
Phase Angle	Calculated[a]		Measured
0°	1.741[b]	1.628[c]	1.678
45°	1.751	1.627	1.671
90°	1.742	1.638	1.631
135°	1.748	1.637	1.676
180°	1.730	1.637	1.635

[a] Moon and Li (1985a).
[b] Based on four smallest log r points in log C versus log r.
[c] Based on seven smallest log r point in log C versus log r.

A comparison of the optical and numerical methods for experimental Poincaré maps for the buckled beam is also shown in Table 7-3. In this set of tests, the phase of the Poincaré map trigger was changed. The optical measurement of fractal dimension confirms the results of the numerical method, namely, that the dimension is independent of the phase of the map. This implies that the dimension of the strange attractor itself is $1 + D$, where D is the planar map dimension.

7.7   FRACTAL BASIN BOUNDARIES

Basins of Attraction

In most physical *linear* systems, there is just one possible motion for a given input. For example, the response of a linear mass–spring–damper system to an initial impulse force is just a decaying response, where the mass eventually comes to rest. Such a system has but one attractor, namely, the equilibrium point. However, in nonlinear systems, it is possible for more than one outcome to occur depending on the input parameters such as force level or initial conditions. For example, the system may have more than one equilibrium position or it may have more than one periodic or nonperiodic motion as in certain self-excited systems.

Equilibrium positions and periodic or limit cycle motions are called *attractors* in the mathematics of dissipative dynamical systems. The range of values of certain input or control parameters for which the motion tends toward a given attractor is called a *basin of attraction* in the space of parameters. If there are two or more attractors, then the transition from one basin of attraction to another is called a *basin boundary* (see Figure 7-29). In classical problems, we expect the basin boundary to be a smooth, continuous line or surface as in Figure 7-29. This implies that when the input parameters are away from the

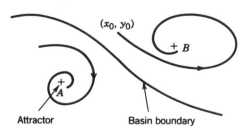

Figure 7-29   Sketch of two dynamic attractors in phase space and the boundary between their basins of attraction in initial condition space.

boundary, small uncertainties in the parameters will not affect the outcome. However, it has been discovered that in many nonlinear systems, this boundary is nonsmooth. In fact it is fractal—hence the term *fractal basin boundary*. The existence of fractal basin boundaries has fundamental implications the behavior of dynamical systems. This is because small uncertainties in initial conditions or other system parameters may lead to uncertainties in the outcome of the system. Thus predictability in such systems is not always possible (see the papers by Grebogi et al., 1983b, 1985a,b, 1986).

Sensitivity to Initial Conditions: Transient Motion in a Two-Well Potential

Before we examine a problem with a fractal basin boundary, it is instructive to look at a case where the basin boundary is smooth, but the outcome is sensitive to initial conditions. This is the case of the *transient* dynamics of a particle with damping. This one-degree-of-freedom example is a simple model for the postbuckling behavior of an elastic beam. The equation of motion for this problem is

$$\ddot{x} + \gamma\dot{x} - \tfrac{1}{2}x(1 - x^2) = 0 \tag{7-7.1}$$

Unlike the related problem with periodic forcing, the complete dynamics can be described in a two-dimensional phase plane $(x, y, = \dot{x})$. The displacement and time have been normalized such that the two stable equilibrium positions in the phase plane are $(\pm 1, 0)$ and the undamped natural frequency is one radian per second. The control parameters are the damping γ and the initial conditions $x(0) = x_0$, $\dot{x}(0) = y_0$. Although there are three equilibrium positions, $x = 0, \pm 1$, only the latter two are stable and thus we will have *two competing basins of attraction*.

Dowell and Pezeshki (1986) have examined the basins of attraction for this problem as illustrated in Figure 7-30. They subdivided the basins into how many times the particle orbits cross the $x = 0$ axis before settling down to $x = \pm 1$. One can see that for large initial conditions there are alternating bands where the particle will eventually go to the left or right attractor. Although these boundaries are smooth, the size of the bands approaches zero as the damping $\gamma \to 0$. Thus, if there is some finite uncertainty in the initial conditions as denoted by the circle of radius ε in Figure 7-30, one has no certainty of which attractor the particle will go toward if $\varepsilon > \varepsilon_0(\gamma)$, where lim

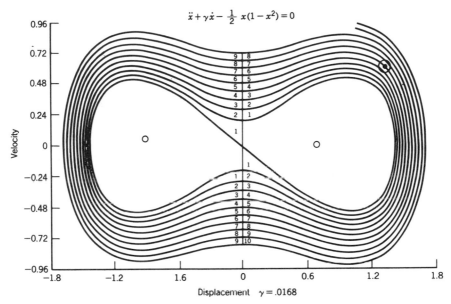

Figure 7-30   Basins of attraction for the unforced, damped motion of a particle in a two-well potential. The numbers indicate the number of times the trajectory crosses $x = 0$ before going to one of the two equilibrium points at $x = \pm 1$. [From Dowell and Pezeshki (1986).]

$\varepsilon_0 \to 0$ as $\gamma \to 0$. For finite damping, we can obtain certainty of the end state only if we have accurate enough information about the initial state.

In the next example, we will show a fractal basin boundary where the outcome is always uncertain no matter how small ε is; that is, $\varepsilon_0 = 0$.

Fractal Basin Boundary: Forced Motion in a Two-Well Potential

In this section, we will examine the periodic forcing of a particle in a two-well potential:

$$\dot{x} = y \qquad\qquad (7\text{-}7.2)$$

$$\dot{y} = -\gamma y + \tfrac{1}{2}x(1 - x^2) + f_0\cos \omega t$$

As discussed in earlier chapters, the dynamics of the particle can be described in a three-dimensional phase space $(x, y, z = \omega t)$. In the earlier discussions, however, we focused on chaotic motions for this system. Here we will only consider motions which are *periodic* about

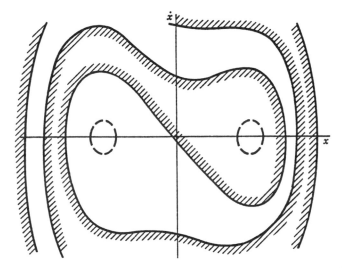

Figure 7-31 Smooth basin boundary for low-amplitude forcing of two-well potential oscillator. The attractors are periodic orbits about left and right equilibrium points. [From Moon and Li (1985b) with permission of the American Physical Society, copyright 1985.]

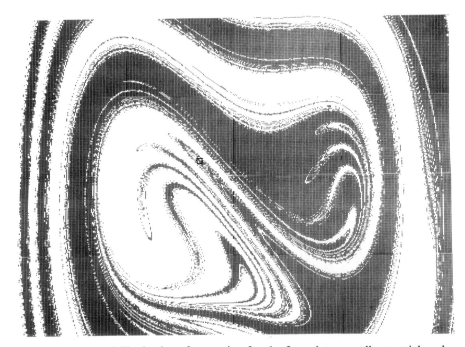

Figure 7-32 Fractal-like basins of attraction for the forced, two-well potential problem for forcing amplitude above the Melnikov criterion (7-7.3). [From Moon and Li (1985b) with permission of the American Physical Society, copyright 1985.]

either the left or right equilibrium positions, $x = \pm 1$. Thus, the attractors in this problem may be considered limit cycles. [If we take a Poincaré map of the asymptotic motion, we will have a finite set of points near one of the equilibrium positions $(\pm 1, 0)$.] Here we do not distinguish between period-1 or period-2 subharmonics. We assume that the forcing f_0 is small enough to avoid chaotic vibrations and high-period subharmonics.

In this example, we fix γ, f_0, and ω and vary the initial conditions. The results are shown in Figures 7-31–7-33 and are obtained from numerical simulation using a fourth-order Runge–Kutta integration algorithm (see Moon and Li, 1985b for details).

The results in Figure 7-31 show that when f_0 is small enough, the basin boundary is smooth, but when f_0 is greater than some critical value, the boundary becomes fractal-looking as shown in Figure 7-32. (This figure is based on integration of 400×400 initial conditions.) To ascertain whether this boundary is fractal, we have taken a small region of initial condition space and have expanded this region. The

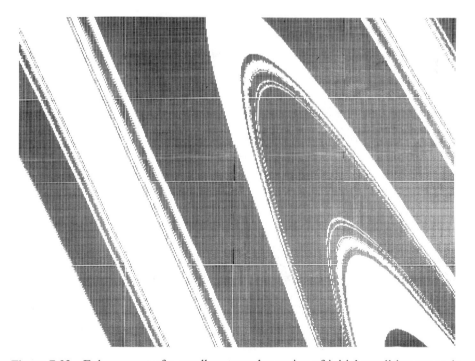

Figure 7-33 Enlargement of a small rectangular region of initial condition space in Figure 7-23 showing fractal-like structure on a finer scale. [From Moon and Li (1985b) with permission of the American Physical Society, copyright 1985.]

results are shown in Figure 7-33. Thus, we see that on a finer and finer scale the boundary shows evidence of fractal structure. These results have important implications for classical dynamics insofar as predictability goes. See also Color Plates 5 and 6.

Two other examples of basin boundary calculations are illustrated in Color Plates 7 and 8 and on the jacket for a particle in a three-well and four-well potential. These problems were described in Chapter 6 (see also Li and Moon, 1990a,b). In the three-well potential problem the particle has one degree of freedom and is excited by a periodic force. The two photos in Color Plates 7 and 8 show the evolution of the basin boundaries as the force level is increased. The four-well problem is a two-dimensional one. The fractal nature of the basins of attraction is shown on the jacket for a force level high enough to produce homoclinic orbits in the Poincaré map.

Homoclinic Orbits: A Criterion for Fractal Basin Boundaries

Although the main theme of this book has been chaotic dynamics, the results of the previous section demonstrate that one of the properties of chaotic dynamics, namely, parameter sensitivity and unpredictability, may also be characteristic of certain nonchaotic motions. This prospect stirs terror in the computers of those engineers involved in numerical simulation of nonlinear systems. In such systems, the output of a calculation may be sensitive to small changes in variables such as initial conditions, control parameters, round-off errors, and numerical algorithm time steps. This lack of robustness may exist even when the problem is a transient one or has a periodic output.

First, we expect that those systems most susceptible to fractal basin boundary behavior will be those with multiple outcomes, such as multiple equilibrium states or periodic motions. For example, if we consider the impact of an elastic–plastic arch (see Symonds and Yu, 1985 and Poddar et al., 1986) or periodic excitation of a rotor or pendulum, there are at least two possible outcomes. In the case of the arch, the end state could be either the arch bend up or down. In the case of the rotor, one could have rotation clockwise or counterclockwise.

The second clue to establishing the possibility of fractal basin boundaries is more subtle and requires more mathematical intuition. We have seen in Chapters 1, 3, and 6 that nonlinear systems which tend to stretch and fold regions of phase space in what are called *horseshoe maps* have a certain element of sensitivity to initial conditions as well as a variety of subharmonics solutions. As discussed in Chapter 6, it was shown that horseshoe map properties result when

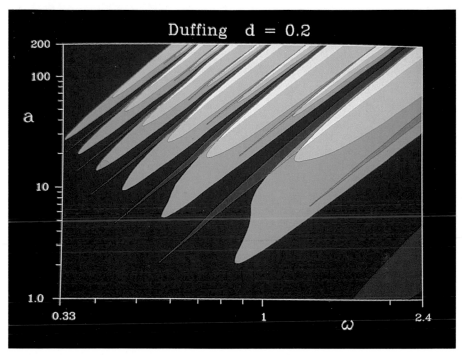

Color Plate 1 Dynamic regimes for the forced motion of a particle in a cubic force field; $\ddot{x} + d\dot{x} + x + x^3 = a\ cos\ \omega t$. Dark blue: symmetric, period -1 motion; lighter blues: two asymmetric period -1 and period -2 motions; red: other period -1 solutions. (Courtesy of Professor W. Lauterborn, Technische Hochschule, Darmstadt, FRG; copyright U. Parlitz C. Scheffczyk, W. Lauterborn.)

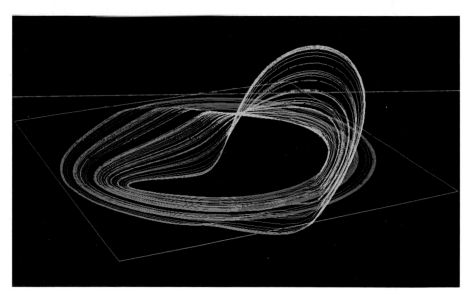

Color Plate 2 Experimental chaotic attractor for periodic forcing of a mass under dry friction. [From Feeny and Moon (1992).]

Color Plate 3 Mandelbrot set (black). Each color represents the same number of iterations to escape to "infinity."

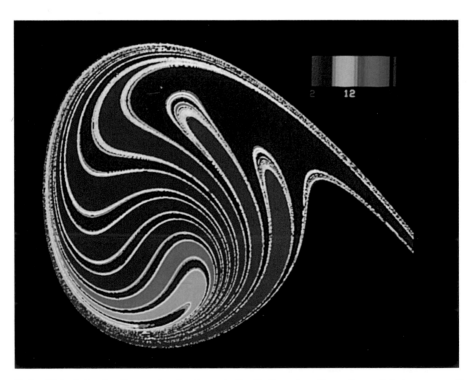

Color Plate 4 Basin of attraction for the periodically forced motions of a particle in a one-well potential with a saddle point (dark purple). Other colors indicate a measure of time to escape out of the well, see Chapter 6 and Thompson (1989b). (Graphic courtesy of Professor J. Cusumano, Pennsylvania State University.)

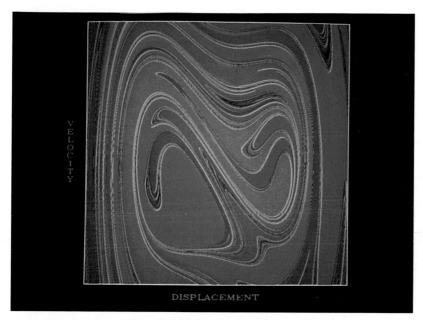

Color Plate 5 Basins of attraction for the periodically forced motions of a particle in a two-well potential. [Based on Moon and Li (1985).]

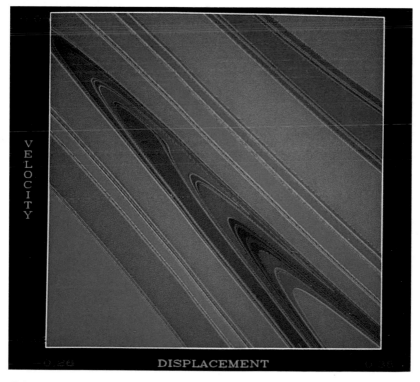

Color Plate 6 Enlargement of the central section of Color Plate 5 for motion in a two-well potential. [Based on Moon and Li (1985).]

Color Plate 7 Basins of attraction for periodically forced motions of a particle in a three-well potential — close to the homoclinic criterion. [From Li and Moon (1990a).]

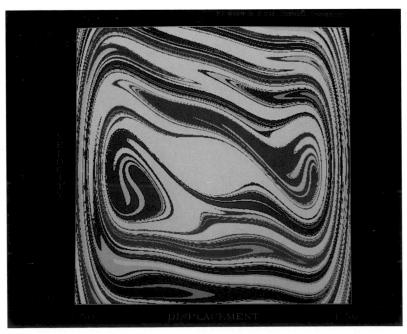

Color Plate 8 Basins of attraction for three-well potential problem in Color Plate 7–force greater than homoclinic criteria. [From Li and Moon (1990a).]

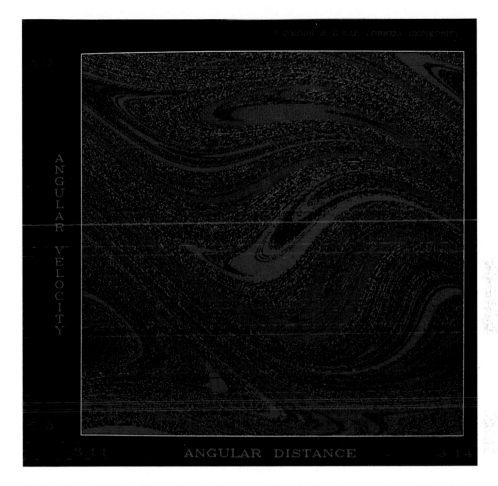

Color Plate 9 Basins of attraction for the parametrically forced magnetic pendulum; blue, clockwise; red counterclockwise. [Color graphic, G.-X. Li, based on Moon, Cusumano, and Holmes (1987).]

Color Plate 10 Lagrangian chaotic mixing of dye in a fluid between two oscillating cylinders. (Courtesy of Professor J. M. Ottino, Northwestern University, Evanston, Illinois.)

Color Plate 11 Lagrangian chaotic mixing of dye in a fluid between two oscillating cylinders. (Courtesy of Professor J. M. Ottino, Northwestern University, Evanston, Illinois.)

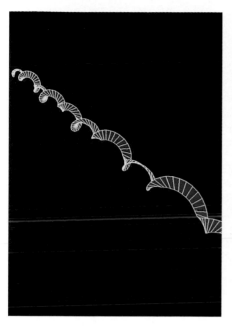

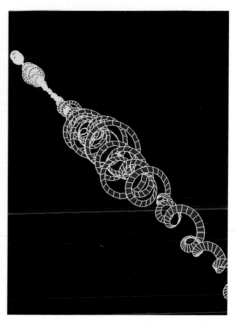

Color Plate 12 Spatially periodic deformation of the elastica. [From Davies and Moon (1992).]

Color Plate 13 Spatially quasiperiodic deformation of the elastica. [From Davies and Moon (1992).]

Color Plate 14 Spatially chaotic deformation of the elastica. [From Davies and Moon (1992).]

Color Plate 15 Alternating torque mixing of elasto-plastic solids. [From Feeny et al. (1992).]

Color Plate 16 Alternating torque mixing of elasto-plastic solids. [From Feeny et al. (1992).]

the Poincaré map associated with the flow in phase space develops homoclinic points in dissipative nonlinear systems. A criterion was derived by Holmes (see Guckenheimer and Holmes, 1983) using a method by Melnikov [Eq. (6-3.20)]. In the case of the forced motion of a particle in a two-well potential, it turns out that this criterion gives a very good indication of fractal basin boundaries even when the motion is *not* chaotic. The criterion for the equation of motion (7-7.2) is given by

$$f_0 > \frac{\gamma\sqrt{2}}{3\pi\omega} \cosh\left(\frac{\pi\omega}{\sqrt{2}}\right) \tag{7-7.3}$$

Evidence for this conclusion is given in Figure 7-34 (e.g., see Moon and Li, 1985b). This figure summarizes the results of many calculations of basin boundaries similar to those in Figures 7-31–7-33. Below the Holmes–Melnikov criterion the numerically calculated basin boundary appears to be smooth, whereas above the criterion curve the boundary appears fractal.

The connection between homoclinic orbits and fractal basin boundaries is not entirely a mystery especially if we examine the results in Figure 7-35. In this figure, we have superimposed two calculations.

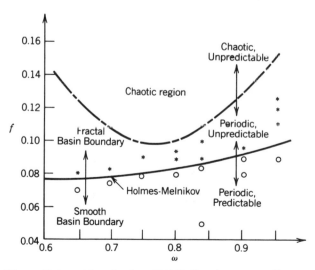

Figure 7-34 Homoclinic orbit criterion (7-7.3) for the two-well potential problem with fractal-like and smooth basin boundary observation fron numerical studies. [From Moon and Li (1985b) with permission of the American Physical Society, copyright 1985.]

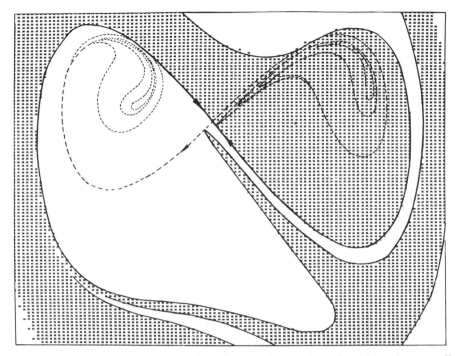

Figure 7-35   Superimposed plots of the basins of attraction of the forced, two-well potential problem and the associated stable and unstable manifolds of the Poincaré map at the critical force level (7-7.3). [From Moon and Li (1985b) with permission of the American Physical Society, copyright 1985.]

The first is the basin boundary for the two-well potential for a force amplitude just below the Holmes–Melnikov curve. We can see that the boundary has developed a long finger as compared with that in Figure 7-31 for a smaller force. The second calculation in Figure 7-35 is the determination of the stable and unstable manifolds of the Poincaré map which emanate out of the saddle point near the origin. The first observation is that the basin boundary is *identical* to the stable manifold of the Poincaré map. The second observation is that the unstable manifolds, shown as the dashed curves, are just touching the stable manifolds. This is to be expected because at the criterion the two manifolds touch and form homoclinic points. In theory, beyond this criterion, the two manifolds of the Poincaré map must touch an infinite number of times, which results in an infinite folding of the stable manifold and hence an infinite folding of the basin boundary and the resulting fractal properties.

The idea that basins of attraction of different motions can become intertwined is not a new concept in nonlinear dynamics, as can be seen in the classic book by Hayashi (1953) on nonlinear oscillations.

Professor Hayashi's book illustrates the intertwining of three basins of attraction, each associated with a particular subharmonic motion of the oscillator. These diagrams were obtained by Hasashi and his co-workers using analog computers, and they showed how a small change in initial conditions could switch the output from one attractor to another. Although this knowledge was available in the 1950s, and perhaps earlier, the relationship of these basin boundary diagrams to fractals and chaos was not made until around 1980. (See also the discussion in Chapter 4 on Hayashi, Ueda, and the "Japanese attractor.")

The above discussion assumes the existence of only two attractors. However, even in the two-well potential problem, within each well there may be two or more attractors; for example, there could be two subharmonic solutions in the vicinity of each well. When there are more than two attractors, it is possible for one basin boundary to become fractal and another basin boundary to remain smooth. This also suggests that there could be multihomoclinic tangencies or homoclinic criteria. A discussion of multiple coexisting attractors and basin boundaries is presented in a paper by Battelino et al. (1988) in which they treat the forced motion of two-coupled Van der Pol oscillators. Two other studies on multiple basin boundaries have been presented in two Cornell University dissertations by G.-X. Li (1984, 1987) (see also Li and Moon 1990a,b). These studies examine a particle in three- and four-well potentials. A brief discussion of multiple homoclinic orbit criteria for these problems was presented in Chapter 6. Color plates of basins of attraction for both the three-well and four-well potential problems are shown in Color Plates 7 and 8 and on the jacket.

In these multiple-well potential problems, mixed fractal and smooth basin boundaries can arise when there are more than one saddle point in the Poincaré map. Thus, the outflow trajectory (unstable manifold) of one saddle can intersect either its own inflow trajectory or that of another saddle. Each such entanglement is bound to generate a horseshoe map structure that in turn produces the fractal basin boundary. However, these multiple entanglements may occur for different values of the control parameter and hence the possibility of multiple homoclinic tangency criteria, each leading to greater and greater sensitivity to initial conditions.

Fractal Basin Boundaries and Robust Design

In the design of most practical engineering devices, the system is usually assumed to operate near one stable dynamic attractor. Thus, a design which is sensitive to either initial conditions or control param-

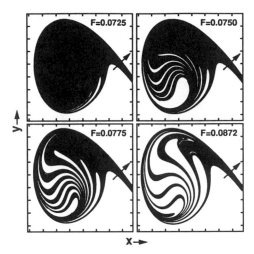

Figure 7-36 A set of basin boundaries for the periodically forced one-well potential oscillator. [From Thompson (1989b).]

eters is not robust. But how does one quantify robustness? Professor J. M. T. Thompson has attempted to answer this question in a series of papers relating to the dynamic capsize of ships (Thompson, 1989a,b; Thompson et al., 1990; Soliman and Thompson, 1989). These studies are based on a model of a particle in a one-well potential with a one-sided escape barrier: (see Figure 6-17 and Color Plate 4)

$$\ddot{x} + B\dot{x} + x - x^2 = F \sin \omega t \qquad (7\text{-}7.4)$$

For zero forcing, $F = 0$, this system has a saddle point at $x = 1$ and a stable spiral attractor at $x = 0$. The basin of attraction for $x = 0$, in fact, is determined by the stable manifold of the saddle point in the phase plane $(x, \dot{x})$. For small enough forcing, $F \neq 0$, there is a saddle point in the Poincaré map, and the inflow curve to this saddle (stable manifold) also defines the basin of attraction. This boundary can be found numerically by iterating the Poincaré map backwards in time for a set of initial conditions lying along the stable eigenvector of the saddle point of the map. A set of four such basin boundaries are shown in Figure 7-36 for the one-well potential for four different forcing levels (Thompson, 1989). As the force increases, the fractal tongues invade the basin. Thompson then defines robustness in terms of the degree of erosion of area of the basin of attraction. This is illustrated in Figure 7-37. One can see that even when F is increased past the homoclinic tangle value calculated from Melnikov's theory (Eq. (6-3.28b)), the

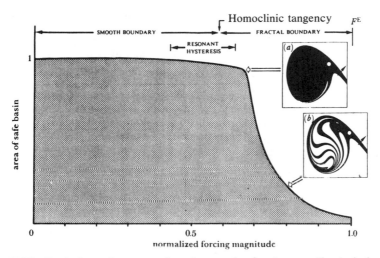

Figure 7-37 Basin boundary area function as the forcing amplitude is increased. *Insets:* (a) $F = 0.0725$; (b) $F = 0.0872$. [From Thompson et al. (1990).]

basin area is robust until a critical point at which the safe area erosion is accelerated by a small increase in F. These ideas and other so-called safety integrity measures for dynamical systems show how the concept of fractal geometry can be used to quantify intuitive features of design of nonlinear engineering devices (see Soliman and Thompson, 1989).

Dimension of Basin Boundaries and Uncertainty

Yorke and co-workers at the University of Maryland have produced numerous studies of basin boundaries, fractals, and chaos. In one study they have shown that the fraction ϕ of uncertain initial conditions in the phase space as a function of the radius of uncertainty ε is related to the fractal dimension of the basin boundary (e.g., see McDonald et al., 1985)

$$\phi \approx \varepsilon^{D - d}$$

where D is the dimension of the phase space and d is the capacity fractal dimension of the basin boundary. When the boundary is smooth, $d = D - 1$ or

$$\phi \sim \varepsilon$$

For example, if the relative uncertainty in initial conditions were $\varepsilon = 0.05$, then the uncertainty of the outcome as a fraction of all initial conditions would be $\delta \approx 22\%$ when $d = 1.5$ and $D = 2$.

A technique for calculating d for basin boundaries is described in a number of the Maryland group papers. The technique differs from that for trajectories because the boundary points are never given but are formed from the set of points that lie in neither of two attracting sets. Such fractal sets have been labeled "fat fractals." [See Grebogi et al. (1985c) for a discussion of fat fractals and their application to basin boundary calculations.]

Transient Decay Times: Sensitivity to Initial Conditions

In the preceding discussion we described how the development of a fractal basin boundary leads to uncertainty about which attractor the system will approach as $t \rightarrow \infty$. However, one may also be interested in how much time it takes to approach the attractor. Pezeshki and Dowell (1987) have calculated an initial-condition–transient-time plot for the two-well potential as shown in Figure 7-38. In this diagram each point is coded in color or shade to represent the transient time to approach a periodic orbit around either the left or right potential well. The two wells are not distinguished, only the transient times. They observed fractal-looking patterns when the forcing amplitude was above the homoclinic orbit criterion (7-5.3). This means that, given some uncertainty in initial conditions, both the transient decay time and the particular attractor are unpredictable for certain nonlinear problems.

Fractal Boundaries for Chaos in Parameter Space

We have seen how small changes in initial conditions can dramatically change the type of output from a dynamical system. It is natural to ask whether a similar sensitivity exists in the other parameters that control the dynamics, such as forcing amplitude or frequency or the damping or resistence in a circuit. One example is discussed here—a fractal experimental boundary between chaotic and periodic motions in a forced one-degree-of-freedom oscillation.

When two or more types of motions are possible in a system, one usually determines the range of parameters for which one or another type of motion will exist. In the case of the forced motion of a particle in a two-well potential (see Chapters 2 and 6), it is of great interest to know when chaotic motions or periodic motions will occur when the

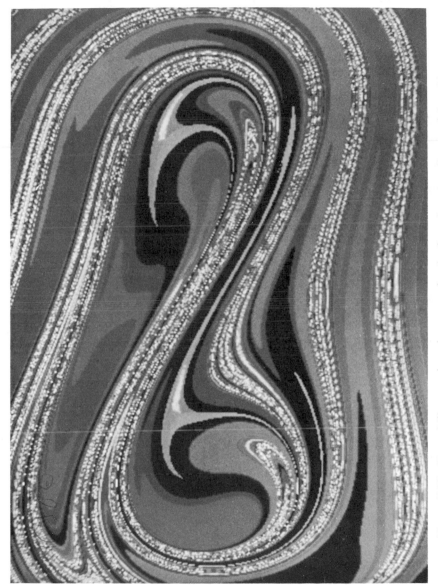

Figure 7-38 Fractal-like transient time maps for the forced, two-well potential problem. Each shade represents a different time for the motion to approach a steady periodic orbit. [From Pezeshki and Dowell (1987).]

input force is periodic. The equation that describes this oscillation is by now familiar to the reader [Eq. (7-7.2)]. In this problem, we have used a nondimensionalization procedure to eliminate all but three parameters (γ, f, ω). As discussed in Chapter 6, both Holmes (1979) and Moon (1980a) derived criteria relating (γ, f, ω) for when chaotic motion would occur. These relations [Eqs. (6-3.27) and (6-3.46)] have the form

$$f > F(\omega, \gamma) \qquad\qquad (7\text{-}7.5)$$

Fixing the nondimensional damping γ, both criteria are smooth curves in the (f, ω) plane as shown in Figure 7-39. When these criteria are compared with experimental data (see Moon, 1984b), however, two differences are obvious: The theoretical criteria are lower bounds, and the experimental criterion looks ragged and may therefore be fractal.

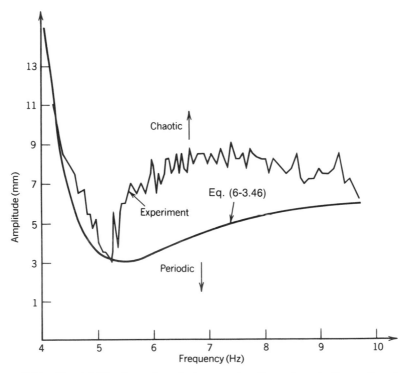

Figure 7-39 Fractal-like boundary between chaotic and periodic motion in the forcing-amplitude–frequency plane. Experimental data are from the vibration of a buckled beam. [From Moon (1984b) with permission of the American Physical Society, copyright 1984.]

The experiments were carried out on the now familiar buckled, steel, cantilever beam placed above two permanent magnets (Figure 2-2*b*). The elastic beam, magnets, and support are placed on an electromagnetic shaker which drives the system at a given amplitude A_0 and frequency ω. The nondimensional force in (7-7.2) is related to this forcing amplitude by

$$f_0 = -A_0\omega^2$$

The experiments were carried out by fixing the forcing frequency and slowly increasing the driving amplitude of the shaker. With the beam vibrating initially with periodic motion about one of the buckled equilibrium positions, the amplitude was increased until the tip of the beam jumped out of the initial potential well.

To determine whether the motion was chaotic or periodic, Poincaré maps were used. To motion was measured by strain gauges attached to the beam at the clamped end, and the strain versus strain rate served as the phase plane. Poincaré maps of these signals were synchronized at the driving frequency. Chaos was determined when the finite set of points of the Poincaré map (as observed on an oscilloscope; see Chapter 5) became unstable and a Cantor-set-like pattern appeared on the screen.

At least five sets of data for chaotic boundaries were taken for different beam-magnet configurations, and all showed a nonsmooth behavior. In the data shown in Figure 7-31, approximately 70 frequencies were sampled between 4 and 9 Hz.

To determine if the boundary between chaotic and periodic motions is fractal, the fractal dimension of the set of experimental points was measured. First, we connected the points with straight line segments. Second, we used the caliper method to measure the length of the boundary as a function of caliper size. This is the same method described by Mandelbrot (1977) to measure the fractal dimension of the coastline of various countries. Thus, we are approximating the experimental boundary by N line segments, each of length ε. As we decrease the caliper size, ε (the number of line segments needed to approximate the curve) increases. The total length is then

$$L = N(\varepsilon)\varepsilon \qquad (7\text{-}7.6)$$

For a nonfractal curve, $N \simeq \varepsilon^{-1}$ or $N = \lambda/\varepsilon$; thus λ becomes a measure of the length of the boundary. However, for fractal curves, such as the Koch curve, $N = \lambda\varepsilon^{-D}$, where ε is small and D is not an integer.

Thus by measuring L versus ε,

$$L = \lambda \varepsilon^{1-D} \qquad (7\text{-}7.7)$$

we can obtain the fractal dimension by measuring the slope of the log L versus log ε curve, or

$$D = \lim_{\varepsilon \to 0} \left(1 + \frac{\log L}{\log(1/\varepsilon)} \right) \qquad (7\text{-}7.8)$$

One can show that this procedure is equivalent to the idea of covering the set of points with small squares as discussed in the definition of the capacity fractal dimension [Eq. 7-1.2].

The results of this series of measurements are shown in Figure 7-40 for two sets of data. The lengths of the boundary curves appear to increase with decrease in caliper size, and they imply a fractal dimension of between 1.24 and 1.28. Thus, there is convincing evidence that the boundary curve between periodic and chaotic regimes in the parameter space of (f, ω) is fractal. It should be noted, however, that while the single-mode description of the chaotic elastic beam [Eq. (7-7.2)] agrees very well with the experimental results insofar as Poincaré maps are concerned, the actual experiment has infinitely many degrees of freedom which one hopes do not influence the low-frequency behavior. However, it may be possible that higher modes

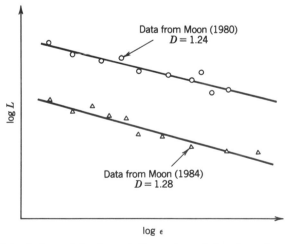

Figure 7-40   Calculation of the fractal dimension of the chaos boundary of Figure 7-39.

could influence or are even essential to the fractal nature of the boundary curve in Figure 7-39. Further research on this question is necessary to provide a clear answer.

In any event, these results suggest that a clear-cut criterion for chaos may not be possible. The apparent fractal nature of the criterion boundary may be inherent in many systems, and one may have to settle for upper or lower bounds on the chaotic regimes.

7.8 APPLICATION OF FRACTALS IN THE PHYSICAL SCIENCES

We have seen in this chapter the way in which fractals enter dynamics and the dynamic basis of creating fractals. However, there are many applications of this modern branch of mathematics where dynamics is not the central issue. This is especially the case of the characterization of geometric forms in the natural and manufactured world by the use of fractal concepts. A very good introduction to some of these applications may be found in the book by Feder (1988). In this section we describe in brief a few of these applications to geology, fluid mechanics, materials characterizations, and fractal mechanics. This section is written to show the broader applications of fractals in physics besides dynamics. But we also present these examples of fractal physical forms to suggest that in each there may have been a dynamic, chaotic process that created them which is yet to be discovered.

Geology

One of the early applications of fractals was the observation that measurement of some geological features such as coastlines or rivers depends on the size of one's measurement instrument (Mandelbrot, 1982); the coastlines of Britain and Norway are clear examples. In this way of thinking, the conventional definition of the length of a natural geological feature may not be applicable because the length, L, depends on the basic unit of the caliper, ε, that is

$$L = N(\varepsilon)\varepsilon$$

As discussed in the previous sections, for fractal-like geological features, the number of caliper lengths $N(\varepsilon)$ may be proportional to ε^{-D}, where for $D = 1$ the curve is not fractal. Thus, the length–caliper size relation takes the form $L = \lambda \, \varepsilon^{1-D}$, where D is the fractal dimension. Parts of the British coastline boasts a fractal dimension of $D = 1.2$, whereas Norway's ragged shoreline has a dimension $D = 1.5$.

In another geological application, Feder (1988) describes the consequences of multiple branching of rivers for which the relation between drainage area A and the length of the longest branch of the river L suggests a fractal scaling $L = \beta A^{1/D}$. In the states of Virginia and Maryland in the United States, Hach (1957) found that $D = 1.2$. Again, we see that there must have been underlying geologic dynamical processes that created these fractal features in the earth's surface topology. A recent discussion of fractals and chaos in the geological sciences may be found in the book by Turcotte (1992).

Viscous Fingering: Hele-Shaw Flow

The case of fluid flow between two plates was studied by Hele-Shaw in 1898. Recent interest in this experiment revolves around the fractal-like interface between a gas under pressure and the fluid in the cell [see Feder (1988) and Chapter 4; also see Homsy (1987) for a review]. A sketch of the geometry is shown in Figure 7-41a, and a typical picture of the fractal viscous fingers that develop is shown in Figure 7-41b. In the theory for this phenomenon, inertial forces are small, and the equilibrium involves a balance of viscous stress, surface tension, and applied pressure. The basic mechanism involves an instability of a gas–fluid interface in which a wrinkled surface becomes more stable than a flat surface, which leads to a dendritic-type growth pattern. These problems are of importance, for example, in understanding fluid flow in porous material such as oil recovery processes.

The physical picture of the development of the fractal figure in the fluid in the Hele-Shaw cell looks like a dynamic model call *DLA-diffusion-limited aggregation*. The *DLA* model involves particles that move in a random way until they reach some surface where they cluster. *DLA* models are used to model many growth phenomena in fractal physics, including electrochemical dissolution processes, dielectric breakdown (Murat and Aharony, 1986), and possibly fracture of metals (Louis et al., 1986).

In experiments that simulate flow through porous media, a viscous fluid such as epoxy is placed between two plates with a monolayer of small glass beds (~1 mm diameter). The flow of air into this cell produces fractal patterns similar to those of *DLA* models (Maloy et al., 1985) as shown in Figure 7-42.

It is interesting that this ostensibly static, deterministic experiment

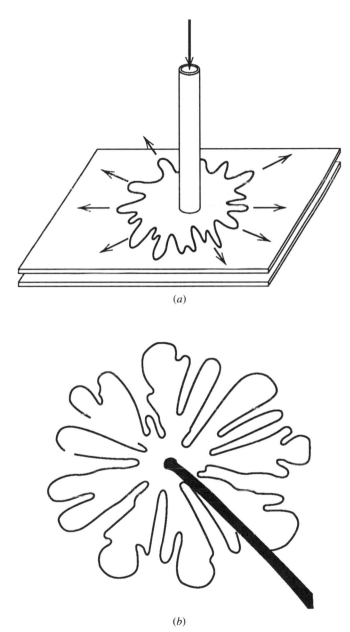

(a)

(b)

Figure 7-41   (a) Sketch of experiment for viscous fingering of a fluid in a Hele-Shaw cell. (b) Fractal-like fingering of a thin fluid layer under pressure in a Hele-Shaw cell. [From Homsy (1987).]

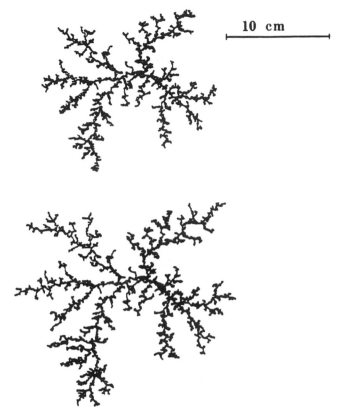

Figure 7-42 Fractal pattern from air displacing liquid epoxy in a glass sphere porous medium monolayer. (From Maloy et al. (1985).)

should be modeled by a dynamic process with underlying randomness. The questions which arise in this phenomena are not unlike those that are surfacing in the discussion of spatiotemporal chaos (see Chapter 8).

Fractal Materials and Surfaces

In classical continuum mechanics one takes for granted certain scaling relations such as area–length relations $A \sim L^2$ or mass–length behavior $M \sim L^3$. However, there are certain classes of materials such as silica gels, polymers, or porous solids which at some length scales do not follow the classic mass–length relation but instead behave as

$$M \sim L^D$$

where $1 \leq D \leq 3$ for chain-like molecular structures, or $2 \leq D \leq 3$ for plate-like structures. The value of the fractal dimension, D, for different materials has been measured in a number of laboratories (e.g., see Schaefer and Keefer, 1984a,b, 1986) using the scattering of light, x-rays, or neutrons.

In the classical theory of linear wave scattering from a geometric object (e.g., a sphere or cylinder), one usually assumes an incident plane wave field $u_I = Ae^{i(\omega t - kx)}$ and then adds a scattered wave field $u_s = e^{i\omega t}S(\mathbf{r}; k)$, where $\mathbf{r}$ is the position vector from the center of the scatterer. For uncorrelated wave sources (e.g., nonlaser light), one can then calculate the energy scattered out of the incident wave field by the scatterer. One of the first such calculations was that by Lord Rayleigh, who, in 1871, showed that the scattering cross section σ for wavelengths $2\pi/k \gg L$ (L is the size of the scatterer) satisfied a power law:

$$\sigma \sim (kL)^{-4}$$

Using a technique called *small-angle x-ray scattering* (SAXS), it has been found that for fractal objects the scattering intensity I depends on a noninteger power law (Schaefer and Keefer, 1984a,b), that is

$$I \sim \kappa^{-x}$$

where $\kappa = (4\pi/\lambda)\sin(\theta/2)$, λ is the wavelength of the incident wave, θ is the scattering angle, and x is called the *Porod exponent*. For mass fractals where $M \sim L^D$, the scattering exponent x equals D. For porous materials, x is found to be a function of the porosity.

Fracture of Solids. If turbulence remains one of the unsolved problems in the physics of fluids, then fracture and fatigue are its counterpart in the physics of solids. However, unlike turbulence, the process of fracture has not received much attention from the new dynamicists except for a few examples. One of these is a study by Mandelbrot et al. (1984) in which fractured surfaces in steel were characterized using fractal measures (see also Feder, 1988). In this experimental work, the fractured surface was coated with a nickel layer and then polished to expose islands of steel surrounded by a nickel sea. In the characterization of the surface, the perimeter of these islands and their area were assumed to be related by a noninteger exponent. Using a series of maraging steel specimens, each heat-treated at different temperatures and fractured by a short time impact force, they found a distribution of fractal dimensions of the surface from 2.10 to 2.28. From these data they found that the impact energy required to fracture the steel was inversely proportional to this fractal dimension of the surface. Explanation of this result in terms of a dynamical model or molecular physics is still wanting.

In a more recent study of dynamic crack propagation in solids by a group at the University of Texas, Fineberg et al. (1991) have observed the dynamics of crack propagation in the brittle plastic polymethyl methacrylate (PMMA). These cracks were observed to propagate at speeds of up to 600 m/s. The computer visualization of surface profilometer data reveals a complex fractal-looking surface, whereas the crack velocity time history shows an erratic chaotic-looking behavior. Such studies may lead to nonlinear dynamic models that may give clues to the often unpredictable dynamics of fracture and fatigue.

Not all fractured surfaces are created under dynamic impact. When a metallic structure is taken through a stress cycle many times (10^2–10^7 cycles), small flaws in the material can sometimes develop first into microcracks and then link up into a catastrophic failure. In these problems, inertial effects are often not very large. However, one may be able to view the advance of a crack as an iterated dynamic map. For example, suppose a crack in a thin plate is described by a position vector $\mathbf{r}_n = (X_n, Y_n)$, where X_n, Y_n are the Cartesian components at the end of the nth stress cycle. While X_n, Y_n may grow without bounds, we can assume that the incremental crack advance displacements are bounded; that is, suppose we define $X_n = X_{n-1} + u_n$, $Y_n = Y_{n-1} + v_n$. Then it might be possible that a function exists which relates (u_n, v_n) to (u_{n+1}, v_{n+1}), that is,

$$u_{n+1} = F(u_n, v_n)$$
$$v_{n+1} = G(u_n, v_n)$$

Of course, this is just speculation at this time. But, careful observation of crack tip advance often exhibits an unpredictable time history. It will remain for future research to see if such models can be found. The fractal nature of fractured surfaces gives credence to the belief in an underlying nonlinear dynamic model [see also Lung (1986), Markworth et al. (1988), and Russel et al. (1991)].

PROBLEMS

7-1 Consider the construction of a Cantor set that starts with a uniformly dense distribution of points on a line and begins by throwing out the middle β percent of the set. Iteration of this process results in a Cantor-type fractal set of points. Show that the box-counting or capacity fractal dimension is given by

$$d = \frac{\ln 2}{\ln [2/(1 - \beta)]}$$

7-2 Define an "iterated function set" of linear transformations that will replicate the Cantor set in Problem 7-1 (see Section 7.3 or Barnsley, 1988).

7-3 Define a fractal-creating operation that takes a line element of length L and replaces it by eight equal segments of length $L/4$ as shown in the figure below. Draw at least four iterations on a large piece of graph paper. Use the four sides of a square as the initial line elements.

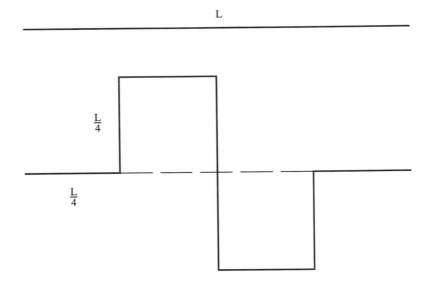

7-4 Consider a fractal-creating operation that starts with a cubic element of side L and removes elements that drill three mutually perpendicular square-shaped holes of side $L/3$. Sketch two iterations in an isometric drawing.

7-5 *Fractal Sponge.* Consider a three-dimensional cube whose lengths are divided into thirds, thus creating a set of 27 subcubes. Imagine a laser cutting or chemical etching process that eliminates the central cubes on each of the faces as well as the central cube (i.e., drill three mutually orthogonal square holes).

Show that iteration of this process will lead to a three-dimensional fractal object with box-counting dimension $d = \log 20/\log 3$. [This set is attributed to K. Menger 1926. See Peitgen et al. (1992) for a discussion of this problem.]

7-6 *Sierpinski Carpet.* In this construction of a two-dimensional fractal set, one starts with a square which is divided into nine equal squares. Then the central square is removed, leaving eight. One repeats this algorithm, dividing each of the eight into nine pieces and removing the central square. Sketch several iterations of this process on a large piece of graph paper. Show that the box-counting or fractal dimension is $D = \log 8/\log 3$.

7-7 *Two-Scale Cantor Set.* One can generate a fractal set with multiple scaling rules. As an example, consider a more generalized Cantor set. Begin with a line of length 1 and replace it with two sets of lengths s_1, s_2, where $s_1 + s_2 < 1$. Then one can show that the fractal dimension D is given by (e.g., see Feder, 1988)

$$s_1^D + s_2^D = 1$$

Choose $s_1 = \frac{1}{4}$, $s_2 = \frac{1}{2}$. Sketch several iterations of this fractal generator. Show that the dimension is given by $D = \log x/\log (1/2)$, where x satisfies $x^2 + x - 1 = 0$ and $D = 0.694$.

7-8 *Fractal Coastlines.* A now classic example of a physical manifestation of fractal geometry is the measurement of the length of coastlines on the Earth as described by Mandelbrot (1977) in his earlier book. The method is described briefly in this book in Section 7.7 [see Eqs. (7-7.6) and (7-7.7)]. Find a very good atlas or book of photographs of the Earth from space and choose a fractal-looking coastline and measure its fractal dimension using the caliper method described in Section 7.7. Plot the length between two points as a function of caliper length and find D. Good choices are Norway, Great Britain, or the west coast of British Columbia, Canada. The results of the first two cases can be compared with those in Feder (1988) (e.g., for Norway, $D = 1.52$).

7-9 *Fractal Contour Maps.* The same technique for coastlines can be applied to contours of the same elevation above sea level. Again, choose a good relief map with fine detail in the contours and plot length versus caliper size to calculate the fractal dimension using Eq. (7-7.7). For examples, try the southern relief contours of the province of Quebec, Canada at elevations of 300–500 meters. Another example is the western relief contour of the Sierra Nevada in California at 600–1500 meters.

7-10 In Chapter 6 we saw that the prediction of chaos or the calculation of Lyapunov exponents for specific physical systems was limited to a few examples. Likewise, the tools for calculation of the fractal dimension of a strange attractor for a specific physical dynamical system are not available at this time. However, one example where one can make an educated estimate of the fractal dimension is the particle is a two-well potential force with damping and periodic forcing (7-7.2). This estimate is based on the use of the relation between Lyapunov exponents and fractal dimension (7-2.14), $d_L = 1 - \lambda_1/\lambda_2$. To estimate λ_1, λ_2, use the linearized equation (7-7.2) near the origin, $\ddot{x} + \gamma\dot{x} - x = f\cos\omega t$, to obtain estimates of the Lyapunov exponents,

$$\lambda_{1,2} = \pm \left[\left(1 + \frac{\gamma^2}{4}\right)^{1/2} \pm \frac{\gamma}{2} \right] t_p$$

where t_p is the Poincaré map time increment, $t_p = 2\pi/\omega$. Derive an expression for $d_L(\gamma)$. Show that as $\gamma \to 0$, $d_L \to 2$ and that as $\gamma \to \infty$, $d_L \to 1$. Compare with results in Figure 7-17 (see also Moon and Li, 1985a).

7-11 Run a 2-D Barnsley map (7-3.5) for the Sierpinski triangle

$$\mathbf{x}_{n+1} = W\mathbf{x}_n$$

by adding a quadratic term to one of the maps in W and see if you get folding of the triangle. Run the map with a random choice of the next map in the set W.

7-12 Barnsley (1988) has many beautiful pictures in his book of a fractal fern-like object. Use the 2-D map (7-3.5) and the following matrices to generate one of these ferns. Choose the next ω_i in (7-3.5) using a random number.

	A	B	C	D	E	F
ω_1	0	0	0	0.2	0	0
ω_2	0.8	−0.04	0.04	0.8	0	2
ω_3	0.2	−0.25	0.2	0.2	0	2
ω_4	−0.15	0.3	0.25	0.25	0	0.4

8

SPATIOTEMPORAL CHAOS

Down beneath the spray, down beneath the whitecaps, that beat them-
selves to pieces against the prow, there were jet-black invisible waves,
twisting and coiling their bodies. They kept repeating their patternless
movements, concealing their incoherent and perilous whims.

Yukio Mishima
The Sound of Waves

8.1 INTRODUCTION

If the histroy of events is written in time, then the history of time is
often written in space: contrails behind a jet, the wake of a passing
boat, fissures after an earthquake, tracks in the sand from a snake or
a worm. These are common experiences of spatial patterns that record
some dynamic events. The phonograph cylinder of Edison is an obvious example of how dynamic data can be stored in spatial patterns
and, of course, in the stroke of an artist's brush. If temporal patterns
are regular and periodic we should expect to see regular spatial patterns. However, if temporal events are chaotic, how does this manifest
itself in space? In some physical systems, all the particles are spatially
coherent even if they behave chaotically in time, whereas in the case
of fluid turbulence, one has both spatial and temporal complexity.

Until recently (circa 1987), most of the research on chaos was
confined to temporal dynamics. In fact, all the previous discussion in
this book has been about temporal dynamics only. But for physical
systems, described by the partial differential equations of physics, one
must deal simultaneously with both space and time. In fact, the lack

of any discussion of spatial patterns so far means that we have made implicit assumptions about the spatial or modal distribution in the physical phenomenon.

The study of spatiotemporal chaotic dynamics is still in the exploratory stage. It has not generated the kind of generic tools and results that can be applied to different physical problems in the same way that temporal nonlinear dynamics can. The field spans a wide range of physical problems, ranging from surface waves in a stationary fluid, electrohydrodynamics instabilities in liquid crystals, and solid-state plasmas in Ge crystals to complex twists and knots in an elastic tape or yarn. A simple experiment in spatiotemporal chaos may be performed by pointing a video camera at the video display terminal (TV) it is connected to (see e.g., Crutchfield, 1988 and Peitgen et al., 1992, p21–27). And, of course, the mother of all spatiotemporal chaos is the fully developed turbulence we are familiar with in everything from weather patterns to flows through jet engines. One of the strategies of many physical scientists with regard to the problem of turbulence, however, is not to tackle it head on (although jet engine designers must deal directly with the problem), but to study the transition from low-dimensional dynamics to high-dimensional phase space behavior by looking at the development of increasingly complex spatial patterns.

At present a clear definition of spatial "chaos" or complexity is not universally accepted. For some it represents an increase in the fractal dimension of the dynamic attractor associated with increasing number of coupled spatial modal functions. For others the measure involves a loss of spatial correlation or an increase in spatial entropy (e.g., see Kaneko, 1990 and Dowell and Virgin, 1990).

An example of spatiotemporal complexity can be seen in the generation of surface wave patterns in a fluid excited by harmonic excitation. This problem, which was described earlier in Chapter 6, has been studied by Gollub and Ramshankar (1990) and involves a shallow layer of fluid in a container under vertical excitation. The problem goes back to Michael Faraday in 1831 (see Gollub and Ramshankar, 1991 for a review). The phenomenon can sometimes be observed in a coffee cup when placed on a vibrating surface. For high enough frequencies, short wavelength wave patterns appear. For certain excitation amplitude and frequency parameters the wave pattern can be regular (Figure 8-1), but for other parameters the patterns can become more complex and can change in time. The surface wave patterns appear to suffer defects similar to those found in solid crystals such as disclinations and dislocations. One of the principal questions that scientists want

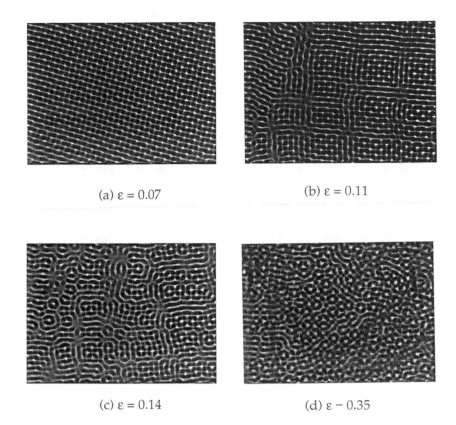

(a) $\varepsilon = 0.07$ (b) $\varepsilon = 0.11$

(c) $\varepsilon = 0.14$ (d) $\varepsilon - 0.35$

Figure 8-1 Photographs of surface wave patterns due to vibration of a fluid in a circular container. (a) $\varepsilon = 0.07$; (b) $\varepsilon = 0.11$; (c) $\varepsilon = 0.14$; (d) 0.35. [From Gollub and Ramshankar (1991). ε is the relative forcing level beyond the flat surface instability.]

to answer in these spatiotemporal studies is how different patterns are selected when many are possible and how they evolve in time.

Similar patterns can be seen in convective flow patterns in a shallow fluid layer heated from below. Another example is a thin layer of nematic liquid crystal with an applied electric field (Ranberg et al., 1989).

Many other experiments are beginning to appear that attempt to quantify spatiotemporal dynamics. For example, an Italian group has studied spatial patterns in a Rayleigh–Benard convection cell using a laser scanning measurement system (Rubio et al., 1989). They observe two types of spatiotemporal regimes representing localized oscillations and traveling wave-type patterns. (See also Figure 1-1.)

In another work, Tam and Swinney (1990) have investigated spatiotemporal patterns in a reaction diffusing system. The system is based on the Belousov–Zhabotinsky reaction in a Couette cylindrically shaped reactor.

Spatial Chaos—The Wave Guide Paradigm

As a simple example of how spatial chaos can arise naturally in physical systems, we consider the model in Figure 8-2a. Here a nonlinear oscillator is connected to a linear nondispersive semi-infinite wave guide such as a taut string or an electrical transmission guide. If the oscillator is weakly connected to the wave guide, then chaotic motions of the oscillator in time $W(t)$ will be spatially stored in the wave guide in the form of linear right, running waves $u(x, t) = f(x - c_0t)$. Boundary conditions between the oscillator and the wave guide can result in the temporal history stored as information in the spatial wave

$$u(x, t) = A \cdot W\left(t - \frac{x}{c_0}\right) \qquad (8\text{-}1.1)$$

Spatial Chaos—The Edison Phonograph Model

Thomas Edison invented a device that stored information on the surface of a cylinder in response to the motion of a needle. By moving the cylinder in the axial direction, a spatial record of the temporal history of the oscillation needle could be recorded. In many mechanical systems a similar mechanism is involved such as the cutting of metal from a cylindrical workpiece on a machine lathe. As we have seen from the work of Grabec (1988) (see Chapter 4), the vibration of the tool can leave a record in the machined surface.

Another example involves roller bearings. This model, however, involves "writing over" the past deformation history on the surface of the bearing (Figure 8-2b). Thus, one can imagine that chaotic motions in the device attached to the bearing can be recorded in the spatial deformations on the surface of the bearing due to plasticity effects.

Aside from physical experiments in *continuous* media, another strategy has been to look at *discrete* mathematical and computational models with a large number of coupled cells. Geometrically, these take the form of either (a) periodic chains of identical oscillators or one-dimensional maps or (b) two- and three-dimensional lattice structures. The models also range from those with discretized space,

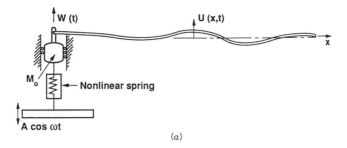

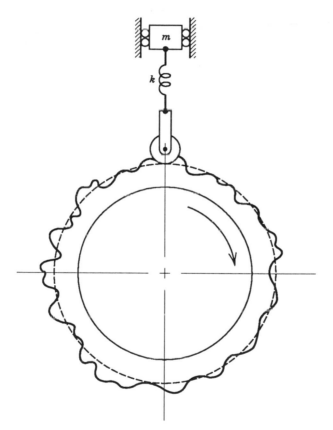

Figure 8-2 (a) The "wave guide" model for storing temporal chaotic dynamics in a spatial pattern. (b) The "write over" model for creating complex spatial patterns from a temporally chaotic system.

time, and state variables (cellular automata) to coupled cell maps with discretized space and time, to coupled differential equations with only space discretized. These models are discussed in Section 8.3.

8.2 SPATIAL COMPLEXITY IN STATIC SYSTEMS

The Twisted Elastica and the Spinning Top

The central idea of this section is that spatially complex patterns can be found in familiar systems in static equilibrium. Our example focuses on a long, thin, flexible, tapelike continuum called in the classical literature the *elastica* (e.g., see Love, 1922). Such stringlike objects are familiar as audio, film, or videotape and as yarn, wire, or measuring tape. Spatial complexity is often familiar in such objects in the form of despooled film tape or in the form of twisted yarn or fishing line. Macromolecular structures may also exhibit such complexity. Such complex static patterns in space may have exact analogs in the chaotic dynamic orbits of a top or pendulum in phase space.

The analogy between the temporal dynamics of a rotating body and the static deformations of a long, thin elastic body goes back more than a century to Kirchhoff (1859) (see also Love, 1922). The simplest version of this analogy is that between a pendulum and a buckled rod. The equations of motion of a planar pendulum (Figure 8-3), written in terms of the angular displacement θ and the angular momentum L, are given by

$$J\frac{d\theta}{dt} = L$$

$$\frac{dL}{dt} = -r_c mg \sin \theta$$

(8-2.1)

where J is the moment of inertia about the point of rotation, r_c is the distance to the center of mass, m is the mass, and g is the gravity constant. The equations of static equilibrium for a buckled elastic rod, written in terms of the slope angle θ and bending moment M, have the same precise form as those of the pendulum, that is,

$$D\frac{d\theta}{ds} = M$$

$$\frac{dM}{ds} = -P \sin \theta$$

(8-2.2)

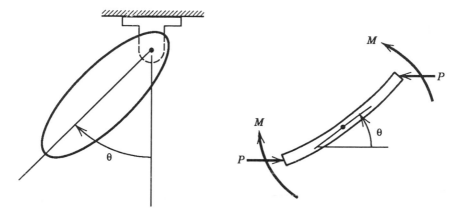

Figure 8-3 Analogy between the temporal dynamics of the pendulum and the spatial deformation of the elastica.

where D is the bending modulus and P is the compressive end load on the rod. If all the parameters in either (8-2.1) or (8-2.2) are constant in time or space, then the solutions can be found in terms of elliptic integrals (Love, 1922), and no chaos exists as in Figure 5-2. However, it is known that a pendulum under external or parametric time periodic forcing may exhibit chaotic dynamics (Koch and Levey, 1985).

This suggests that if the bending modulus D in (8-2.2) were to vary periodically in space (e.g., $D = D_0 + D_1 \cos ks$), then spatially chaotic equilibrium solutions may be found for the buckled elastica.

This idea has been studied by several authors since the first edition of this book, including Mielke and Holmes (1988), Thompson and Virgin (1988), and El Naschie (1990), and El Naschie and Kapitaniak (1990). The first authors presented a detailed mathematical study, whereas the second and third authors presented numerical and qualitative experimental evidence for spatial chaos in the buckled elastica.

The extension of this analogy to a rigid body spinning in three dimensions and a thin elastic tape twisted in space is straightforward and may be found in the classic text on elasticity by Love (1922).

To sketch the analogy we note that the angular momentum of a rigid body referred to its center of mass may be written in terms of the components of its angular velocity vector $\boldsymbol{\omega} = (\omega_1, \omega_2, \omega_3)$;

$$\mathbf{L} = J_1\omega_1\mathbf{e}_1 + J_2\omega_2\mathbf{e}_2 + J_3\omega_3\mathbf{e}_3 \qquad (8\text{-}2.3)$$

where $\{\mathbf{e}_1, \mathbf{e}_2, \mathbf{e}_3\}$ are an orthogonal triad of principal axes of the inertial matrix and $\{J_1, J_2, J_3\}$ are three principal inertias. Under applied moments (which may vary in time) $\{M_1, M_2, M_3\}$, the equations of motion take the form

$$\frac{d}{dt}(J_1\omega_1) = (J_2 - J_3)\omega_2\omega_3 + M_1$$

$$\frac{d}{dt}(J_2\omega_2) = (J_3 - J_1)\omega_1\omega_3 + M_2 \qquad (8\text{-}2.4)$$

$$\frac{d}{dt}(J_3\omega_3) = (J_1 - J_2)\omega_1\omega_2 + M_3$$

These equations are called *Euler's equations (e.g., see Goldstein, 1988). When the J_i are constants* and $M_i = 0$, the solution is known in terms of three constants of the motion. Two of these are the kinetic energy and the angular momentum. However, if either M_i is periodic in time or one of the principal inertias J_i varies in time (e.g., $J_2 = J_0 + A \cos \theta t$), then chaotic motions are possible as in the suspected tumbling of one of the moon's of Jupiter, Hyperion (see Chapter 4).

The analogous equations for the spatial deformation of a long, thin elastic rod or tape (Figure 8-4) are governed by equations for the internal bending moment, **G**, produced by bending stresses on the cross section and its relation to the curvatures of the centerline of the rod (κ_1, κ_2, τ). The curvatures (κ_1, κ_2), as one recalls from analytic

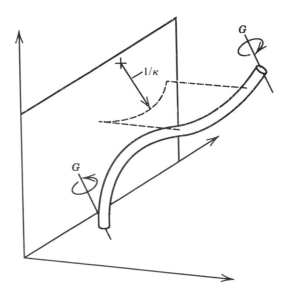

Figure 8-4   Three-dimensional deformation patterns in elastica-type structure.

geometry, are inversely proportional to the radius of bending, while the torsion τ is a measure of the twist about the centerline of the rod.

In analogy to the spinning top, the bending moment is written in components of the three principal geometric axes of the cross section, one of which lies along the centerline $\{e_1, e_2, e_3\}$, that is,

$$\mathbf{G} = A(\kappa_1 - \kappa_{10})\mathbf{e}_1 + B(\kappa_2 - \kappa_{20})\mathbf{e}_2 + C(\tau - \tau_0)\mathbf{e}_3 \quad (8\text{-}2.5)$$

where A, B are bending moduli, C is the torsion modulus, and $(\kappa_{10}, \kappa_{20}, \tau_0)$ are the initial curvatures when the elastica is moment-free. The resulting equations of equilibrium are simplified under the assumption of zero net force at each cross section:

$$\frac{d}{ds} A(\kappa_1 - \kappa_{10}) = (B - C)\kappa_2\tau$$

$$\frac{d}{ds} B(\kappa_2 - \kappa_{20}) = (C - A)\kappa_1\tau \quad (8\text{-}2.6)$$

$$\frac{d}{ds} C(\tau - \tau_0) = (A - B)\kappa_1\kappa_2$$

One can see that these equations have the same structure as those of the spinning top. Thus, several possibilities for spatial chaos follow the analogy to (8-2.4).

First, one of the bending moduli could be periodic in space, that is, $B = B_0 + b \cos ks$. Or, one could have an initial periodically varying curvature, that is, $\kappa_{20} = \kappa_0 \cos ks$. In either case, $2\pi/k$ is the wavelength of the disturbance and is assumed to be larger than the geometric scale of the cross section.

In order to get a description of deformation in space, however, one must add to these equations the so-called Frenet–Seret equations of differential geometry:

$$\frac{d\mathbf{e}_i}{ds} = \mathbf{\Omega} \times \mathbf{e}_i, \qquad \mathbf{\Omega} = (\kappa_1, \kappa_2, \tau) \quad (8\text{-}2.7)$$

Finally, to get the position of the centerline of the elastica $\mathbf{r}$ in its deformed shape, another relation is required:

$$\frac{d\mathbf{r}}{ds} = \mathbf{e}_3 \quad (8\text{-}2.8)$$

With a periodic disturbance in space, these equations constitute a four-dimensional phase-space system $(\kappa_1, \kappa_2, \tau, ks)$. Thus, it is natural to use a Poincaré section synchronized with the spatial disturbance $ks_i = 2\pi i + \phi_0$. The resulting three-dimensional map is still difficult to visualize. However, this system is identical to that of a Hamiltonian dynamical system. In our case the conserved quantity is the moment vector, **G** along the rod. This relation can be used to eliminate one of the variables in the Poincaré map, so that a two-dimensional map is possible.

An example of the numerical integration of these equations is shown in Figures 8-5 and 8-6 (see Davies and Moon, 1992). Figure 8-5 shows a Poincaré map where several different solutions are possible for the same bending moment. One can see that both a quasiperiodic and a chaotic spatial solution (diffuse set of points) can exist. This case is for a spatially varying bending modulus. Spatial twisting deformations of the elastica are shown in Figure 8-6 and in the color plates (CP-12–14). These show that incredible spatial complexity is possible in the elastica.

One point should be noted here. These numerical solutions require some attention to computational errors that can arise. Namely, one has to adjust the integration step size in order to keep the bending moment **G** · **G** constant along the rod.

To complete the analogy with the spinning top, one should imagine

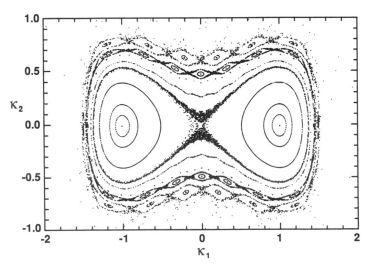

Figure 8-5 Spatial Poincaré map of the deformation of a long elastica with spatially periodic change in cross section.

Figure 8-6 Spatial complex twisting deformations of the elastica.

oneself riding on a small cart that travels on the twisted tape with constant velocity. Then the chaotic rotations of the cart will be precisely those of a spinning top when one of its inertias varies in time.

These chaotic solutions are thought to be related to the homoclinic tangling of stable and unstable manifolds that emanate form saddle points of the Poincaré maps (e.g., see Mielke and Holmes, 1988; also see Chapter 6). Again, as we have tried to emphasize from the beginning in Chapter 1, evolutionary laws that create horseshoes in the phase space seem to be one of the principal mechanisms for creating spatial as well as temporal chaos in physical systems. Another example of the role of horseshoe maps and spatial chaos is discussed in Section 8.4 (see subsection entitled "Chaotic Mixing of Fluids").

8.3 COUPLED CELL MODELS

A classic model of approximating a spatially continuous medium is the use of coupled cells or lattice models (e.g., see Brillouin, 1964).

In inertial models a periodic array of masses is assumed to interact with neighboring masses and one constructs an infinite set of coupled ordinary differential equations. In linear models one derives a relationship between frequency and wavelength, the so-called dispersion relations, and a superposition principal is valid.

Nonlinear coupled lattice models have been used to describe such nonlinear phenomena as solitons and shock waves (e.g., see Tasi, 1990 or Toda, 1989), and as models for nonlinear electrical transmission lines. Interest has recently been revived in such models to explore spatiotemporal chaos.

There are three basic mathematical types of coupled cell models:

(1) Cellular automata (Wolfram, 1984, 1986)
(2) Coupled maps (Crutchfield and Kaneko, 1987)
(3) Coupled differential equations (Umberger et al., 1989; Moon et al., 1991)

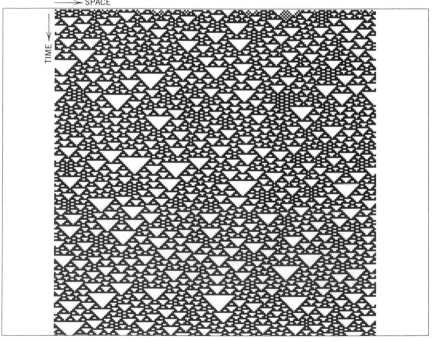

Figure 8-7 Coupled cell model with discretized space and time variables and a state space of a finite set of symbols (0, 1) using the rule following (8-3.1) and randomly chosen initial conditions.

Coupled Automata

This model is depicted in Figure 8-7, for a one-dimensional chain of cells with nearest neighbor interaction. Space and time are discretized, as is the state variable. In fact, the state variable is only allowed to take on a finite set of states represented by a set of symbols.

For example, in a binary symbol set, the state variable could take on either 0 or 1, black or white, L or R. To effect a dynamic on the lattice, a rule must be chosen governing the new state variable at time $n + 1$ at lattice site α, in terms of the old state variable at time n. We can see at once that two sets of integers are needed for space and time. A nearest neighbor law might take the form

$$A_{n+1}^{\alpha} = F(A_n^{\alpha}, A_n^{\alpha+1}, A_n^{\alpha-1}) \qquad (8\text{-}3.1)$$

For example, for a binary symbol pair (R, L) one could adopt a rule given in the following table called the 122 rule in Wolfram (1986).

	nth state		
$\alpha - 1$	α	$\alpha + 1$	$n + 1$
R	R	R	R
L	R	R	L
R	R	L	L
L	R	L	L
L	L	L	R
R	L	L	L
L	L	R	L
R	L	R	R

Wolfram (1986) and others have shown that such symbol dynamics on a lattice can exhibit very complex patterns and behavior as illustrated in Figure 8-7. For example, a spatially simple initial row of symbols can generate a chaotic-looking pattern. Or a randomly chosen initial row can generate a regular pattern after several iterations. While simple in concept, the major drawback to these models is a lack of rigorous connection between physical laws in the continuum one is trying to model (such as a turbulent fluid) and the symbol rule that generates the iterated coupled symbol map.

Coupled Maps

In this model, one discretizes space and time, but allows the state variable to take on continuous values:

$$x_{n+1}^{\alpha} = F(x_n^{\alpha}, x_n^{\alpha+1}, x_n^{\alpha-1}) \tag{8-3.2}$$

For example, a popular model is to assume that each cell is governed by a simple one-dimensional map such as the logistic map with coupling to nearest neighbors:

$$x_{n+1}^{\alpha} = (1 - G)f(x_n^{\alpha}) + \frac{\varepsilon}{z}[f(x_n^{\alpha+1}) + f(x_n^{\alpha-1})] \tag{8-3.3}$$

where $f(x) = 1 - ax^2$. This so-called coupled map lattice law has a diffusive interaction between cells (Kaneko, 1989). Using such models one is able to define Lyapunov exponents, entropy measures, correlation functions, mutual information, and other thermodynamic quantifies of spatiotemporal chaos (Kaneko, 1989).

Coupled Differential Equations

These models are identical to the classical studies of Brillouin (1946) or Toda (1981), but new tools have been used to try to characterize spatiotemporal dynamics, especially chaotic-looking spatial patterns. One such study is the work of the group at the University of Maryland (Umberger et al., 1989). The model is shown in Figure 8-8 and is similar to the earthquake model of Carlson and Langer (1989) in Chapter 4. The Maryland group uses a chain of Duffing oscillators, similar to a set of buckled beams, or a set of two-well potential oscillators,

$$\ddot{x}_{\alpha} = -\gamma\dot{x}_{\alpha} + \frac{\delta}{z}x_{\alpha}[a - x_{\alpha}^2] + f\cos\omega t + \varepsilon D[x_{\alpha}] \tag{8-3.4}$$

where the coupling operator is defined as

$$D[x_{\alpha}] = x_{\alpha+1} - 2x_{\alpha} + x_{\alpha-1}$$

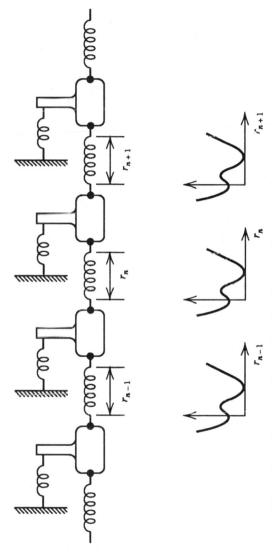

Figure 8-8 Coupled cell model with discretized space and continuous time and state variables.

(Note we are using the subscript α to denote spatial position in the chain.)

Chain of Toda Oscillators

Another example of a chain of coupled oscillators with a nonlinear force interaction between masses is a model used to describe anharmonic intermolecular forces in a crystal lattice (Toda, 1981). This model has received both numerical and analytical study by Geist and Lauterborn (1988). The equations take the form

$$\dot{d}_i = v_i - v_{i+1}$$
$$M\dot{v}_i = \exp(d_{i-1}) - \exp(d_i) + \gamma[v_{i+1} - 2v_i + v_{i-1}] + F_i(t)$$

$$(8\text{-}3.5)$$

where $F_i = 0$ for $i \neq i_0$ and $F_{i_0} = A \sin \omega_0 t$. In this model, d_i represents the distance between neighboring masses, v_i is the velocity of the ith mass, and γ is a damping constant. A periodic force is applied to one of the masses in the chain. Numerical integration of these equations for 15 masses is shown in Figure 8-9 from the paper of Geist and Lauterborn (1988). The spatial complexity in Figure 8-9 is clearly evident.

Experiments on Coupled Lattice Chaos

A few experiments are beginning to emerge using coupled cell lattices. In electrical circuits one should look at the papers of Purwins et al. (1987, 1988). Another two-dimensional lattice based on cellular neural networks has been studied by Chua and Yang (1988).

In the following we describe a mechanical experiment based on coupled masses on a taut string (Moon et al., 1991). The physical problem is shown in Figure 8-10. Eight small aluminum spheres sit on a string under tension. The string is fixed at one end and periodically excited at the other end with an electromagnetic shaker. The nonlinearity in this case is represented by an amplitude constraint, so that if the masses exceed a certain amplitude they will impact with a fairly rigid wall. This problem is not entirely academic, because in high-speed printers a chain of masses with typeface characters is moved across the paper.

The unique character of this experiment is the signal processing. The format of the data output was designed to provide a link to cellular automata. That is, if the kth mass does not hit the constraint within a certain time period, a 0 would be stored in the kth register, whereas

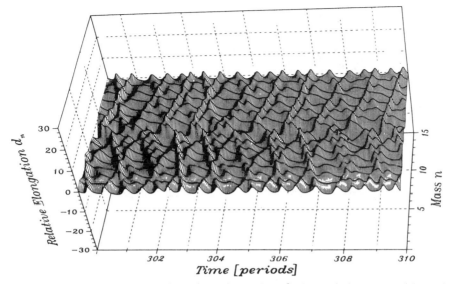

Figure 8-9 Numerical integration of the dynamics of 15 coupled masses with Toda potential forces (8-3.5). [From Geist and Lauterborn (1988).]

if it hits the wall a 1 would be stored. The time interval chosen was the first quarter cycle of each forcing cycle. Thus, this represented a finite Poincaré window.

Finally, eight masses were chosen so that the eight symbols of 0's or 1's could be coded into a binary number. Thus, at the end of each Poincaré time window, a binary number S_n from 0 to 255 would code the spatial impact pattern. In this way a large number of statistical data could be obtained for both space and time.

Spatial Return Maps

One of the features of this experiment was the use of a spatial pattern return map. After each Poincaré window the spatial pattern number

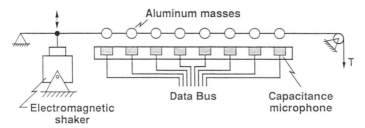

Figure 8-10 Sketch of experiment with eight coupled masses on a string and an amplitude constraint.

S_n was plotted versus the previous pattern number S_{n-1}. In this way one could track the changes in the impact pattern as one varied, say, the forcing amplitude or frequency.

The results of these experiments are shown in Figures 8-11 and 8-12. Figure 8-11a shows the name of the bead that hits. Figure 8-11b shows the pattern history. The next figure (Figure 8-12) shows the pattern number return map. For example, for only two masses hitting, one can see that there are two different spatial patterns which alternate. For more masses hitting, one can also see that many spatial patterns are involved.

Along with the measurements, a numerical simulation was performed (Figure 8-13) (see Moon et al., 1991). Using the numerical data an entropy measure was used (e.g., see Crutchfield and Kaneko, 1987). The entropy is based on probability measures $\{P_i\}$, that is,

$$S = -\sum P_i \log P_i \qquad (8\text{-}3.6)$$

where P_i measures the probability that one of the 256 configuration patterns would occur. In the numerical experiment, 4300 cycles of data were taken and the first 600 were discarded. Of the remaining 3700 cycles, 330 groups of 400 cycles were used to calculate P_i, each averaged over 330 sets of spatiotemporal data.

The resulting entropy was then plotted as a function of the mass-wall gap to driving amplitude ratio (Figure 8-13). One can see that entropy increases as the gap is made smaller, which seems to measure the increase in complexity of the spatiotemporal impact patterns.

These experiments and others are still exploratory. We are still looking for a new "Feigenbaum number" that will relate these observations in one experiment to some universal mathematical model. In Kuhn's theory of scientific revolutions, we are still looking for the right spatiotemporal paradigm that will unite these otherwise disparate experiments.

8.4   LAGRANGIAN CHAOS

Chaotic Mixing of Fluids

In looking at the beautiful color pictures of fractal basin boundaries (see color plates), one is struck with the similarity to mixing of paints of different colors. Whereas the bending and stretching formations that are responsible for temporal, fractal dynamics can only be seen

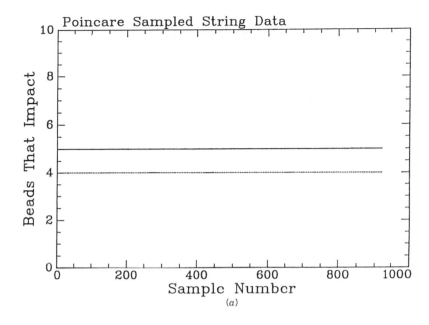

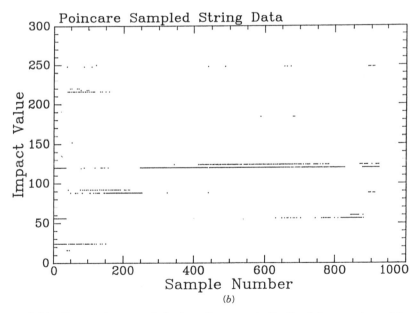

Figure 8-11 Space–time symbol plots for a periodically driven string with eight masses (Figure 8-10). (*a*) Two-bead impact. (*b*) Multiple-bead impact. [From Moon et al. (1991).]

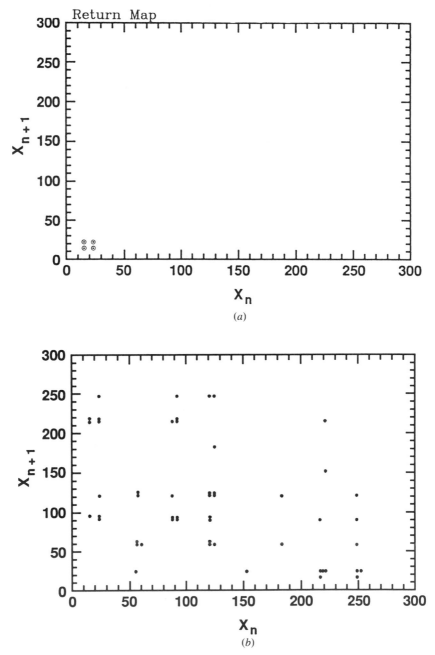

Figure 8-12   Spatial pattern number return map for the driven string with eight masses (Figures 8-10 and 8-11). (*a*) Periodic impact. (*b*) Chaotic impact. [From Moon et al. (1991).]

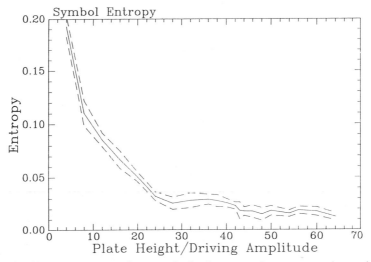

Figure 8-13 Entropy measure for complexity in space–time patterns in a string with eight masses (Figure 8-10). [From Moon et al. (1991).]

in abstract phase space, in mixing of fluids, the folding and stretching can be seen in physical space directly.

Flow patterns in fluids can be visualized in two ways. One can fix attention to one location $\mathbf{r}$ and then describe the velocity as it changes in time $\mathbf{v}(\mathbf{r}; t)$. Alternatively, one can fix one's eye on a single fluid particle and follow its position in space $\mathbf{r}(t)$, with velocity $\mathbf{V} = (\dot{x}, \dot{y}, \dot{z})$. The fixed spatial reference flow description is called *Eulerian*, whereas the particle-based description is called *Lagrangian*. In general, one obtains three equations relating the particle velocity to the spatial velocity functions:

$$\dot{x} = V_x(x, y, z, t)$$

$$\dot{y} = V_y(x, y, z, t) \qquad (8\text{-}4.1)$$

$$\dot{z} = V_z(x, y, z, t)$$

The similarities of these equations to the Lorenz equations (1-3.9) are clear, and chaotic trajectories are known to exist as can be seen in any turbulent flow. However, in most turbulent flows the spatial (Eulerian) patterns often change in time with as much complexity as the chaotic particle trajectories. What is remarkable, however, is that it is possible for the Eulerian velocity patterns to be regular (i.e., either stationary in 3-D or periodically time-varying in 2-D) whereas the individual particle trajectories are chaotic.

These sets of problems are important in chemical and other related technologies where mixing, stirring, or advection are important. Also, in fluid or gaseous combustion, mixing of fuel and oxygen is important (e.g., see Gouldin, 1987). In fact, these problems constitute examples where chaos is desirable. Modern studies of chaotic mixing using dynamical systems ideas have been done by Aref and Balachandar (1986), Ottino (1989a), and Chaiken et al. (1987), to name a few of the principal researchers. A very readable description can be found in an article by Ottino (1989b), and a review of chaotic advection of fluids may be found in Aref (1990).

To illustrate the basic ideas, we describe the phenomenon as it occurs in 2-D physical fluid flow (as contrasted with "flows" in phase space). In particular we consider the case of incompressible fluid (such as water or oil) where the divergence of the velocity field is zero, that is,

$$\nabla \cdot \mathbf{v} = 0 \tag{8-4.2}$$

In this case the velocity in two dimensions can be described by a scalar function called a *stream function,* $\psi(x, y, t)$:

$$\mathbf{v} = \nabla \times \psi \mathbf{e}_z \quad \text{or} \quad v_x = \frac{\partial \psi}{\partial y}, \quad v_y = -\frac{\partial \psi}{\partial x} \tag{8-4.3}$$

ψ is then found by solving the momentum equation of fluid mechanics. When ψ is independent of time, we have a stationary flow pattern in space. Equations (8-4.3) are then used to describe the particle paths, that is,

$$\dot{x} = \frac{\partial \psi(x, y, t)}{\partial y}$$
$$\dot{y} = -\frac{\partial \psi(x, y, t)}{\partial x} \tag{8-4.4}$$

These equations are precisely the same as those for a single particle with position $q \equiv x$ and momentum $p \equiv y$ and an energy or Hamiltonian function $H(q, p) \equiv \psi$. (See also Eq. (6-3.18).)

Of course, a steady 2-D flow cannot produce chaos, so one introduces a time disturbance by slowly varying the flow pattern periodi-

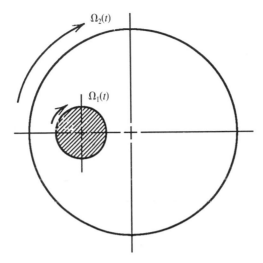

Figure 8-14 Sketch of geometry of Stokes flow experiment in a circular container.

cally in time, that is, $\psi(x, y, t + T) = \psi(x, y, t)$. One now has the analog of a periodically forced oscillator without dissipation. Two examples have received extensive study in the literature, namely Stokes' flow (Figure 8-14) and flow in a rectangular cavity (Figure 8-15).

Stokes Flow

In this problem, an incompressible, viscous fluid flows between two rotating cylinders whose centers are displaced (Figure 8-14). It can be shown that when the rotation rates of the cylinders are steady, then an exact solution can be found when viscous forces dominate the

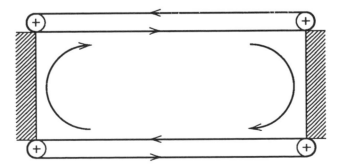

Figure 8-15 Sketch of geometry of fluid flow experiment in a rectangular cavity with moving walls. [From Leong and Ottino (1989).]

inertial forces. In these solutions, particles travel in closed orbits in the plane.

However, if one allows one of the cylinders to have a slow time-periodic change in rotation, then chaotic orbits of the particles can occur (see Color Plates 10, 11).

Analysis of this periodic cylinder rotation problem has been given by Aref and Balachandar (1986) and Chaiken et al. (1987). A beautiful experimental study of the Stokes problem chaotic mixing has been presented by Chaiken et al. (1986). [See also Tabor (1989, Section 4.8) for a description of this experiment.]

In either numerical or experimental studies, one can see the effect of folding and stretching in chaotic mixing. An example from the paper by Chaiken et al. (1987) for the Stokes flow problem is shown in Figure 8-16. An initially short straight segment of fluid particles is stretched and folded after several cycles of the periodic rotation of the inner cylinder.

Rectangular Cavity

A fine experimental study of fluid mixing in a rectangular cavity with moving walls has been reported by Leong and Ottino (1989) (see

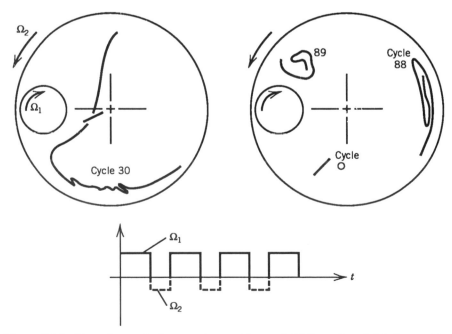

Figure 8-16   Experimental picture of stretching and folding of an initially straight segment of fluid particles in a Stokes flow field. [From Chaiken et al. (1987).]

Figure 8-15). In this study a viscous fluid is confined in a rectangular cavity in which two of the walls can move. If the velocity of the walls is steady, then the streamlines form time independent closed orbits. However, in the experiments the authors slowly varied periodically the wall velocity resulting in complex particle motions which exhibit the stretching and folding shown in the Stokes flow problem. In these experiments, one can tag an initial group of fluid particles with a different color. After several cycles of wall motions, one can observe complex folded patterns in the tracer particles as shown in Figure 8-17. A discussion of the role of symmetry and chaotic mixing for this problem is discussed in the paper by Franjione et al. (1989).

Three-Dimensional Problems and Turbulence

If the Eulerian flow field $V(x, y, z)$ has a three-dimensional character, then the particle trajectory can become chaotic without the need for time-periodic boundary conditions. One example of this is the flow in a twisted pipe, which has been studied by Jones et al. (1989). In this problem a viscous fluid flows in a circular pipe, whose centerline is bent in a semicircular arc. However, the next planar section of the pipe is twisted out of the plane relative to the previous section. Iterating this idea in space, these authors obtained a two-dimensional mapping of the fluid particle position as it passes from one circularly bent section to another. Iteration of this 2-D mapping then leads to stochastic or chaotic orbits of the fluid particle. A finite number of such twisted pipe segments may be useful as a practical device for mixing fluids.

It should be noted, however, that these examples of Langrangian

Figure 8-17 Complex mixing patterns for an oscillatory fluid flow process. [From Leong and Ottino (1989).]

fluid chaos constitute a limited class of fluid motions and cannot be understood as "solving" the general problem of fluid turbulence (Aref, 1990) which remains still a distant but slightly closer target.

Chaotic Mixing in Plastic Material

Mixing is usually associated with fluids. But one can also discover mixing problems in more solid materials. Solids are distinguished from fluids because they can resist shear. However, some solid materials will yield or flow when the shear stress exceeds some critical value. In the somewhat academic example discussed here, it is shown how chaotic trajectories of particles under alternating shear deformation in a plastic material can lead to spatial complexity that is similar to the folding and stretching processes in the mixing of fluids. (Such elasto–plastic mixing may have taken place in the earth's mantle over a geological time scale.)

Consider a sheet of plastic material in which we apply a body tangential torque distribution centered about point A in Figure 8-18 (Feeny et al., 1992). The body force is such that the shear stress is a constant value: $\tau_{r\theta} = \kappa$.

In the theory of plasticity the material will flow when the shear stress reaches a yield value τ_0. The strain rate $\dot{\gamma}$ is directed related to the shear stress so that $\dot{\gamma} = \lambda \tau_0$. It can also be shown that the strain

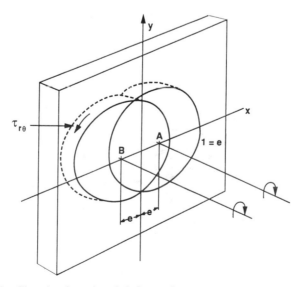

Figure 8-18 Sketch of torsional deformation mechanics of a thin plastic sheet.

rate is related to the velocity field so that, if v is the circumferential deformation rate, under circular symmetry

$$\dot{\gamma} = \frac{dv}{dr} - \frac{v}{r} = \lambda\tau_0 \qquad (8\text{-}4.5)$$

We assume that the shearing body force is applied within a radius normalized to unity, so that $v(r = 1) = 0$. Then the velocity field is given by

$$v(r) = \lambda\tau_0 r \ln r \qquad (8\text{-}4.6)$$

In terms of the angular deformation, one can then write

$$\theta = \tau \ln r, \qquad 0 \le r \le 1$$
$$\theta = 0, \qquad 1 < r < \infty \qquad (8\text{-}4.7)$$

where the parameter τ is a measure of both the yield stress and the time of application of the body torque.

To obtain a chaotic mixing deformation, we use an adaptation of the "blinking vortex" model of Aref (1984). We first apply the torque at point A and then apply it at B and so on, alternating the deformation process in a periodic way. This process leads to an iterated map

$$\theta_\alpha^{n+1} = \theta_\alpha^n + \tau \ln r_\alpha^n$$
$$r_\alpha^{n+1} = r_\alpha^n \qquad (8\text{-}4.8)$$

where $\alpha = $ A or B and the superscripts n, $n + 1$ indicate the value of the variable at the nth and $(n + 1)$st cycles. This map can be shown to be area-preserving.

In the fluid blinking vortex model, one has two co-rotating vortices in which one turns on while the other turns off in a periodic manner.

Three types of graphic results are presented. In the first we show what happens to a line element in the plastic sheet after 8 or 16 iterations of the cycles (Figure 8-19). It is clear that there are stretching and folding operations in the two-dimensional space which in temporal dynamics would lead to horseshoes and homoclinic tangles.

The second graphic looks at a few initial points in the plane after many iterations of the map (Figure 8-20). This is effectively a Poincaré map of the alternating plastic deformation process. Here we see struc-

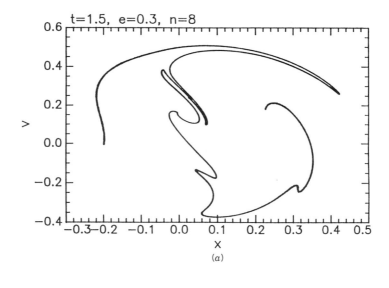

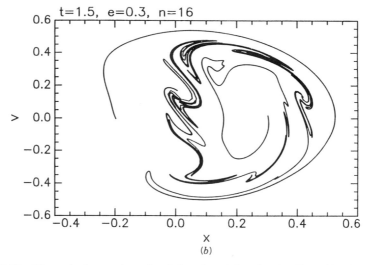

Figure 8-19   Numerical experiments of chaotic mixing due to alternating torques on a plastic material showing the deformation of a line element after 8 and 16 iterations (8-4.8). [From Feeny et al. (1992).]

tures that remind one of Hamiltonian temporal dynamics with quasiperiodic orbits and stochastic (diffuse) orbits that give evidence for chaos. (Similar to Figure 8-5 or Figure 3-35.)

The final graphics are some color plates (CP-15,16). Here we assign different colors to four quadrants of the plastic sheet and look at the patterns after several deformation cycles. These pictures graphically

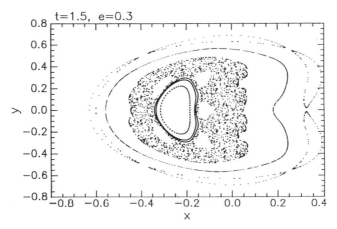

Figure 8-20 Poincaré map for several initial conditions for the alternating torque plasticity mixing model (8-4.8) showing quasiperiodic and stochastic orbits. [From Feeny et al. (1992).]

show the folding and stretching processes that occur in these mixing problems and are similar to those observed experimentally in fluid mixing problems.

Although his example is artificial, the availability of an explicit map allows one to look at the spatial mixing in a straightforward way. It also shows how simple velocity fields can lead to spatially complex particle trajectories. Furthermore, it suggests that processes such as forging of metals, kneading of baker's dough, geomechanical deformations, and deformation of clay in the making of pottery may have in them mechanisms for chaotic mixing.

PROBLEMS

8-1 Give six examples of how the history of temporal events is written in spatial patterns.

8-2 *Periodic Chain Structures.* As a prelude to understanding non-linear dynamics in periodic chain oscillator structures, examine the propagation of harmonic waves in a linear chain of equal masses m and springs of stiffness k. If ω is the frequency and κ is the wave number ($\lambda = 2\pi/\kappa$ is the wavelength), then show that in a linear chain ω and κ must be related by a dispersion relation

$$\omega^2 = 2\omega_0^2(1 - \cos \kappa)$$

where $\omega_0^2 = k/m$.

8-3 Show that the phase velocity of harmonic waves in the uniform linear lattice of Problem 8-2 is given by

$$v = \pm\omega_0 \frac{\sin(\kappa/2)}{\kappa/2}$$

where v is the number of lattice cells per time.

8-4 In a linear one-dimensional semi-infinite lattice, what is the nature of the motion when the end mass is excited at a frequency $\omega > 2\omega_0$, where $\omega_0^2 = k/m$, for a series chain of alternating masses and springs and $\omega_0^2 = 1/LC$ for a chain of linear inductors L alternating with linear capacitors C connected to ground.

8-5 Nonlinear lattice models have been used to model many systems, ranging from the dynamics of a long train of railroad cars to the dynamics of macromolecules such as DNA. The equations can often be written in terms of a nonlinear potential function $V(r_n)$, where $r_n = x_{n+1} - x_n$ is the relative displacement between neighboring cells. Write the equations of motion for a system with a double-well potential. Can you find more than one static solution?

8-6 *Solitons*. One nonlinear potential lattice model uses an exponential force potential (Toda, 1989) $V(r) = (a/b)e^{-br} + ar$ $(ab > 0)$. A lattice with this potential is known to admit so-called *solitary wave* or *soliton* solutions where a given deformation pattern can propagate without distortion and where two such waves can interact and preserve their identity similar to linear wave systems. Show that a solitary wave solution for the Toda lattice is given by

$$x_n = \frac{1}{b}\ln\frac{1 + e^{2(\kappa n - \kappa \pm \beta t)}}{1 + e^{2(\kappa n \pm \beta t)}} + \text{constant}$$

Here $\beta = (ab/m)^{1/2}\sinh\kappa$, and the width of the wave is proportional to $1/\kappa$. Also show that the solitary wave for the Toda lattice above has a speed $C = \beta/\kappa$.

8-7 *Coupled Map Lattice.* When both time and space are discretized, one can sometimes use coupled map lattices (CML) to model spatiotemporal dynamics. CML systems have been extensively studied by the Japanese physicist K. Kaneko at the University of Tokyo (e.g., see Kaneko, 1990). In one CML model the equations take the form

$$x_{n+1}(i) = (1 - \varepsilon)f[x_n(i)] + \tfrac{1}{2}\varepsilon\{f[x_n(i + 1)] + f[x_n(i - 1)]\}$$

This is known as the diffusive coupling model. Here the integers n, $n + 1$ represent time and $i - 1$, i, $i + 1$ represent spatial positions. A standard paradigm is to choose $f[x]$ as the logistic map, that is, $f = 1 - ax^2$. As a computer exercise, choose $1 \leq i \leq N$ where $N = 8$, 32, or 64 depending on the size of your computer. First, choose a set of initial conditions for $n = 1$ and plot a space–time diagram with dimensions $N \times n$ (e.g., $n = 100$). Use course-graining to visualize the space–time patterns. For example, if $|x_n(i + 1) - x_n(i)| > 0.3$, plot a black dot or other symbol. Otherwise leave the space blank. Kaneko often starts with a random set of initial conditions to see if organized spatial patterns develop after sufficient iteration time.

8-8 *The Elastica and the Pendulum.* A special case of the Kirchhoff analogy between the spatial deformation of a thin elastica and the temporal dynamics of a rigid body concerns planar deformation under a compressive load. Let Θ represent the slope of the elastica and let s be the distance along the filament. Assume that on each cross section a compressive load P acts parallel to the x axis. Furthermore, assume that the bending moment at each cross section is proportional to the change in curvature, that is, $M = EI(\kappa - \kappa_0)$, where $\kappa = d\Theta/ds$ and $\kappa_0(s)$ is the initially zero force configuration. (E is Young's modulus and I is the second moment of area.) Then show that the equation of deformation is similar to that of a forced pendulum

$$EI\Theta'' + P \sin \Theta = EI\kappa_0'(s)$$

8-9 Suppose the initial curvature of the planar elastica is zero, that is, $\kappa_0(s) = 0$. Then use the analogy with the pendulum to draw the deformation solution corresponding to the separatrix or heteroclinic orbit for the pendulum.

8-10   *Blinking Vortices.* A vortex flow field in an incompressible fluid
has zero radial velocity and a circular velocity component given
by $v_r = 0$, $v_\theta = \Gamma/2\pi r$, where Γ is called the *circulation*. Consider now two centers of vortical flow along the x axis at $x =$
$\pm a$. Assume that while one vortex operates for time T, the other
is quiet. (Hence the term "blinking vortices.") Derive a map

$$(x_{n+1}, y_{n+1}) = (F(x_n, y_n), G(x_n, y_n))$$

where "$(n + 1)$" denotes the new position of the vector (x_n, y_n)
after one full cycle. (See also Avef, 1984.)

8-11   Divide the plane into four quadrants. Sketch the deformation
of the quadrants after one cycle of the blinking vortex flow. Can
you sketch two cycles?

8-12   It is natural to use thermodynamic ideas such as entropy when
discussing spatiotemporal dynamics (e.g., see Kaneko, 1989).
In classical statistical thermodynamics the entropy is a measure
of the disorder in a system. Entropy measures are easily defined
when both space and time are discretized. For example, for a
one-dimensional, nearest neighbor, coupled cell system each of
which has a two-symbol state space of either A or B, one can
calculate the probability P_i that a certain pattern i of A's and
B's will occur. If there are M such patterns possible, then an
entropy S can be defined as

$$S = -\sum_{i=1}^{M} P_i \log P_i$$

Choose a set of eight cells. Show that the number of possible
configurations is $M = 2^8$. Compare S for two configurations: (a)
an alternating AB pattern and (b) a pattern where A or B are
chosen at random for all eight cells.

APPENDIX A

GLOSSARY OF TERMS IN CHAOTIC AND NONLINEAR VIBRATIONS

Almost periodic: A time history made up of a number of discrete incommensurate frequencies.

Arnold tongues: Refers to the motion of coupled nonlinear oscillators for which the ratio of frequencies may become locked at some value p/q (p, q integers). The tongue refers to the shape of the locked region in some parameter space, and the name refers to the Soviet dynamicist V. I. Arnold who discovered them.

Attractor: A set of points or a subspace in phase space toward which a time history approaches after transients die out. For example, equilibrium points or fixed points in maps, limit cycles, or a toroidal surface for quasiperiodic motions are all classical dynamical attractors.

Baker's map: A transformation of the plane (a mapping from the plane to the plane), which takes a rectangular area, streches it in one direction, shrinks it in a transverse direction, cuts it in half, and places it back over the original area. Similar to horseshoe map. Repeated iterations of the map transform the original set of points into a fractal structure. Named after the operations of a cook or baker, who repeatedly forms and reforms a piece of pie dough.

Basin of attraction: A set of initial conditions in phase space which leads to a particular long-time motion or attractor. Usually this set of points is connected and forms a continuous subspace in phase space. However, the boundary between different basins of attraction may or may not be smooth.

Bifurcation: Denotes the change in the type of long-time dynamical motion when some parameter or set of parameters is varied (for example, as when a rod under a compressive load buckles—one equilibrium state changes to two stable equilibrium states).

Cantor set: Formally a set of points obtained on a unit interval by throwing out the middle third and iterating this operation on the remaining intervals. This operation, when carried to the limit, leads to a fractal set of points on the line with a dimension between 0 and 1 (ln 2/ln 3).

Capacity: One of the many definitions of the fractal dimension of a set of points. The basic idea is to count the minimum number of cubes of size ε needed to cover a set of points. If this number behaves as ε^{-d} as $\varepsilon \to 0$, the exponent d is called the *capacity fractal dimension*.

Catastrophe theory: In many physical systems, the equilibrium points are derived from a potential function by setting the derivatives of this potential with respect to the generalized coordinates equal to zero. Catastrophe theory has to do with the dependence of the number of equilibrium points on the parameters in the problem such as force loads in elastic systems. Near certain critical values of these parameters, this theory predicts that the number of equilibrium points will change in prescribed ways and that these changes are universal for certain classes of potential functions. The roots of the theory are attributed to the French mathematician Rene Thom. In engineering mechanics, a special version of the theory was developed independently and deals with the sensitivity of critical loads to imperfections in the structure.

Center manifold: In dynamical systems theory, the motions in the neighborhood of an equilibrium point can be classified according to whether the eigensolutions are stable, unstable, or oscillatory. The subspace of phase space which is spanned by the purely oscillatory solutions is sometimes called the *center manifold*.

Chaotic: Denotes a type of motion that is sensitive to changes in initial conditions. A motion for which trajectories starting from slightly different initial conditions diverge exponentially. A motion with positive Lyapunov exponent.

Circle map: This is a map or difference equation that maps points on a circle onto the original circle. In the theory of two coupled oscillators, some motions in phase space can be viewed as motion on a toroidal surface. A Poincaré section that intersects the smaller diameter of the torus constitutes a circle map.

Combination tones: (See also *Quasiperiodic*.) In vibrations and acoustics, frequencies that appear as the sum or difference of two fundamental frequencies. More generally, frequencies of the form ($n\omega + m\omega_2$), where n and m are positive or negative integers.

Deterministic: Refers to a dynamic system whose equations of motion, parameters, and initial conditions are known and are not stochastic or random. However, deterministic systems may have motions that appear random.

Duffing's equation: A second-order differential equation with a cubic nonlinearity and harmonic forcing $\ddot{x} + c\dot{x} + bx + ax^3 = f_0\cos\omega t$. Named after G. Duffing (circa 1918).

Equilibrium point: In a continuous dynamical system, a point in phase space toward which a solution may approach as transients decay ($t \to \infty$). In mechanical systems, this usually means a state of zero acceleration and velocity. For maps, equilibrium points may come in a finite set where the system visits each point in a sequential manner as the map or difference equation is iterated. (Also called a *fixed point*.)

Ergodic theory: In Hamiltonian mechanics (no dissipation), it refers to the randomlike motions of coupled nonlinear systems of particles and the evolution of collective properties of the total system.

Feigenbaum number: A property of a dynamical system related to the period-doubling sequence. The ratio of successive differences between period-doubling bifurcation parameters approaches the number 4.669 ... . This property and the Feigenbaum number have been discovered in many physical systems in the prechaotic regime.

Fixed point: See *Equilibrium point*.

Fractal: A geometric property of a set of points in an n-dimensional space having the quality of self-similarity at different length scales and having a noninteger fractal dimension less than n.

Fractal dimension: The fractal dimension is a quantitative property of a set of points in an n-dimensional space which measures the extent to which the points fill a subspace as the number of points becomes very large. (See *Capacity*.)

Global/local motions: Local motions refer to solutions to dynamical systems that do not wander far from equilibrium points. Global solutions concern motion between and among equilibrium points or solutions that are not confined to a small region of phase space.

Hamiltonian mechanics: Formally, a method to derive the equations of motion of an N-degree-of-freedom dynamical system in terms of $2N$ first-order differential equations (Hamilton, 1805–1865). In practice, a Hamiltonian problem often refers to a nondissipative system in which the forces can be derived from a scalar potential.

Hausdorff dimension: A mathematical definition of fractal properties related to the capacity dimension.

Henon map: A set of two coupled difference equations with one quadratic nonlinearity. When one parameter is set to zero, the equations resemble the logistic or quadratic map. Named after a French astronomer.

Heteroclinic orbit: An orbit in a map or difference equation that occurs when stable and unstable orbits from different saddle points intersect.

Homoclinic orbit: An orbit in a map that occurs when stable and unstable manifolds of a saddle point intersect.

Hopf bifurcation: The emergence of a limit cycle oscillation from an equilibrium state as some system parameter is varied. Named after a mathematician who gave precise conditions for its existence in a dynamical system.

Horseshoe map: A map of the plane onto the plane. Points in the lower half of a rectangular domain are stretched and contracted and mapped into a vertical strip in a section of the left half-plane, while points in the upper half are stretched and contracted and mapped onto a vertical strip in the right half-plane. The process is like transforming a rectangular domain into a horseshoe-shaped set of points—hence the name. Similar to the baker's transformation. Repeated iterations can yield a fractal-like set of points.

Hyperchaos: A dynamic system where the phase space is stretched in two or more directions (i.e., two or more positive Lyapunov exponents).

Intermittency: A type of chaotic motion in which long time intervals of regular, periodic, or stationary dynamical motion are followed by short bursts of randomlike motion. The time interval between bursts is not fixed but is unpredictable.

Invariant measure: A distribution function that describes the long-time probability of finding the motion of a system in a particular region of phase space.

KAM theory: The initials stand for the theorists Kolmogorov, Arnold, and Moser, who developed a theory regarding the existence of periodic or quasiperiodic motions in nonlinear Hamiltonian systems (i.e., systems that have no dissipation and in which the forces can be derived from a potential). This theory states that if small nonlinearities are added to a linear system, the regular motions will continue to exist.

Lagrangian chaos: The state of motion of a fluidlike substance in which the spatial patterns of flow are regular, but the fluid particle motions are chaotic.

Limit cycle: In the engineering literature, a periodic motion that arises from a self-excited or autonomous system as in aeroelastic flutter or electrical oscillations. In the dynamical systems literature, it also includes forced periodic motions. (See also *Hopf bifurcation*.)

Linear operator: Any mathematical operation (e.g., differentiation, multiplication by a constant) in which the action on the sum of two functions is the sum of the action of the operation on each function. Akin to the principle of superposition.

Lorenz equations: A set of three first-order autonomous differential equations that exhibit chaotic solutions. The equations were derived and studied by E. N. Lorenz of M.I.T. in 1963 as a model for atmospheric convection. This set of equations is one of the principal paradigms for chaotic dynamics.

Lyapunov exponents: Numbers that measure the exponential attraction or separation in time of two adjacent trajectories in phase space with different initial conditions. A positive Lyapunov exponent indicates a chaotic motion in a dynamical system with bounded trajectories. Named after the dynamicist Lyapunov (1857–1918) (in some books spelled Liapunov).

Mandelbrot set: If z is a complex variable, the quadratic map $z \rightarrow z^2 + c$ has more than one attractor. Fixing the initial conditions, one can vary the complex parameter c to determine the basin of attraction as a function of c. The basin boundary that results is fractal, and the basin is known as the *Mandelbrot set* after a mathematician at IBM.

Manifold: A subspace of phase space in which solutions with initial conditions in the manifold stay in the manifold or subspace, under the action of the differential or difference equations.

Map, mapping: A mathematical rule that takes a collection of points in some n-dimensional space and maps them into another set of points. When this rule is iterated, a map is similar to a set of difference equations.

Melnikov function: One theory of chaotic motions focuses on the saddle points of Poincaré maps of continuous phase-space flows.

Near such points there are subspaces where trajectories are swept into the point (stable manifolds) and subspaces where trajectories are swept away from the point (unstable manifolds). The Melnikov function provides a measure of the distance between these stable and unstable manifolds. One theory contends that chaos is possible when these two manifolds intersect or when the Melnikov function has a simple zero. (Named after a Russian mathematician circa 1962.)

Multifractal: A set of geometric patterns with multiple scaling relationships. A set with a distribution of fractal dimensions.

Navier–Stokes equations: A set of three partial differential equations governing the velocity field in the flow of an incompressible, linear, viscous fluid (Navier, 1785–1836; Stokes, 1819–1903).

Noise: In experiments, noise usually denotes the small random background disturbance of either mechanical, thermal, or electrical origin.

Nonlinear: A property of an input–output system or mathematical operation for which the output is not linearly proportional to the input. For example, $y = cx^n$ ($n \neq 1$), or $y = x\, dx/dt$, or $y = c(dx/dt)^2$.

Period doubling: Refers to a sequence of periodic vibrations in which the period doubles as some parameter in the problem is varied. In the classic model, these frequency-halving bifurcations occur at smaller and smaller intervals of the control parameter. Beyond a critical accumulation parameter value, chaotic vibrations occur. This scenario to chaos has been observed in many physical systems but is not the only route to chaos. (See *Feigenbaum number*.)

Phase space: In mechanics, phase space is an abstract mathematical space whose coordinates are generalized coordinates and generalized momentum. In dynamical systems, governed by a set of first-order evolution equations, the coordinates are the state variables or components of the state vector.

Poincaré section (map): The sequence of points in phase space generated by the penetration of a continuous evolution trajectory through a generalized surface or plane in the space. For a periodically forced, second-order nonlinear oscillator, a Poincaré map can be obtained by stroboscopically observing the position and velocity at a particular phase of the forcing function (H. Poincaré, 1854–1912).

Quantum chaos: In quantum theory, every classical dynamical system has a quantum counterpart. The question of quantum chaos is a search for the dynamical measures of chaotic Newtonian systems taken to their quantum limits. This question has not been resolved to date.

Quasiperiodic: A vibration motion consisting of two or more incommensurate frequencies.

Rayleigh–Benard convection: Circulatory patterns in a fluid produced by a thermal gradient and gravitational forces. The chaos model of Lorenz attempted to simulate some of the dynamics of thermal convection.

Renormalization: A mathematical theory in functional analysis in which properties of some mathematical set of equations at one scale can be related to those at another scale by a suitable change of variables. Developed by the Nobel-prize-winning physicist K. Wilson (Cornell University). Used in the theory of quadratic maps to derive the Feigenbaum number.

Reynolds number: A nondimensional group in fluid mechanics proportional to a velocity parameter and a characteristic length and inversely proportional to the kinematic viscosity. The transition from laminar to turbulent flow in many fluid problems occurs at a critical value of the Reynolds number (O. Reynolds, 1842–1912).

Rotation number: (Also Winding number.) When a system has two oscillators with frequencies ω_1 and ω_2, the rotation number, based on ω_1, measures the average number of orbits of frequency ω_2 in an orbit of ω_1.

Saddle point: In the geometric theory of ordinary differential equations, an equilibrium point with real eigenvalues with at least one positive and one negative eigenvalue.

Self-similarity: A property of a set of points in which geometric structure on one length scale is similar to that at another length scale. (See also *Fractal, Renormalization*.)

Shil'nikov chaos: A mathematical model for chaotic dynamics proposed by the Soviet scientist, Shil'nikov. The model is based on a phase space flow around a single unstable equilibrium point with either an unstable spiral outflow and stable inflow or an unstable outflow and a stable spiral inflow.

Spatiotemporal chaos: The dynamics of physical systems with loss of correlation in both space and time.

Stochastic process: Often refers to a type of chaotic motion found in conservative or nondissipative dynamical systems.

Strange attractor: Refers to the attracting set in phase space on which chaotic orbits move. An attractor that is not an equilibrium point nor a limit cycle, nor a quasiperiodic attractor. An attractor in phase space with fractal dimension.

Surface of section: See *Poincaré section (map)*.

Symbolic dynamics: Refers to a dynamic model in which not only time is discretized but the state variables take on a finite set of values, for example, $(-1, 0, 1)$. Because the set of values is finite, one is free to use any set of symbols, say (L, C, R). A dynamic trajectory then consists of a sequence of symbols. Related also to cellular autonoma.

Taylor–Couette flow: The flow of fluid between two rotating concentric cylinders.

Torus (invariant): The coupled motion of two undamped oscillators is imagined to take place on the surface of a torus, with circular motion around the small radius representing the oscillatory vibration of one oscillator and motion around the large radius direction representing the other oscillator. If the motion is periodic, a closed helical trajectory will wind around the torus. If the motion is quasiperiodic, the orbit will come close to all points on the torus.

Transient chaos: A term describing motion that looks chaotic during a finite time; that is, it appears to move on the strange attractor, but eventually settles into a periodic or quasiperiodic motion.

Unfolding: In the mathematical theory of stability, a term that describes a set of problems which are close to some idealized problem, as when a small amount of asymmetry is introduced into a problem with symmetry or when small damping is added to a nondissipative dynamical problem. The change in stability or dynamical properties of the idealized problem as some nonidealized terms are added is called an *unfolding*.

Universal property: A property of a dynamical system that remains unchanged for a certain class of nonlinear problems. For example, the Feigenbaum number relating the sequence of bifurcation parameters in period doubling is the same for a certain class of nonlinear, noninvertible, one-dimensional maps.

Van der Pol equation: A second-order differential equation with linear restoring force and nonlinear damping which exhibits a limit cycle behavior. The classic mathematical paradigm for self-excited oscillations. (Named after B. Van der Pol, circa 1927.)

Winding numbers: see *Rotation number*.

APPENDIX B

NUMERICAL EXPERIMENTS
IN CHAOS

The spirit of the approach to chaotic vibration in this book has been an empirical one of exploring the range of physical phenomena in which chaotic dynamics play a role. While not all readers will have access to a laboratory or have the inclination to do experiments, most readers have some access to digital computers. Thus, this appendix contains a number of numerical experiments using either a personal computer or minicomputer in which the reader can explore the dynamics of the now classic paradigms of chaos. Other numerical exercises may be found in some of the problems at the end of each chapter.

B.1 LOGISTIC EQUATION—PERIOD DOUBLING

Perhaps the easiest problem with which to begin study in the new dynamics is the population growth model or logistic equation

$$x_{n+1} = \lambda x_n(1 - x_n)$$

Period-doubling phenomena were observed by a number of researchers (e.g., see May, 1976) and, of course, Feigenbaum (1978), who discovered the famous parameter scaling laws (see Chapters 1 and 3). Two numerical experiments on a desktop computer are fairly easy to perform. In the first plot, we have x_{n+1} versus x_n with a range of $0 \leq x \leq 1$. The period-doubling regime is below $\lambda = 3.57$. Start with $\lambda < 3.0$ to see a period-1 orbit. To see the long-term orbit, plot the first

441

30–50 iterates with dots and plot the later iteration with a different symbol. Of course, one can also plot x_n versus n to see both the transient and steady-state behavior. Chaotic orbits may be found for $3.57 < \lambda \le 4.0$. A period-3 window may be found around $\lambda = 3.83$ (see May, 1976).

The next experiment involves generating a bifurcation diagram. In this picture, the long-term values of the orbit are plotted as a function of the control parameter. Start with some initial condition (e.g., $x_0 = 0.1$) and iterate the map for say 100 steps. Then plot x_n for another 50 steps on the vertical axis with the λ value on the horizontal axis (or vice versa). Take step sizes for λ to be around 0.01 and look at the range $2.5 < \lambda < 4.0$. The diagram should produce the classic pitchfork bifurcation at the period-doubling points. Can you calculate the Feigenbaum number from this experiment? (See Figure 3-18.)

May (1976) also lists other experiments with one-dimensional maps, for example,

$$x_{n+1} = x_n \exp[\lambda(1 - x_n)]$$

He described this as a model for growth of a single species population which is regulated by an epidemic disease. Look at the region $2.0 < r < 4.0$. The period-doubling accumulation point and chaos begin at $r = 2.6824$ (May, 1976). This article also lists data for several other computer experiments.

B.2   LORENZ EQUATIONS

A fascinating numerical experiment worth trying is the one in Lorenz's original 1963 paper. In this paper, Lorenz simplifies equations derived by Saltzman (1962) based on the fluid convection equations of mechanics (see Chapter 4). Lorenz acknowledges Saltzman's discovery of nonperiodic solutions of the convection equation. Lorenz chose the now classical parameters to study chaotic motions: $\sigma = 10$, $b = \frac{8}{3}$, $r = 28$ for equations

$$\dot{x} = \sigma(y - x)$$
$$\dot{y} = rx - y - xz$$
$$\dot{z} = -bz + xy$$

His data in Figures 1 and 2 of the 1963 paper may be reproduced by choosing initial conditions $(x, y, z) = (0, 1, 0)$ and a time step of $\Delta t = 0.01$ and projecting the solution on either the z–x or z–y planes.

To derive a one-dimensional map based on this flow, Lorenz chose to look at *successive maxima* of the variable z which he called M_n. A plot of M_{n+1} versus M_n reveals a map shaped like a tent. Lorenz then went on to study a simplified version of the map called a tent map, which is a bilinear version of the logistic equation (See Figure 3-29)

$$M_{n+1} = \begin{cases} 2M_n & \text{if } M_n < \frac{1}{2} \\ 2 - 2M_n & \text{if } M_n > \frac{1}{2} \end{cases}$$

B.3 INTERMITTENCY AND THE LORENZ EQUATION

An illustration of intermittency may be seen on the computer by numerically integrating the Lorenz equations with a Runge–Kutta algorithm,

$$\dot{x} = \sigma(y - x)$$

$$\dot{y} = -xz + rx - y$$

$$\dot{z} = xy - bz$$

using the parameters $\sigma = 10$, $b = \frac{8}{3}$, and $166 \leq r \leq 167$. For $r = 166$, a periodic time history of say $z(t)$ will be obtained, but for $r = 166.1$ or larger, "bursts" or chaotic noise will appear (e.g., see Manneville and Pomeau, 1980). By measuring the average number of periodic cycles between bursts, N (the laminar phase), one should obtain the scaling

$$N \sim \frac{1}{(r - r_c)^{1/2}}$$

where $r_c = 166.07$.

B.4 HENON ATTRACTOR

An extension of the quadratic map on the line to a map on the plane was proposed by the French astronomer Henon:

$$x_{n+1} = 1 + y_n - ax_n^2$$

$$y_{n+1} = bx_n$$

When $b = 0$, one obtains the logistic map studied by May and Feigenbaum. Values of a and b for which one will get a strange attractor include $a = 1.4$ and $b = 0.3$. Plot this map on the x–y plane with graph limits $-2 \leq x \leq 2$ and $0.5 \leq y \leq 0.5$. After obtaining the attractor, rescale your graph to focus on one small area of the attractor. Run the map for a much longer time and look for fine-scale fractal structure. If you have the patience or a fast computer, rescale and run again for an even smaller area of the plane. (See Figure 1-24.)

If you have a program to calculate Lyapunov exponents, the reported Lyapunov exponent is $\lambda = 0.2$ and the fractal dimension for this attractor is $d_L = 1.264$. One can also vary a and b to see where the attractor exists and to find period-doubling regions of the (a, b) plane [see Guckenheimer and Holmes (1983, p. 268) and Ott (1981)].

B.5 DUFFING'S EQUATION: UEDA ATTRACTOR

This model for an electric circuit with a nonlinear inductor was discussed in Chapter 4. The equations for the model in first-order form are

$$\dot{x} = y$$
$$\dot{y} = -ky - x^3 + B \cos t$$

Chaotic oscillations were studied quite extensively by Ueda (1979). Use a standard numerical integration algorithm such as a fourth-order Runge–Kutta and examine the case $k = 0.1$, $9.8 \leq B \leq 13.4$. For $B = 9.8$, one should get a period 3. (Take a Poincaré map when $t = 2\pi n$, $n = 1, 2, \ldots$.) The period-3 motion should bifurcate to chaos around $B = 10$. Beyond $B = 13.3$, the motion should become periodic again with a transient chaos regime. (See Figure 3-33.)

Also, compare the fractal nature of the attractor as damping is decreased for $B = 12.0$ and $k = 0.2, 0.1, 0.05$. Note also that for $k = 0.3$ only a small piece of the attractor remains, while for $k = 0.32$ the motion has become periodic.

B.6 TWO-WELL POTENTIAL
DUFFING–HOLMES ATTRACTOR

This example has been discussed throughout the book. Several numerical experiments are worth trying. The nondimensional equations are

$$\dot{x} = y$$

$$\dot{y} = -\delta y + \tfrac{1}{2}x(1 - x^2) + f\cos\omega t$$

(This can be put into a third-order autonomous system by setting $z = \omega t$ and writing $\dot{z} = \omega$.) The factor of $\tfrac{1}{2}$ makes the small-amplitude natural frequency in each well equal to unity. The criterion for chaos for fixed damping $\delta = 0.15$ and variable f, ω has been discussed in Chapter 6. An interesting region to explore is $\omega = 0.8, 0.1 \leq f \leq 0.3$. In this regime, one should go from periodic to chaotic to periodic windows in the chaotic region and out of the chaotic region at $f = 0.3$. Another interesting region is $\delta = 0.15$, $\omega = 0.3$, and $f > 0.2$. In all studies, the reader is encouraged to use a Poincaré map. In using a small desktop computer, one can achieve reasonable computing speeds if the program is run in a compiled form. (See Figure 6-3.)

Another interesting experiment is to fix the parameters, say $f = 0.16$, $\omega = 0.833$, and $\delta - 0.15$, and vary the phase of the Poincaré map; that is, plot (x, y) when $t_n = (2\pi/\omega)n + \varphi_0$ and vary φ_0 from 0 to π. One should see an inversion of the map for $\varphi_0 = 0, \pi$. Is this related to the symmetry of the equations? (See Figure 5-7.)

B.7   CUBIC MAP (HOLMES)

We have illustrated many of the concepts of chaotic vibrations with the model of the two-well potential attractor. The dynamics are described by a nonlinear second-order differential equation (see Chapters 2 and 3), but an explicit formula for the Poincaré map of this attractor has not been found. Holmes (1979) has suggested a two-dimensional cubic map which has some of the features of a negative stiffness Duffing oscillator

$$x_{n+1} = y_n$$

$$y_{n+1} = -bx_n + dy_n - y_n^3$$

A chaotic attractor may be found near the parameter values $b \approx 0.2$ and $d = 2.77$ (see also Problem 3.1).

B.8   BOUNCING BALL MAP (STANDARD MAP)

[See Holmes (1982) and Lichtenberg and Lieberman (1983).] As discussed in Chapter 3, a Poincaré map for a ball bouncing on a vibrating table can be obtained exactly in terms of the nondimensional impact velocity v_n and phase of the table motion $\varphi_n = \omega t_n \pmod{2\pi}$:

$$v_{n+1} = (1 - \varepsilon)v_n + K \sin \varphi_n \qquad \varphi_n \pmod{2\pi}$$

$$\varphi_{n+1} = \varphi_n + v_{n+1}$$

where ε represents energy lost during impact.

Case 1: $\varepsilon = 0$ Conservative Chaos

This case is studied in Lichtenberg and Lieberman (1983) as a model for acceleration of electrons in electromagnetic fields. Iterate the map and plot points on the (v_n, φ_n) plane. To obtain $\varphi \pmod{2\pi}$, one can use

$$\frac{\varphi}{2\pi} - ABS\left(\frac{\varphi}{2\pi}\right)$$

in advanced BASIC. To get a good picture, you must vary the initial conditions. For example, choose $\varphi = 0.1$ and run the map for several hundred iterations for different values of v between $-\pi < v < \pi$.

The interesting cases are for $0 < K < 1.5$. For $K \ll 1$, one can see quasiperiodic closed orbits around the periodic fixed points of the map. For $K \approx 1$, one should see regions of conservative chaos near the separatrix points. (See Figure 6-26.)

Case 2: $0 < \varepsilon < 1$

This case corresponds to a dissipative map where energy is lost at each impact. First try $K \approx 1.2$ and $\varepsilon \approx 0.1$. Note that although the early iterations look chaotic, as in Case 1, the motion settles into a periodic orbit. To get fractal-like chaos, raise K to ~ 5.8–6.9. To get a more fractal-looking strange attractor, use $\varepsilon \approx 0.3$–0.4 and $K \approx 6.0$.

B.9   CIRCLE MAP: MODE-LOCKING, WINDING NUMBERS, AND FAIREY TREES

A point moving on the surface of a torus is a conceptual model for the dynamics of two coupled oscillators. The amplitudes of each motion are represented by minor and major radii of the torus and are often assumed to be fixed. The phases of each oscillator are represented by two angles describing positions along the major and minor circumferences. A Poincaré section of the minor circumference of the torus produces a one-dimensional difference equation called the *circle map:*

$$\varphi_{n+1} = \varphi_n + f(\varphi_n)$$

where $f(\varphi)$ is a periodic function.

Each iteration of the map represents an orbit of one oscillator around the large circumference of the torus. A popular example for study is the so-called *standard circle map* (normalized by 2π):

$$x_{n+1} = x_n + \Omega - \frac{K}{2\pi} \sin 2\pi x_n \quad (\text{mod } 1)$$

Possible motions observed in this map are periodic, quasiperiodic, and chaotic. To see periodic cycles, plot the points on a circle with rectangular coordinates $u_n = \cos 2\pi x_n$ and $v_n = \sin 2\pi x_n$.

When $K = 0$, Ω represents a *winding number*—the ratio of two frequencies of the uncoupled oscillators. When $K \neq 0$, the map may be periodic when Ω is an irrational number. The oscillators are then said to be *mode-locked*. When $0 < K < 1$, one can observe mode-locked or periodic motions in finite-width regions of the Ω axis, $0 \leq \Omega_k \leq \Omega \leq \Omega_{k+1} < 1$, which, of course, includes nonrational values of Ω. For example, for $K = 0.8$, a two-cycle can be found for $0.48 < \Omega < 0.52$ and a three-cycle can be found in the region $0.65 < \Omega < 0.66$. To find these regions for $0.7 < K < 1.0$, calculate the winding number W as a function of Ω, $0 \leq \Omega \leq 1$. The winding number is calculated by suspending the mod 1 action and using

$$W = \lim_{N \to \infty} \frac{x_N - x_0}{N}$$

In practice, one has to choose $N > 500$ to get good data. Plotting W versus Ω, one should see a series of plateaus of mode-locked regions. To see more mode-locked regions, one should choose a small $\Delta\Omega$ region and plot W for many points in this small domain.

Each mode-locked plateau in the $W(\Omega)$ plot corresponds to a p/q rational number representing p cycles of one oscillator to q cycles of the other. These p/q ratios are ordered in a sequence called a *Fairey tree*. Given two mode-locked regions r/s and p/q at Ω_1 and Ω_2, respectively, a new mode-locked region will exist somewhere between these two, $\Omega_1 < \Omega < \Omega_2$, with a winding number given by

$$W = \frac{r + p}{s + q}$$

Starting from $0/1$ at $\Omega = 0$ and $1/1$ at $\Omega = 1$, one can begin to generate the whole infinite sequence of mode-locked regions. Most, however, are very narrow. Note that the size $\Delta\Omega$ of these mode-locked regions approaches zero as $K \to 0$ and gets wider as $K \to l$. The shapes of mode-locked regions in the K–Ω plane are sometimes called *Arnold tongues*. See Figure 2-23 for an example.

B.10 RÖSSLER ATTRACTOR: CHEMICAL REACTIONS, RETURN MAPS

Thus far, each of the principal fields of classical physics has developed a simple paradigm for chaotic dynamics: fluid mechanics—Lorenz equations; structural mechanics—Duffing–Holmes two-well attractor; electrical science—Duffing–Ueda attractor. Another simple model motivated by the dynamics of chemical reactions in a stirred tank is the following proposed by Rössler (1976b):

$$\dot{x} = -(y + z)$$
$$\dot{y} = x + \alpha y$$
$$\dot{z} = \alpha + z(x - \mu)$$

The system often studied is the case $\alpha = \frac{1}{5}$. Period-1, -2, and -4 motions may be found for $\mu = 2.6, 3.5,$ and 4.1. Chaotic motions may be found for $\mu > 4.23$. This model has the properties of a linear oscillator with negative damping and feedback,

$$\ddot{y} - a\dot{y} + y = -z$$

This example is also illustrative of higher-dimensional systems whose dynamics are approximated by a one-dimensional map. Take the Poincaré section for $y = 0$ and plot the x_n values on the x–z plane in the form of a one-dimensional map, that is, x_{n+1} versus x_n. Note the resemblance to the quadratic or logistic map. It should be no surprise that period doubling is observed in this system.

B.11   FRACTAL BASIN BOUNDARIES: KAPLAN–YORKE MAP

An example of a two-dimensional map with a fractal basin boundary is one studied by Kaplan and Yorke (1978) and McDonald et al. (1985):

$$x_{n+1} = \lambda_x x_n \quad (\text{mod } 1)$$

$$y_{n+1} = \lambda_y y_n + \cos 2\pi x_n$$

where λ_x is an integer. When $\lambda_x = 2$ and $|\lambda_y| < 1$, this map possesses a strange attractor which the reader can explore on the computer. [For $\lambda_y = 0.2$, the fractal dimension is reported to be 1.43 (Russel et al., 1980).]

However, to look at a simple numerical experiment in basin boundaries, try the case $\lambda_x = 3$, $\lambda_y = 1.5$. For this case, there are two attractors $y \pm \infty$. Set the scale for $0 \leq x \leq 1$, $-2.0 \leq y \leq 2.0$. To get the boundary, choose some initial x value and scan a set of initial y values. For each set of initial conditions, iterate the map until $|y| > 10$ or some other large value. If $y \rightarrow +\infty$, leave a blank at the (x_0, y_0) point; if $y \rightarrow -\infty$, print a dot. If one scans y_0 from bottom to top once the boundary is crossed, one can omit further y values and choose another x_0.

This example is one of the few for which an explicit formula for the boundary may be obtained:

$$y = - \sum_{j=1}^{\infty} \lambda_y^{-j} \cos(2\pi \lambda_x^{j-1} x)$$

(see McDonald et al., 1985).

The capacity dimension of this boundary derived by McDonald et al. is $d = 2 - (\ln \lambda_y / \ln \lambda_x) = 1.63 \ldots$. This boundary is continuous, of infinite length, and nowhere differentiable.

B.12   TORUS MAPS

The motion of coupled nonliner oscillators is sometimes imagined to occur on the surface of a torus. When the number of oscillators is two, a Poincaré section of the torus yields a circle map. However, when the number of oscillators is *three*, the dynamic interaction of the phase of each oscillator takes place on some abstract torus. The Poincaré section of this three-torus yields a two-dimensional map on a two-torus. Grebogi et al. (1985a) have studied such maps and have produced *beautiful* pictures of chaotic attractors. The set of equations takes the form

$$\theta_{n+1} = \left[\theta_n + \omega_1 + \frac{\varepsilon}{2\pi} P_1(\theta_n, \psi_n) \right] \quad (\text{mod } 1)$$

$$\psi_{n+1} = \left[\psi_n + \omega_2 + \frac{\varepsilon}{2\pi} P_2(\theta_n, \psi_n) \right] \quad (\text{mod } 1)$$

The functions P_1 and P_2 are periodic functions of the form

$$P_\sigma(\theta, \psi) = \sum_{r, s} A_{r, s}^{(\sigma)} \sin \left[2\pi(r\theta + s\psi + B_{r, s}^{(\sigma)}) \right]$$

where $\sigma = 1, 2$ and (r, s) take on combinations of $(1, 0)$, $(0, 1)$, $(1, 1)$, and $(1, -1)$. The values of $A_{r, s}^{(1)}$, $A_{r, s}^{(2)}$, $B_{r, s}^{(1)}$ and $B_{r, s}^{(2)}$ were chosen randomly by Grebogi et al. The details are listed in their paper in Table 1. Iterations of this map produce spectacular pictures of this strange attractor in the torus, which with high resolution are suitable for framing (see Figures 7, 9, 10, and 11 of Grebogi et al., 1985a).

Maps of this kind are also related to the Newhouse–Ruelle–Takens theory of the quasiperiodicity route to chaos.

APPENDIX C

CHAOTIC TOYS

During many lectures given on chaos, I have demonstrated chaotic vibrations with a simple, inexpensive vibrating beam. This chaotic toy has many times made a believer out of a doubting Thomas and has provided motivation to study the often difficult mathematical theory behind chaotic phenomena. In this appendix, I describe several chaotic toys or desktop experiments and also provide some detailed description of the buckled beam experiment (which has been mentioned many times in this book) for the more serious experimenter. During one of these lectures, a physicist (with tongue in cheek) dubbed this experiment the "chaotic Moon-beam." This experiment has had great success in providing both qualitative and quantitative verification of many of the theoretical ideas about chaos.

Another chaotic toy is a version of the forced pendulum sometimes seen in adult toy shops under names such as the "Space ball." A description of this experiment is also given.

A short description of a simple neon bulb circuit with chaotic flashing lights is also provided. For those interested in a simple circuit that exhibits period-doubling phenomena, I recommend the circuit described in Matsumoto et al. (1985) called the "double scroll attractor," which was discovered by L. Chua of University of California at Berkeley. This circuit is described in Chapter 4. Several other demonstrations of chaos are also described, including one shown to me by a Soviet physicist which I call ping-pong ball chaos. Coupled pendulums provide nice desktop chaotic toys, and Professor N. Rott's paper on how to design them is described. Finally, we mention the possible role of chaos and nonlinear dynamics in the kinetic sculpture of Calder and Tinguely.

C.1   THE CHAOTIC ELASTICA: A DESKTOP CHAOTIC VIBRATION EXPERIMENT

This mechanical toy is very inexpensive to build and can demonstrate three different chaotic phenomena:

1. The two-well potential attractor (or buckled beam)
2. Bilinear oscillator
3. Out-of-plane chaotic vibration of a thin beam

A sketch of the device is shown in Figure C-1 for the buckled beam problem. It consists of a small, hobby-type, battery-run motor with an eccentric weight as a source of forced vibrations and a thin steel cantilevered beam with two magnets near the free end of the beam to

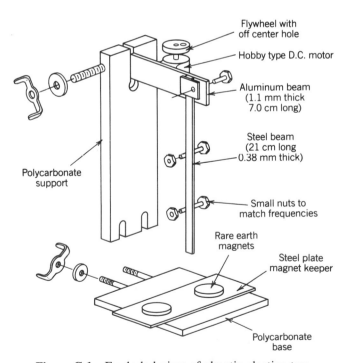

Flywheel with off center hole

Hobby type D.C. motor

Aluminum beam (1.1 mm thick 7.0 cm long)

Steel beam (21 cm long 0.38 mm thick)

Polycarbonate support

Small nuts to match frequencies

Rare earth magnets

Steel plate magnet keeper

Polycarbonate base

Figure C-1   Exploded view of chaotic elastica toy.

provide nonlinear buckling forces. Two masses are attached to the thin beam to match dynamically the driving frequency (4–8 Hz) to one or two of the natural modes of the beam. A strong polycarbonate plastic acts as a supporting frame, and the base can be secured to a table or desktop with double-sided poster adhesive pads. The whole device can be disassembled and carried in a thin box to fit in one's briefcase for travel.

The device works as follows. With a low voltage applied by two or three D-cell batteries across a potentiometer, the aluminum beam at the top is excited by the rotation of the eccentric weight on the motor. With two rare earth magnets below the beam, the beam will vibrate periodically about one of the two stable equilibrium positions (of course, one can use more than two magnets). With the two masses attached, the steel beam will resonate near the second mode so that the beam tip undergoes large deflection. As the motor speed is increased further, the beam will jump from one equilibrium position to the other. Under the right conditions (e.g., magnet spacing, motor speed, mass positions), which usually take about 5 minutes to search for, the beam will perform chaotic motions.

To achieve a more theatrical effect, I have glued a small mirror on the beam and projected a laser beam on a wall or ceiling with spectacular effects as the motion makes the transition from periodic to chaotic motion.

If the magnets are replaced by a thin metallic channel one can demonstrate chaotic vibrations of a beam with nonlinear boundary conditions (see Chapter 4). If the metal end constraint is very thin, the audience can hear the nonperiodic or periodic tapping of the beam against the constraints.

Another elastica chaos demonstration toy is shown in Figure C-2 and is based on the work of Cusumano and Moon (1990). In this experiment, in-plane vibrations at the second or third resonant frequency become unstable and the beam performs out of plane twisting and bending chaotic oscillations.

C.2   THE "MOON BEAM" OR CHAOTIC BUCKLING EXPERIMENT

As described in Chapters 2, 3, and 4, the forced motion of a buckled cantilevered beam in the field of two strong magnets can be described quite adequately by a nonlinear differential equation of the Duffing

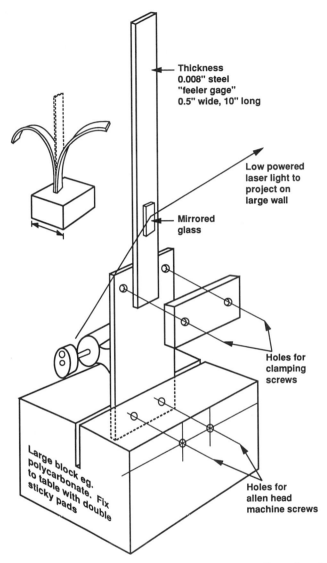

Figure C-2 Sketch of experiment for chaotic torsional bending vibrations of a thin steel "feeler" gauge strip (length 10 inches, thickness 0.008 inch).

type:

$$\ddot{x} + \dot{x} - ax + \beta x^3 = f \cos \omega t$$

Successful experiments in chaotic vibration have been carried out with two different beams by shaking the beam clamp and magnets

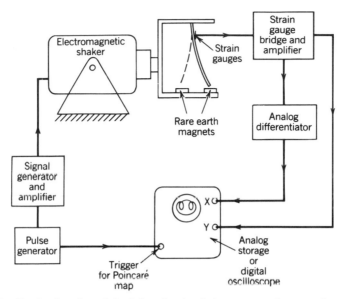

Figure C-3   Professional model of chaotic elastic beam experiment using an electromagnetic shaker.

with an electromagnetic shaker as shown in Figure C-3. Standard electromagnetic shakers can cost from $3000 to $5000 in 1992 prices. However, the resourceful experimenter can improvise one from a $200 audiospeaker by using the magnet and drive coil in the speaker.

A list of specifications for two elastic beams is given in Table C-1. The best magnets to use are rare earth permanent magnets 2.5-cm circular diameter.

With this setup, one can obtain Poincaré maps of chaotic motions (Chapter 5), measure the critical force for chaotic motion as a function of frequency (Chapter 6), or measure the fractal dimension of the motion using time series data (Chapter 7).

To obtain electrical signals proportional to the motion, we used two strain gauges glued near the clamped end of the beam. One gauge is placed on each side and the two resistors (i.e., the gages) are connected to two legs of a Wheatstone bridge.

The output of the bridge is amplified and electronically filtered. One can also design an inexpensive circuit such as a Bessel filter to differentiate the filtered signal. In both devices, one should use care to achieve minimum distortion in both amplitude and phase for frequencies up to at least twice the natural frequency of the beam near one of the two magnets.

TABLE C-1   Parameters for Chaotic Buckled Beam Experiments[a]

	Model A	Model B
Dimensions of elastic steel beam		
Length	18.0 cm (7.4 in.)	12.4 cm (4.9 in.)
Width	9.5 mm (0.375 in.)	12.7 mm (0.5 in.)
Thickness	0.23 mm (0.009 in.)	0.38 mm (0.015 in.)
Constrained layer (for damping)	0.05 mm (0.002 in.)	0.025 mm (0.001 in.)
Natural frequencies		
Cantilevered—no magnets	4.6, 26.6, 73.6 Hz	18 Hz
Cantilevered–with magnets	9.3 Hz	38 Hz
Damping		
Without constrained layer	0.0033	
With steel shim and		
double-sided tape layer	0.017	
Buckled displacement with magnets	±20 mm	±15 mm
Frequency range for test data	6–12 Hz	21–35 Hz
Driving amplitude range	±2–5 mm	± 5 mm
Magnets, rare earth, 2.5-cm		
diameter, 0.2 tesla on one face		

[a] See also Moon (1980a).

Damping is a critical property in this experiment. Most metallic structures have very low damping, and the Poincaré maps will look more like Hamiltonian or conservative chaos than fractal or dissipative chaos. In our experiments, we used constrained layer damping to increase dissipation. A simple way to do this is to put double-sided sticky cellophane tape along the beam and to put a thin shim-type metal layer (0.1 mm) on top of this. When constrained layer damping is placed on each side of the beam, a significant increase in damping can be achieved and some very beautiful fractal-looking Poincaré maps can be obtained.

The reader should see Chapter 5 for other suggestions about experiments in chaotic vibrations.

C.3   A CHAOTIC DOUBLE PENDULUM OR "SPACE BALL"

This toy has several variations, two of which are shown in Figure C-4. The commercial versions are well made (from Taiwan), but I could not find the name of any manufacturer (nor for that matter any patent numbers) on the devices. The basic principles involve the forced motion of a pendulum that interacts with a magnetic circuit in the base. Attached to the primary pendulum is another rotating arm.

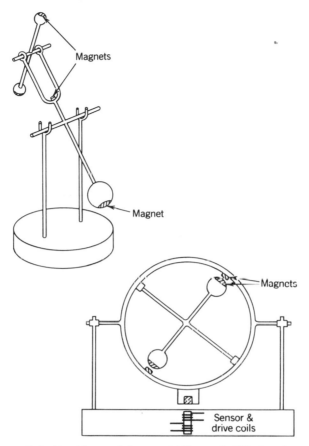

Figure C-4   Double pendulum and "Space ball" chaotic toys.

Several configurations are possible as shown in Figure C-4. In all cases, the pivot point of the second arm is forced by the motion of the driven pendulum. In some versions of this toy, small magnets on both arms interact when the second arm rotates past the primary arm.

A simple but clever driving circuit is used to provide current impulses to a driving magnet as shown in Figure C-5. When the lower pendulum oscillates, the magnetic field in the attached magnet generates a voltage in a coil in the base circuit. This voltage is applied to a transistor which begins to conduct when this motion-induced voltage reaches a critical value. During the conduction phase, current can flow out of the 9-V battery into a second coil wrapped around the magnet, thus providing a pulsed torque to the pendulum. In most cases, the motion of the driven pendulum is almost periodic, whereas the second arm performs chaotic rotations. Professor Alan Wolf of Cooper Union,

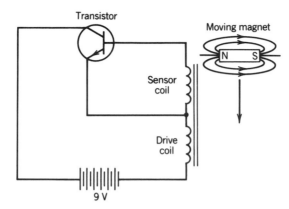

Figure C-5 Pulsed torque circuit for the chaotic pendulum toy.

New York City, and colleagues have analyzed this toy and have shown the motion to be chaotic.

C.4 NEON BULB CHAOTIC TOY

For those with an electrical bent, another toy is one described by Rolf Landauer of IBM, Yorktown Heights, New York, in 1977 in an internal IBM memorandum entitled ''Poor Man's Chaos.'' A similar study was published by Gollub et al. (1978). The circuit is shown in Figure C-6 and consists of two neon bulb circuits coupled together. A single circuit can perform relaxation oscillations (e.g., see Minorsky, 1962). When coupled together, the two circuits can exhibit stationary, periodic, or chaotic dynamics that are made visible to the observer by the flashing neon bulb. In the Gollub et al. paper, however, tunnel diodes were used in place of the neon bulbs and inductances were added to the circuits.

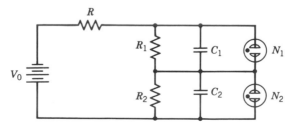

Figure C-6 Circuit for neon bulb experiment. [After R. Landauer (1977).]

Landauer was inspired to build this chaotic toy by memories of a similar exercise in a U.S. Navy electronic technician training program in 1945. This gives further evidence to my claim that chaotic vibrations were observed in the past but were seen as curiosities because there were no theoretical foundations for their serious study.

C.5 ROLLER COASTER CHAOS

This demonstration was sent to me by Professor Lawrence Virgin of Duke University, North Carolina. The experiment mimics the dynamics of a double-well potential Duffing oscillator

$$m\ddot{x} + c\dot{x} - ax + bx^3 = F_0 \cos \Omega t$$

In this device, a small cart rides on a track with one hill and two valleys (Figure C-7). The excitation is provided by moving the whole track harmonically by a motor and a scotch yoke mechanism.

To obtain an output, a wheel on the cart is coupled to a potentiometer by means of a chain linkage. Care was taken to prevent slippage

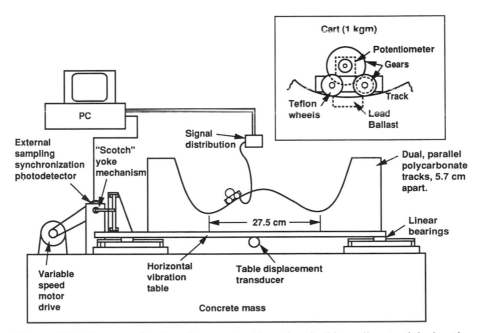

Figure C-7 Diagram showing the construction of a double-well potential chaotic oscillator with motorized cart. (Courtesy, Prof. L. Virgin, Duke University)

between the cart wheels and the track. Output from the potentiometer, which was proportional to the motion along the curved path, was sent to a laboratory computer for analysis or displayed on an oscilloscope.

A cheap version of this device for qualitative demonstration is to buy a flexible tape draftsman's curve and shape it into a double-well configuration. Then place the tape between two flat thin plastic plates and place a ball bearing or toy marble between the plates. Hand excitation will then show how the ball jumps from one side to the next in a somewhat erratic way. This is not precise, but it is inexpensive!

C.6   DUFFING OSCILLATOR DEMONSTRATOR

This design was sent to me by Professors Edwin Kreuzer and C. Wilmers of Technische Universität Hamburg, Federal Republic of Germany. It is based on a form of the Duffing equation which generates the Japanese attractor of Ueda (1979):

$$\ddot{x} + k\dot{x} + x^3 = B \cos t$$

The small analog computer diagram is shown in Figure C-8. The excitation frequency is 400 Hz. The controls allow a range of parameters, $0 < k < 1.0$, $0 < B < 20$, which should take one into and out of chaos. The output is displayed on an oscilloscope. A Poincaré map output has also been designed into this device. An audio speaker output is provided to hear the difference between periodic and chaotic oscillations.

C.7   WATER-TAP, PING-PONG BALL CHAOS

This idea was given to me by Professor M. A. Gol'dshtik of the Institute of Thermophysics, Novosibirsh, USSR, during a visit to my laboratory. As shown in Figure C-9, one places a ping-pong ball or other light-weight sphere on a flat plate and then places the ball under a jet of flowing water, most easily obtained from a water tap. Under the right flow, the ball will perform chaotic-looking oscillations on the plate. I do not know of any analysis of this clever experiment, however. It is a very simple demonstration of flow-induced chaotic vibrations as discussed in Chapter 4.

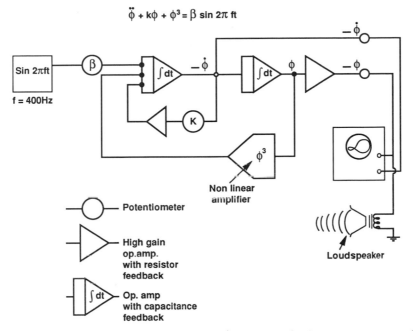

$$\ddot{\phi} + k\phi + \phi^3 = \beta \sin 2\pi \, ft$$

Figure C-8 Sketch of analog computer circuit to create the Japanese attractor (Figure 3-33).

C.8 ROTT'S COUPLED PENDULUMS

A chaotic toy that I have seen on many a dynamicists desk is a two- or three-arm coupled pendula device. I suspect that this demonstration has been around a long time, however. Professor N. Rott of Stanford University some time ago published a paper describing the mechanics and construction of these toys (Rott, 1970). The analysis is based on nonlinear resonance and, in particular, the 2:1 resonance. For example, in the two-link device shown in Figure C-10, two angles describe the configuration: α, γ.

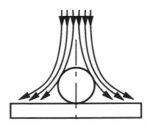

Figure C-9 Sketch of water-jet, ping-pong-ball chaos experiment.

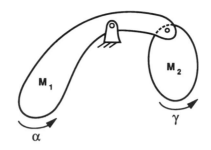

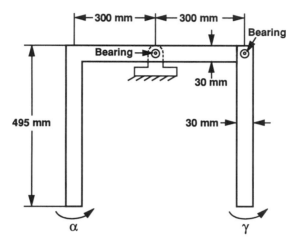

Figure C-10   Sketch of a 2:1 resonance double-pendulum demonstrator.

For small motions about the stable equilibrium position, Rott derives equations of motion of the form

$$\ddot{\alpha} + \omega_1^2\alpha = F(\alpha, \gamma, \dot{\gamma}, \ddot{\gamma})$$

$$\ddot{\gamma} + \omega_2^2\gamma = G(\alpha, \gamma, \dot{\alpha}, \ddot{\alpha})$$

where F, G are nonlinear coupling terms and the constants ω_1^2, ω_2^2 are related to the geometry. Neglecting the coupling, one has two oscillators with frequencies ω_1, ω_2. Rott adjusts the geometry of his pendulums so that $\omega_1/\omega_2 = 1:2$ and optimizes the design so that the relative amplitudes of the coupled device are equal. This allows energy to flow easily from one pendulum to the other and can result in some transient chaotic behavior. He also has designs for two other devices, one of which is a three-pendulum device that looks like a puppet (Figure C-11). Needless to say, the choice of good bearings in these devices

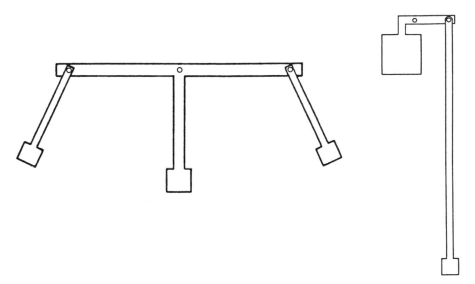

Figure C-11   Multiple-pendulum toys.

is essential, because they are excited by hand and eventually damp out depending on the aerodynamic and bearing damping. Professor Rott has recently marketed an executive toy based on his paper called "Pendemonium."

C.9   KINETIC ART

Sculptural art has evolved from an essentially static medium in the 19th century to a dynamic art form in the 20th century, especially in the works of Alexander Calder, Jean Tinguely, and George Rickey. Part of the fascination with kinetic art is the changing variety of forms or patterns so that designing dynamic sculpture to produce chaotic behavior is often essential to the success of the piece. Calder's mobiles may be seen as generalizations of Rott's coupled pendulums, though it is not clear whether Calder had any analytical insight into his designs such as nonlinear resonance. His multiple-pendulum mobiles are often excited by natural air currents in the installation space or the main support rotates with constant speed, such as the one in the National Gallery of Art in Washington, D.C. (e.g., see Lipman, 1976; also see Figure C-12).

Tinguely, a Swiss-born artist, created dynamic machines with coupled wheels, cogs, gears, and other mechanisms. In 1975 he completed

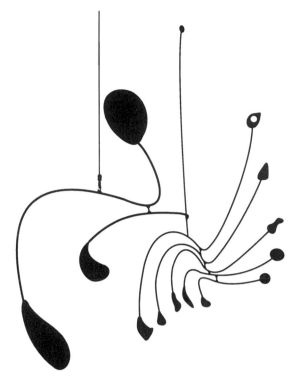

Figure C-12 "Hanging spider" multiple-pendulum mobile of A. Calder, 1940. [From Lipman (1976).]

a piece called *Chaos I* which was created for the small town of Columbus, Indiana, appropriately the headquarters of a machine maker, Cummins Diesel. In a retrospective on Tinguely's life work, the author, Hulten (1987), described *Chaos I* in terms a modern dynamicist would identify with:

> Mechanically, *Chaos I* is a very sophisticated construction, with movements that change rhythm and forms that can alter themselves: from a quiet, calm and graceful tempo, parts of the sculpture will break into a frenzy, gesticulating aggressively.

Sounds like intermittent chaos. The input to Tinguely's chaotic-like machines is often a motor running at constant speed.

George Rickey, an American-born sculptor, creates pieces that are simpler than those of Tinguely. Of special note here are his multiple-pendulum, long needle-like arms, weighted at the bottom. It is not clear, however, if the pendulums are strongly coupled, except through the flexibility of the base (as in the manner of Huygen's pendulum clocks). Usually installed out of doors, they are excited by slight breezes in the air and certainly look unpredictable.

A study of kinetic art and chaos has been published by Viet et al. (1983).

APPENDIX D

BOOKS ON NONLINEAR DYNAMICS, CHAOS, AND FRACTALS

D. K. Arrowsmith and C. M. Place, *An Introduction to Dynamical Systems,* Cambridge University Press, 1990.
This mathematical text is less formal than others. A readable graduate level book.

G. L. Baker and J. P. Gollub, *Chaotic Dynamics,* Cambridge University Press, 1990.
This elementary paperback uses the pendulum to present basic notions of nonlinear and chaotic dynamics

M. Barnsley, *Fractals Everywhere,* Academic Press, 1988.
This is an introductory mathematical book on fractals at the upperclass undergraduate level. Very readable with many computer experiments and color graphics.

P. Bergé, Y. Pomeau, and C. Vidal, *Order Within Chaos: Towards a Deterministic Approach to Turbulence,* John Wiley and Sons, 1984.
This readable book has a fluid mechanics orientation and a good physical discussion of the basic mathematical ideas of chaos.

Robert L. Devaney, *An Introduction to Chaotic Dynamical Systems,* 2nd ed., Benjamin/Cummings, 1986, 1989.
This is an introductory mathematical text with virtually no physical examples. However, it is readable at the upperclass undergraduate level.

Robert L. Devaney, *Chaos, Fractals, and Dynamics,* Addison–Wesley, 1990.
This is a companion to the previous book and presents computer experiments of dynamical systems.

M. S. El Naschie, *Stress, Stability and Chaos,* McGraw-Hill, 1990.
This book is written for structural engineers and introduces modern concepts of catastrophe theory and nonlinear dynamics.

K. Falconer, *Fractal Geometry: Mathematical Foundations and Application,* John Wiley & Sons, 1990.
Even though this book is mathematically based it is readable, though with fewer physical examples than the Feder book (see below).

Jens Feder, *Fractals,* Plenum Press, 1988.
This book introduces concepts of fractals for physical scientists especially in the context of material characterization.

John Guckenheimer and Philip Holmes, *Nonlinear Oscillations, Dynamical Systems, and Bifurcations of Vector Fields,* Vol. 42, Springer-Verlag, 1983.
One of the classic texts in the field. Although a few physical examples are used as motivation the authors assume a high level of mathematical sophistication of the reader. Belongs on every dynamicists bookshelf, however.

Peter Hagedorn, *Non-Linear Oscillations,* 2nd ed., Oxford University Press, 1988.
This is an introductory text on nonlinear vibration in mechanics, but does not discuss chaos.

Chihiro Hayashi, *Nonlinear Oscillations in Physical Systems,* Princeton University Press, 1985.
Hayashi is one of the pioneers of nonlinear oscillation in electrical systems. This is a reprint of his earlier work. Contains early reference to basin boundaries and subharmonic motion.

E. Atlee Jackson, *Perspectives of Nonlinear Dynamics,* Vols. 1 and 2, Cambridge University Press.
This is a wide ranging, mathematical physics text with both physical and mathematical discussion.

A. J. Lichtenberg and M. A. Lieberman, *Regular and Stochastic Motion,* Vol. 38, Springer-Verlag, 1983.
This is now another classical text with a more mathematical physics orientation. Treats conservative chaos as well as dissipative systems.

Ali Hasan Nayfeh and Dean T. Mook, *Nonlinear Oscillations,* John Wiley & Sons, 1979.
This text predates the chaos revolution. It presents a wide survey of classical perturbation methods in the context of various physical problems. Another required text for the dynamicist bookshelf.

G. Nicolis and I. Prigogine, *Exploring Complexity, an Introduction,* W. H. Freeman, 1989.
This is a philosophical text with some calculus level mathematics. An interesting scientific discussion from a Nobel laureate.

H. O. Peitgen, H. Jürgens, and D. Saupe, *Fractals in the Classroom: Part One, Introduction to Fractals and Chaos,* Springer-Verlag, New York, 1992.
This is a very introductory text, with many examples and good graphics. It is less formal than the Barnsley book. Few physical examples, however.

S. N. Rasband, *Chaotic Dynamics of Nonlinear Systems,* John Wiley & Sons, 1990.
This is a very readable mathematical introduction to chaotic dynamics for those with a modest calculus background. Less formal than the Devaney book.

M. Schroeder, *Fractals, Chaos and Power Laws: Minutes from an Infinite Paradise,* W. H. Freeman and Company, New York, 1991.
Another philosophical book written by a physicist with some mathematical notes and physical examples.

H. G. Schuster, *Deterministic Chaos,* 2nd ed., VHC Publishers, Weinheim, Federal Republic of Germany, 1988.
This is very nice physics-oriented introductory book on chaos. Very readable.

I. Stewart, *Les Fractals* (in French), Berling Press, 1982.
This is a clever cartoon book written for the child in everyone. A high school student with elementary French can enjoy this book.

Wanda Szemplinska-Stupnicka, *The Behavior of Nonlinear Vibrating Systems,* Vol. I,II, Kluwer Academic Publ. Dordrecht, The Netherlands.
Volume I treats single degree of freedom problems, while Volume II deals with multidegree of freedom problems. The methods presented involve classical perturbation and stability analysis in a modern context.

J. M. T. Thompson and H. B. Stewart, *Nonlinear Dynamics and Chaos,* John Wiley & Sons, 1986.
Introductory text on bifurcations in dynamical systems with some physical examples. Less emphasis on chaos per se.

S. Wiggins, *Introduction to Applied Nonlinear Dynamical Systems and Chaos,* Springer-Verlag, 1990.
Despite claims to being introductory, this excellent text requires some mathematical sophistication. Few physical examples.

REFERENCES

Abarbanel, H. D. I., Brown, R., and Kennel, M. B. (1991). "Lyapunov Exponents in Chaotic Systems: Their Importance and Their Evaluation Using Observed Data," invited review for *Mod. Phys. Lett. B.*

Abdullaev, S. S., and Zaslavsky, G. M. (1988). "Fractals and Ray Dynamics in a Longitudinally Inhomogeneous Medium," *Sov. Phys. Acoust.* **34**(4), 334–336.

Abraham, N. B., Albano, A. M., Das, B., DeGuzman, G., Yang, S., Gioggia, R. S., Puccioni, G. P., and Tredicce, J. R. (1986). "Calculating the Dimension of Attractors from Small Data Sets," *Phys. Lett.* **114A**(5), 217–221.

Abraham, R. H., and Shaw, C. D. (1983). *Dynamics: The Geometry of Behavior,* Aerial Press, Santa Cruz, CA.

Aref, H. (1984). "Stirring by Chaotic Advection," *J. Fluid Mech.* **143,** 1–21.

Aref, H. (1990). "Chaotic Advection of Fluid Particles," *Philos. Trans. R. Soc. London A* **333,** 273–288.

Aref, H., and Balachandar, S. (1986). "Chaotic Advection in a Stokes Flow," *Phys. Fluids* **29**(11), 3515–3521.

Argoul, F., Arneodo, A., and Richetti, P. (1987a). "Experimental Evidence for Homoclinic Chaos in the Belousov–Zhabotinski Reaction," *Phys. Lett. A* **120**(6), 269–275.

Argoul, F., Arneodo, A., Richetti, P., Roux, J. C., and Swinney, H. L. (1987b). "Chemical Chaos: From Hints to Confirmation," *Acc. Chem. Res.* **20,** 436–442.

Argoul, F., Arneodo, A., Grasseau, G., Gagne, Y., Hopfinger, E. J., and Frisch, U. (1989) "Wavelet analysis of turbulence reveals the multi-fractal nature of the Richardson cascade," *Nature* **338,** March 2, 51–53.

Arnold, V. I. (1978). *Ordinary Differential Equations,* MIT Press, Cambridge, MA.

Arnold, V. I. (1991). "Cardiac Arrhythmias and Circle Mappings," *Chaos* **1,** 20–24.

Arecchi, F. T, Meucci, R., and Gadomski, W. (1987). *Phys. Rev. Lett.* **58,** 2205.

Aubry, N., Holmes, P., Lumley, J. L., and Stone, E. (1988). "The Dynamics of Coherent Structures in the Wall Region of a Turbulent Boundary Layer," *J. Fluid. Mech.* **192,** 115–173.

Babitsky, V. I., Landa, P. S., Ol'khovoi, A. F., and Perminov, S. M. (1982). "Stochastic Behavior of Auto-Oscillating Systems with Inertial Self-Excitation," *Z. Angew. Math. Mech.* **66**(2), 73–81.

Baillieul, J., Brockett, R. W., and Washburn, R. B. (1980). "Chaotic Motion in Nonlinear Feedback Systems," *IEEE Trans. Circuits Syst.* **CAS-27,** (11), 990–997.

Bajaj, A. K., and Sethna, P. R. (1984). "Flow Induced Bifurcations to Three-dimensional Oscillatory Motions in Continuous Tubes," *SIAM J. Appl. Math.* **44**(2), 270–286.

Baker, N. H., Moore, P. W., and Spiegel, E. A. (1971). "Aperiodic Behavior of a Nonlinear Oscillator," *Q. J. Mech. Appl. Math.,* **24**(4), 391–422.

Bapat, C. N., Sankar, S., and Popplewell, N. (1986). "Repeated Impacts on a Sinusoidally Vibrating Table Reappraised," *J. Sound Vib.* **108**(1), 99–115.

Barnsley, M. (1988). *Fractals Everywhere,* Academic Press, New York.

Baryshnikova, Yu. S., Zaslavsky, G. M., Lupyan, E. A., Moiseyev, S. S., and Sharkov, E. A. (1989). "Fractal Analysis of the Pre-hurricane Atmosphere from Satellite Data," *Adv. Space Res.* **9**(7), 405–408.

Bassett, M. R., and Hudson, J. L. (1988). "Shil'nikov Chaos During Copper Electrodissolution," *J. Phys. Chem.* **92,** 6963–6966.

Battelino, P. M., Grebogi, C., Ott, E., and Yorke, J. A. (1988). "Multiple Coexisting Attractors, Basin Boundaries and Basic Sets," *Physica D* **32,** 296–305.

Bau, H. H., and Torrance, K. E. (1981). "On the Stability and Flow Reversal of an Asymmetrically Heated Open Convection Loop," *J. Fluid Mech.* **106,** 412–433.

Beletzky, V. V. (1990). "Nonlinear Effects in Dynamics of Controlled Two-Legged Walking," in *Nonlinear Dynamics in Engineering Systems,* edited by W. Schielen, Springer-Verlag, Berlin, 17–26.

Bender, C. M., and Orszag, S. A. (1978). *Advanced Mathematical Methods for Scientists and Engineers,* McGraw–Hill, New York.

Benettin, G., Cercignani, C., Galgani, L., and Giogilli, A. (1980a). "Universal Properties in Conservative Dynamical Systems," *Lett. Nuovo Cimento* **28,** 1–4.

Benettin, G., Galgani, L., Giogilli, A., and Strelcyn, J. M. (1980b). "Lyapunov Characteristic Exponents for Smooth Dynamical Systems and for Hamiltonian Systems; A Method for Computing All of Them. Part 2: Numerical Application," *Meccanica* **15**, 21–30.

Bennedettini, F., and Moon, F. C. (1992). "Experiments on the Chaotic Dynamics of Two Masses on a Hanging Cable," Cornell University Report.

Bergé, P. (1982). "Study of the Phase Space Diagrams Through Experimental Poincaré Sections in Prechaotic and Chaotic Regimes," *Phys. Scr.* **Tl,** 71–72.

Bergé, P., Dubois, M., Manneville, P., and Pomeau, P. (1980). "Intermittency Rayleigh–Bénard Convection," *J. Phys. (Paris) Lett.* **41**(15), L341–L345.

Bergé, P., Pomeau, Y. and Vidal C. (1984). *Order Within Chaos,* Hermann, Paris.

Bergé, P., Pomeau, Y., and Vidal, C. (1985). *L'Ordre dans le Chaos,* Hermann, Paris.

Bleher, S., Grebogi, C., and Ott, E. (1990). "Bifurcation to Chaotic Scattering," *Physica D* **46,** 87–121.

Brandstater, A., and Swinney, H. L. (1987). "Strange Attractors in Weakly Turbulent Couette–Taylor Flow," *Phys. Rev. A* **35**(5), 2207–2220.

Brandstater, A., Swift, J., Swinney, H. L., Wolf, A., Farmer, J. D., Jen, E., and Crutchfield, J. P. (1983). "Low-Dimensional Chaos in a Hydrodynamics System," *Phys. Rev. Lett.* **51**(6), 1442–1445.

Brandstater, A., Swift, J., Swinney, H. L., and Wolf, A. (1984). "A Strange Attractor in a Couette–Taylor Experiment," in *Turbulence and Chaotic Phenomena Fluids,* Proceedings of the IUTAM Symposium (Kyoto), edited by T. Tatsumi, North–Holland, Amsterdam.

Brillouin, L. (1946). *Wave Propagation in Periodic Structures,* Dover Publ., New York.

Brillouin, L. (1964). *Scientific Uncertainty and Information,* Academic Press, New York.

Brorson, S. D., Dewey, D., and Linsay, P. S. (1983). "The Self-Replicating Attractor of a Driven Semiconductor Oscillator," *Phys. Rev. A Rapid Commun.*

Bryant, P., and Jeffries, C. (1984a). "Bifurcations of a Forced Magnetic Oscillator Near Points of Resonance," *Phys. Rev. Lett.* **53**(3), 250–253.

Bryant, P., and Jeffries, C. (1984b). "Experimental Study of Driven Nonlinear Oscillator Exhibiting Hopf Bifurcations, Strong Resonances, Homoclinic Bifurcations and Chaotic Behavior," Lawrence Berkeley Laboratory report, LBL-16949, January.

Bryant, P., Brown, R., and Abarbanel, H. D. I. (1990). "Lyapunov Exponents from Observed Time Series," *Phys. Rev. Lett.* **65**(13), 1523–1526.

Bucko, M. F., Douglass, D. H., and Frutch, H. H. (1984). "Bounded Regions Chaotic Behavior in the Control Parameter Space of a Driven Non-linear Resonator," *Phys. Lett.* **104**(8), 388–390.

Campbell, D. K., and Rose, H. A. (eds.) (1982). "Order in Chaos," in *Proceedings of a Conference Held at Los Alamos National Lab,* May, North–Holland, Amsterdam.

Carlson, J. M., and Langer, J. S. (1989). "Properties of Earthquakes Generated by Fault Dynamics," *Phys. Rev. Lett.* **62**(22), 2632–2635.

Chaiken, J., Chevray, R., Tabor, M., and Tan, Q. M. (1986). *Proc. Roy. Soc. A* **408**, 165.

Chaiken, J., Chu, C. K., Tabor, M., and Tan, Q. M. (1987). "Lagrangian Turbulence and Spatial Complexity in a Stokes Flow," *Phys. Fluids* **30**(3), 687–694.

Chen, S.-S. (1983). "Instability Mechanisms and Stability Criteria of a Group of Circular Cylinders Subjected to Cross-flow. Part 1: Theory," *J. Vib. Acoust., Stress, Reliability in Design,* **105**, 51–58.

Chen, S.-S. and Jendrzejczyk, J. A. (1985). "General Characteristics, Transition, and Control of Instability of Tubes Conveying Fluid," *J. Acoust. Soc. Am.* **77**, 887–895.

Chirikov, B. V. (1979). "A Universal Instability of Many-Dimensional Oscillator Systems," *Phys. Rep.* **52**, 265.

Chirikov, B. V., and Vecheslavov, V. V. (1989). "Chaotic Dynamics of Comet Halley," *Astron. Astrophys.* **221**, 146–154.

Chua, L. O. and Yang, L. (1988). "Cellular Neural Networks: Theory," *IEEE Trans. Circuits Systems* **CS-35**, 1257–1290.

Chua, L. O., Komuro, M., and Matsumoto, T. (1986). "The Double Scroll Family," *IEEE Trans. Circuits Syst.* **CAS-33**(11), 1073–1097.

Ciliberto, S., and Gollub, J. P. (1985). "Chaotic Mode Competition in Parametrically Forced Surface Waves," *J. Fluid Mech.* **158**, 381–398.

Clemens, H., and Wauer, J. (1981). "Free and Forced Vibrations of a Snap-Through Oscillator," Report of the Institute für Technische Mechanik, Universität Karlsruhe.

Comparin, R. J., and Singh, R. (1990). "An Analytical Study of Automotive Neutral Gear Rattle," *J. Mech. Design* **112**, 237–245.

Copeland, G. S. (1991). "Flow-Induced Vibration and Chaotic Motion of a Slender Tube Conveying Fluid," Doctoral Dissertation, Theoretical and Applied Mechanics, Cornell University, Ithaca, NY.

Copeland, G. S. and Moon, F. C. (1992). "Chaotic Flow-Induced Vibrations of a Flexible Tube with End End Mass," *J. Fluids Struct.* (to appear).

Creveling, H. F., dePaz, J. F., Baladi, J. Y., and Schoenhals, R. J. (1975). "Stability Characteristics of a Single-Phase Free Convection Loop," *J. Fluid Mech.* **67**, 65–84.

Croquette, V., and Poitou, C. (1981). "Cascade of Period Doubling Bifurcations and Large Stochasticity in the Motions of a Compass," *J. Phys. (Paris) Lett.* **42,** L537–L539.

Crutchfield, J. P. (1988). "Spatio-temporal Complexity in Nonlinear Image Processing," *IEEE Trans. Circuits. Syst.* **35**(7), 770–780.

Crutchfield, J. P., Farmer, J. D., and Huberman, B. A. (1982). "Fluctuations and Simple Chaotic Dynamics," *Phys. Rep.* **92**(2), 45–82.

Crutchfield, J. P. and Kaneko, K. (1987). "Phenomenology of Spatio-Temporal Chaos," in *Directions in Chaos,* Vol. 1, edited by Hao Bai-lin, World Scientific, Singapore.

Crutchfield, J. P., and Packard, N. H. (1982). "Symbolic Dynamics of One-Dimensional Maps: Entropies, Finite Precursor, and Noise," *Int. J. Theor. Phys.* **21**(6/7), 433–465.

Cusumano, J. P. (1990), "Low-Dimensional, Chaotic, Nonplanar Motions of the Elastica: Experiment and Theory," Doctoral Dissertation, Department of Theoretical and Applied Mechanics, Cornell University, Ithaca, NY.

Cusumano, J. P., and Moon, F. C. (1990). "Low Dimensional Behavior in Chaotic Nonplanar Motions of a Forced Elastic Rod: Experiment and Theory," in *IUTAM Symposium on Nonlinear Dynamics in Engineering Systems,* edited by W. Schiehlen, Springer-Verlag, Berlin, 59–66.

Cvitanovic, P., and Predrag, P. (1984). *Universality in Chaos,* Heyden, Philadelphia.

Davies, M. A., and Moon, F.C. (1992). "3-D Spatial Chaos in the Elastica and the Spinning Top: Kirchhoff Analog," Cornell University Report, April 1992.

Degn, H., Holder, A. V., and Folsen, L. (1987). *Chaos in Biological Systems,* Plenum Press, New York.

Den Hartog, J. P. (1940). *Mechanical Vibrations,* McGraw–Hill, New York.

Devaney, R. L. (1989). *An Introduction to Chaotic Dynamical Systems,* 2nd ed. Benjamin/Cummings Publishers.

Devaney, R. L. (1990). *Chaos, Fractals, and Dynamics,* Addison–Wesley, Reading, MA.

Ding, M., Grebogi, C., and Ott, E. (1989a). "Dimensions of Strange Nonchaotic Attractors," *Phys. Lett.* A **137**(4,5), 167–172.

Ding, M., Grebogi, C., and Ott, E. (1989b). "Evolution of Attractors in Quasiperiodically Forced Systems: From Quasiperiodic to Strange Nonchaotic to Chaotic," *Phys. Rev.* A **39**(5), 2593–2598.

Ditto, W. L., Spano, M. L., Savage, H. T., Rauseo, S. N., Heagy, J., and Ott, E. (1990). "Experimental Observation of a Strange Nonchaotic Attractor," *Phys. Rev. Lett.* **65**(5), 533–536.

Dowell, E. H. (1975). *Aeroelasticity of Plates and Shells,* Noordhoff International, Groningen.

Dowell, E. H. (1982). "Flutter of a Buckled Plate as an Example of Chaotic Motion of a Deterministic Autonomous System," *J. Sound Vib.* **85**(3), 333–344.

Dowell, E. H. (1984). "Observation and Evolution of Chaos for an Autonomous System," *J. Appl. Mech.*, paper no. 84-WA/APM-15.

Dowell, E. H., and Ilgamova, M. (1988). *Studies in Nonlinear Aeroelasticity,* Springer-Verlag, New York.

Dowell, E. H., and Pezeshki, C. (1986). "On the Understanding of Chaos in Duffing's Equation Including a Comparison with Experiment," *J. Appl. Mech.* **53**(1), 5–9.

Dowell, E. H., and Virgin, L. N. (1990). "On Spatial Chaos, Asymptotic Modal Analysis, and Turbulence," *Trans. ASME* **57**, 1094–1097.

Dubois, M., Bergé, P., and Croquette, V. (1982). "Study of Nonsteady Convective Regimes Using Poincaré Sections," *J. Phys. (Paris) Lett.* **43**, L295–L298.

Eckmann, J. P. (1981). "Roads to Turbulence in Dissipative Dynamical Systems," *Rev. Mod. Phys.* **53**(4), part 1, 643–654.

El Naschie, M. S. (1990). *Stress, Stability and Chaos in Structural Engineering: An Energy Approach,* McGraw–Hill, UK.

El Naschie, M. S., and Kapitaniak, T. (1990). "Soliton Chaos Models for Mechanical and Biological Elastic Chains," *Phys. Lett. A* **147**(5,6), 275–281.

Evenson, D. A. (1967). *Nonlinear Flexural Vibrations of Thin-Walled Circular Cylinders.* NASA Technical Note, NASA TN D-4090, August.

Everson, R. M. (1986). "Chaotic Dynamics of a Bouncing Ball," *Physica* **19D**, 355–383.

Falconer, K. (1990). *Fractal Geometry,* John Wiley & Sons, Chichester, England.

Farmer, J. D., Ott, E., and Yorke, J. A. (1983). "The Dimension of Chaotic Attractors," *Physica* **7D**, 153–170.

Feder, J. (1988). *Fractals,* Plenum Press.

Feeny, B. F., and Moon, F. C. (1989). "Autocorrelation on Symbol Dynamics for a Chaotic Dry-Friction Oscillator," *Phys. Lett. A* **141**(8,9), 397–400.

Feeny, B. F., and Moon, F. C. (1992a). "Chaos in a Forced Dry-Friction Oscillator: Experiments and Numerical Modeling," *J. Sound Vib.* (to appear).

Feeny, B. F., and Moon, F. C. (1992b). "Bifurcation Sequences of a Coulomb Friction Oscillator," *Nonlin. Dynam.* (accepted).

Feeny, B. F., Moon, F. C., Chen, P.-Y., and Mukherjee, S. (1992). "Chaotic Mixing in Rigid Perfectly Plastic Materials," *Bifurcation and Chaos* (submitted).

Feigenbaum, M. J. (1978). "Qualitative Universality for a Class of Nonlinear Transformations," *J. Stat. Phys.* **19**(1), 25–52.

Feigenbaum, M. J. (1980). "Universal Behavior in Nonlinear Systems," *Los Alamos Sci.* (summer), 4–27.

Fineberg, J., Gross, S. P., Marder, M., and Swinney, H. (1991). "Instability in Dynamic Fractures," *Phys. Rev. Lett.* **67**(4), 457–460.

Ford, J. (1988). "Quantum Chaos, Is There Any?" in *Directions in Chaos,* Vol. 2, edited by Hao Bai-lin, World Scientific, Singapore.

Franjione, J. G., Leong, C.-W., and Ottino, J. M. (1989). "Symmetries Within Chaos: A Route to Effective Mixing," *Phys. Fluids A* **1**(11), 1772–1783.

Fung, Y. C. (1958). "On Two-Dimensional Panel Flutter," *J. Aerospace Sci.* **25**(3), 145–159.

Gardiner, C.W. (1985). *Handbook of Stochastic Methods for Physics, Chemistry and the Natural Sciences,* Springer-Verlag, New York.

Geist, K., and Lauterborn, W. (1988). "The Nonlinear Dynamics of the Damped and Driven Toda Chain," *Physica D* **31,** 103–116.

Geist, K., Parlitz, U., and Lauterborn, W. (1990). "Comparison of Different Methods for Computing Lyapunov Exponents," *Prog. Theor. Phys.* **83**(5), 875–893.

Gibiat, V. (1988). "Phase Space Representations of Acoustical Musical Signals," *J. Sound Vib.* **123**(3), 529–536.

Glass, L. (1991). "Cardiac Arrhythmias and Circle Maps—A Classical Problem," *Chaos* **1,** 13–19.

Glass, L. and Mackey, M. C. (1988). *From Clocks to Chaos: The Rhythms of Life,* Princeton University Press, Princeton, NJ.

Glass, L., Guevau, X., and Shrier, A. (1983). "Bifurcation and Chaos in Periodically Stimulated Cardiac Oscillator," *Physica* **7D,** 89–101.

Glazier, J. A., and Libchaber, A. (1988). "Quasi-Periodicity and Dynamical Systems: An Experimentalist's View," *IEEE Trans. Circuits Systems* **35**(7), 790–809.

Gleick, J. (1987). *CHAOS: Making a New Science,* Viking Penguin, New York.

Goldberger, A. L., and West, B. J. (1987a). "Applications of Nonlinear Dynamics to Clinical Cardiology," *Ann. N.Y. Acad. Sci.* **504,** 195–211.

Goldberger, A. L., and West, B. J. (1987b). "Fractals in Physiology and Medicine," *Yale J. Biol. Med.* **60,** 421–435.

Goldberger, A. L., Bhargava, V., West, B. J., and Mandell, A. J. (1986). "Some Observations on the Question: Is Ventricular Fibrillation 'Chaos'?" *Physica* **19D,** 282–289.

Goldberger, A. L., Rigney, D. R., and West, B. J. (1990). "Chaos and Fractals in Human Physiology," *Sci. Am.* **262**(2), 42–49.

Goldstein, H. (1980). *Classical Mechanics,* 2nd ed., Addison–Wesley, Reading, MA.

Gollub, J. P., and Benson, S. V. (1980). "Many Routes to Turbulent Convection," *J. Fluid Mech.* **100**(3), 449–470.

Gollub, J. P., Brunner, T. O., and Danly, B. G. (1978). "Periodicity and Chaos in Coupled Nonlinear Oscillators," *Science* **200,** 48–50.

Gollub, J. P., and Ramshankar, R. (1991). "Spatiotemporal Chaos in Interfacial Waves," in *New Perspectives in Turbulence,* edited by S. Orszag and L. Sirovich, Springer-Verlag, New York, 165–194.

Gollub, J. P., Romer, E. J., and Socalar, J. E. (1980). "Trajectory Divergence for Coupled Relaxation Oscillators: Measurements and Models," *J. Stat. Phys.* **23**(3), 321–333.

Golnaraghi, M. F. and Moon, F. C. (1991). "Experimental Evidence for Chaotic Response in a Feedback System," *J. Dynam. Syst. Meas. Control* **113,** 183–187.

Gorman, M., Widmann, P. J., and Robbins, K. A. (1984). "Chaotic Flow Regimes in a Convection Loop," *Phys. Rev. Lett.* **52**(25), 2241–2244.

Gorman, M., Widmann, P. J., and Robbins, K. A. (1986). "Nonlinear Dynamics of a Convection Loop: A Quantitative Comparison of Experiment with Theory," *Physica D* **19,** 255–267.

Gouldin, F.C. (1987). "An Application of Fractals to Modeling Premixed Turbulent Flames," *Combust. Flame* **68,** 249–266.

Grabec, I. (1986). "Chaos Generated by the Cutting Process," *Phys. Lett. A* **117**(8), 384–386.

Grabec, I. (1988). "Chaotic Dynamics of the Cutting Process," *Int. J. Mach. Tools Manufact.* **28**(1), 19–32.

Grassberger, P., and Proccacia, I. (1983). "Characterization of Strange Attractors," *Phys. Rev. Lett.* **50,** 346–349.

Grassberger, P., and Proccacia, I. (1984). "Dimensions and Entropies of Strange Attractors from a Fluctuating Dynamics Approach," *Physica* **13D,** 34–54.

Grebogi, C., Ott, E., Pelikan, S., and Yorke, J. A. (1984). "Strange Attractors That Are Not Chaotic," *Physica* **13D,** 261–268.

Grebogi, C., Ott, E., and Yorke, J. A. (1983a). "Crises, Sudden Changes in Chaotic Attractors and Transient Chaos," *Physica* **7D,** 181–200.

Grebogi, C., Ott, E., and Yorke, J. A. (1983b). "Fractal Basin Boundaries, Long Lived Chaotic Transients and Unstable-Unstable Pair Bifurcation," *Phys. Rev. Lett.* **50**(13), 935–938.

Grebogi, C., Ott, E., and Yorke, J. A. (1985a). "Attractors on an *N*-Torus: Quasiperiodicity Versus Chaos," *Physica* **15D,** 354–373.

Grebogi, C., Ott, E., and Yorke, J. A. (1985b). "Superpersistent Chaotic Transients," *Ergod. Theor. Dynam. Syst.* **5,** 341–372.

Grebogi, C., McDonald, S. W., Ott, E., and Yorke, J. A., (1985c). "Exterior Dimension of Fat Fractals," *Phys. Lett.* **110A**(l), 1–4.

Grebogi, C., Ott, E., and Yorke, J. A. (1986). "Metamorphoses of Basin Boundaries in Nonlinear Dynamical Systems," *Phys. Rev. Lett.* **56**(10), 1011–1014.

Guckenheimer, J. (1982). "Noise in Chaotic Systems," *Nature* **298,** 358.

Guckenheimer, J., and Holmes, P. J. (1983). *Nonlinear Oscillations, Dynamical Systems and Bifurcations of Vector Fields,* Springer-Verlag, New York.

Guevara, M. R., Shrier, A., and Glass, L. (1990). "Chaotic and Complex Cardiac Rhythms," in *Cardiac Electrophysiology,* edited by D. P. Zipes and J. Jalife, W. B. Saunders, Philadelphia.

Gwinn, E. G., and Westervelt, R. M. (1985). "Intermittent Chaos and Low-Frequency Noise in the Driven Damped Pendulum," *Phys. Rev. Lett.* **54**(15), 1613–1616.

Gwinn, E. G., and Westervelt, R. M. (1986). "Frequency Locking, Quasiperiodicity, and Chaos in Extrinsic Ge," *Phys. Rev. Lett.* **57**(8), 1060–1063.

Gwinn, E. G., and Westervelt, R. M. (1987). "Scaling Structure of Attractors at the Transition from Quasiperiodicity to Chaos in Electronic Transport in Ge," *Phys. Rev. Lett.* **59**(2), 157–160.

Hack, J. T. (1957). "Studies of Longitudinal Stream Profiles in Virginia and Maryland," *US. Geol. Surv. Prof. Pap.* **294–B,** 45–97.

Hagedorn, P. (1988). *Non-linear Oscillations,* 2nd ed. Oxford University Press, New York.

Haken, H. (1985). *Light, Vol. 2: Laser Light Dynamics,* North–Holland, Amsterdam.

Halsey, T. C., Jensen, M. H., Kadanoff, L. P., Procaccia, I., and Shraiman, B. I. (1986). "Fractal Measures and Their Singularities: The Characterization of Strange Sets," *Phys. Rev. A* **33**(2), 1141–1151.

Hao, B.-L. (1984). *Chaos,* World Scientific Publishers, Singapore.

Harrison, R. G., and Biswas, D. J. (1986). "Chaos in Light," *Nature* **22,** 394–401.

Hart, J. E. (1984). "A New Analysis of the Closed Loop Thermosyphon," *Int. J. Heat Mass Transfer* **27**(1), 125–136.

Haucke, H., and Ecke, R. E. (1987). "Mode-Locking and Chaos in Rayleigh–Benard Convection," *Physica D.*

Hayashi, C. (1953). *Forced Oscillations in Nonlinear Systems,* Nippon Printing and Publishing Co., Osaka, Japan.

Hayashi, C. (1985). *Nonlinear Oscillations in Physical Systems.* Princeton University Press, Princeton, NJ.

Hayashi, H., Ishizuka, S., Ohta, M., and Hirakawa, J. (1982). "Chaotic

Behavior in the 'Onchidium' Giant Neuron under Sinusoidal Stimulation,'' *Phys. Lett.* **88A**(8), 435–438.

Held, G. A., and Jeffries, C. (1986). ''Quasiperiodic Transitions to Chaos of Instabilities in an Electron-Hole Plasma Excited by ac Perturbations at One and at Two Frequencies,'' *Phys. Rev. Lett.* **56**(11), 1183–1186.

Helleman, R. H. G. (1980a). ''Self-Generated Chaotic Behavior in Nonlinear Mechanics,'' *Fund. Probl. Stat. Mech.* **5**, 165–233. Reprinted in *Universality in Chaos,* edited by B. P. Cvitanovic, Adam Hilger Publishing Ltd., Briston, 1984, 420–488.

Helleman, R. H. G. (ed.) (1980b). ''Nonlinear Dynamics,'' *Ann. N.Y. Acad. Sci.* **357.**

Hendriks, F. (1983). ''Bounce and Chaotic Motion in Print Hammers,'' *IBM J. Res. Dev.* **27**(3), 273–280.

Hénon, M. (1969). ''Numerical Study of Quadratic Area-Preserving Mappings,'' *Q. Appl. Math.* **27**(3), 291–312.

Hénon, M. (1976). ''A Two-Dimensional Map with a Strange Attractor,'' *Commun. Math. Phys.* **50,** 69.

Hénon, M. (1982). ''On the Numerical Computation of Poincaré Map,'' *Physica* **5D**, 412–414.

Hénon, M., and Heiles, C. (1964). ''The Applicability of the Third Integral of Motion: Some Numerical Experiments,'' *Astron. J.* **69**(l), 73–79.

Hentschel, H. G. E. and Proccacia, I. (1983). ''The Infinite Number of Generalized Dimensions of Fractals and Strange Attractors,'' *Physica* **8,** 435–444.

Hockett, K., and Holmes, P. J. (1986). ''Josephson Junction, Annulus Maps, Birkhoff Attractors, Horseshoes and Rotation Sets,'' *Ergod. Th. and Dynam. Sys.* **6,** 205–239.

Holmes, P. J. (1979). ''A Nonlinear Oscillator With a Strange Attractor,'' *Philos. Trans. R. Soc. London A* **292,** 419–448.

Holmes, P. J. (ed.) (1981). *New Approaches to Nonlinear Problems in Dynamics,* SIAM, Philadelphia.

Holmes, P. J. (1982). ''The Dynamics of Repeated Impacts with a Sinusoidally Vibrating Table,'' *J. Sound Vib.* **84,** 173–189.

Holmes, P. J. (1984). ''Bifurcation Sequences in Horseshoe Maps: Infinitely Many Routes to Chaos,'' *Phys. Lett. A* **104**(6, 7), 299–302.

Holmes, P. J. (1985). ''Dynamics of a Nonlinear Oscillator with Feedback Control,'' *J. Dynam. Syst. Meas. Control* **107,** 159–165.

Holmes, P. J. (1986). ''Chaotic Motions in a Weakly Nonlinear Model for Surface Waves,'' *J. Fluid Mech.* **162,** 365–388.

Holmes, P. J. (1990a). ''Nonlinear Dynamics, Chaos, and Mechanics,'' *Appl. Mech. Rev.* **43**(5), S23–S39.

Holmes, P. J. (1990b). "Poincaré Celestial Mechanics, Dynamical-Systems Theory and 'Chaos'," *Phys. Rep.* **193**(3), 138–163.

Holmes, P. J., and Moon, F. C. (1983). "Strange Attractors and Chaos in Nonlinear Mechanics," *J. Appl. Mech.* **50**, 1021–1032.

Holzfuss, J., and Lauterborn, W. (1989). "Liapunov Exponents from a Time Series of Acoustic Chaos," *Phys. Rev. A* **39**(4), 2146–2152.

Holzfuss, J., and Mayer-Kress, G. (1986). "An Approach to Error-Estimation in the Application of Dimension Algorithms," *Dimensions and Entropies in Chaotic Systems,* edited by G. Mayer-Kress, Springer-Verlag, Berlin.

Homsy, G. M. (1987). "Viscous Fingering in Porous Media," *Annu. Rev. Fluid Mech.* **19**, 271–311.

Horowitz, F. G., and Ruina, A. (1989). "Slip Patterns in a Spatially Homogeneous Fault Model," *J. Geophys. Res.* **94**(B8), 10,279–10,298.

Hsu, C. S. (1981). "A Generalized Theory of Cell-to-Cell Mapping for Nonlinear Dynamical Systems," *J. Appl. Mech.* **48**, 634–642.

Hsu, C. S. (1987). *Cell to Cell Mapping,* Springer-Verlag, New York.

Hsu, C. S., and Kim, M. C. (1985). "Statistics for Strange Attractors by Generalized Cell Mapping," *J. Stat. Phys.*

Huang, J., and Turcotte, D. L. (1990). *Geophys. Res. Lett.* **17**, 223.

Hudson, J. L., Mankin, J. C., and Rössler, O. E. (1984). "Chaos in Continuous Stirred Chemical Reaction," in *Stochastic Phenomena and Chaotic Behavior in Complex Systems,* edited by P. Schuster, Springer-Verlag, Berlin, pp. 98–105.

Hulten, P. (1987). *Jean Tinguely: A Magic Stronger than Death,* Abbeville Press, New York, pp. 220–225.

Hunt, E. R. (1991). "High-Period Orbits Stabilized in the Diode Resonator," in *Proceedings of the Experimental Chaos Conference* (1–3 October 1991, Arlington, VA), World Scientific Publishers, Singapore.

Hunt, E.R., and Rollins, R. W. (1984). "Exactly Solvable Model of a Physical System Exhibiting Multidimensional Chaotic Behavior," *Phys. Rev. A* **29**(2), 1000–1002.

Iansiti, M., Hu, Q., Westervelt, R. M., and Tinkham, M. (1985). "Noise and Chaos in a Fractal Basin Boundary Regime of a Josephson Junction," *Phys. Rev. Lett.* **55**(7), 746–749.

Iooss, G., and Joseph, D. D. (1980). *Elementary Stability and Bifurcation Theory,* Springer-Verlag, New York.

Isomaki, H. M., von Boehm, J., and Raty, R. (1985). "Devil's Attractors and Chaos of a Driven Impact Oscillator," *Phys. Lett.* **107A**(8), 343–346.

Jackson, E. A. (1990). *Perspectives of Nonlinear Dynamics,* Vols. 1 and 2. Cambridge University Press, Cambridge, England.

James, G. E., Harrell, E. M., II, and Roy, R. (1990). "Intermittency and Chaos in Intracavity Doubled Lasers. II," *Phys. Rev. A* **41**(5), 2778–2790.

Jensen, M. H., Kadanoff, L. P., Libchaber, A., Procaccia, I., and Stavans, J. (1985). "Global Universality at the Onset of Chaos: Results of a Forced Rayleigh–Bénard Experiment," *Phys. Rev. Lett.* **55**(25), 2798–2801.

Jensen, R. V. (1989). "Quantum Chaos," in *The Ubiguity of Chaos*, edited by S. Krasner, pp. 98–104.

Jones, S. W., Thomas, O. M., and Aref, H. (1989). "Chaotic Advection by Laminar Flow in a Twisted Pipe," *J. Fluid Mech.* **209**, 335–357.

Kadanoff, L. P. (1983). "Roads to Chaos," *Phys. Today* (Dec.), 46–53.

Kaneko, K. (1989). "Towards Thermodynamics of Spatiotemporal Chaos," *Prog. Theor. Phys.* (*Suppl.*) **99**, 263–287.

Kaneko, K. (1990). "Supertransients, Spatiotemporal Intermittency and Stability of Fully Developed Spatiotemporal Chaos," *Phys. Lett. A* **149**(2,3), 105–112.

Kapitaniak, T. and Steeb, W.-H. (1991). "Transition to Hyperchaos in Coupled Generalized van der Pol Equations," *Phys. Lett. A* **152**(1,2), 33–36.

Kapitaniak, T., Ponce, E., and Wojewoda, J. (1990). "Route to Chaos via Strange Nonchaotic Attractors," *J. Phys. A: Math. Gen.* **23**, L383–L387.

Kaplan, J., and Yorke, J. A. (1978). *Springer Lecture Notes in Mathematics*, No. 730, p. 228.

Karagiannis, K., and Pfeiffer, F. (1991). "Theoretical and Experimental Investigations of Gear-Rattling," *Nonlin. Dyn.* **2**, 367–387.

Keolian, R., Turkevich, L. A., Putterman, S. J., Rudnick, I., and Rudnick, J. A. (1981). "Subharmonic Sequences in the Faraday Experiment: Departures from Period Doubling," *Phys. Rev. Lett.* **47**(16), 1133–1511.

Kirchhoff, G. (1859). "Über das Gleichgewicht und die Bewegung eines unendlich dünnen elastischen Stabcs," *J. Mathematik* (*Crelle*) **56**, 285–313.

Kittel, A., Peinke, J., Klein, M., Baier, G., Parisi, J., and Rössler, O. E. (1990). "Existence of Nowhere Differentiable Boundaries in a Realistic Map," *Z. Naturforsch.* **45a**, 1377–1379.

Klinker, T., Meyer-Ilse, X., and Lauterborn, W. (1984). "Period Doubling and Chaotic Behavior in a Driven Toda Oscillator," *Phys. Lett. A* **101**(8), 371–375.

Kobayashi, S. (1962). "Two-dimensional Panel Flutter 1. Simply Supported Panel," *Trans. Jpn. Soc. Aeronaut. Space Sci.* **5**(8), 90–102.

Koch, B. P. and Levey, R. W. (1985). "Subharmonic and Homoclinic Bifurcation in a Parametrically Forced Pendulum," *Physica D*, 1–13.

Koch, P. M. (1990). "Microwave Excitation and Ionization of Excited Hydrogen Atoms," in *The Ubiguity of Chaos*, edited by S. Krasner, American Association for the Advanced Science, Washington, DC, pp. 75–97.

Kostelich, E. J., Grebogi, C., Ott, E., and Yorke, J. A. (1987). "The Double Rotor Chaotic Attractor," *Physica D*.

Kostelich, E. J., and Yorke, J. A. (1985). "Lorenz Cross Sections and Dimension of the Double Rotor Attractor," in *Dimensions and Entropies in Chaotic Systems,* edited by G. Mayer-Kress, Springer-Verlag, Berlin, pp. 62–66.

Kreuzer, E. J. (1985). "Analysis of Chaotic Systems Using the Cell Mapping Approach," *Ingenieur-Archiv* **55**, 285–294.

Kuhn, T. (1962), *The Structure of Scientific Revolutions,* The University of Chicago Press, Chicago.

Kunert, A. and Pfeiffer, F. (1991). "Description of Chaotic Motion by an Invariant Distribution at the Example of the Driven Duffing Oscillator," *Int. Series Num. Math.* **97,** 225–230.

Landau, L. D. (1944). "On the Problem of Turbulence," *Akad. Nauk Doklady* **44** 339; Russian original reprinted in *Chaos,* Vol. 107, edited by Bai-Lin Hao, World Scientific Publishers, Singapore.

Landauer, R. (1977). "Poor Man's Chaos," Internal Memo IBM Corp.

Lauterborn, W., and Cramer, E. (1981). "Subharmonic Route to Chaos Observed in Acoustics," *Phys. Rev. Lett.* **47**(20), 1145.

Lauterborn, W., and Holzfuss, J. (1991). "Acoustic Chaos," *Int. J. Bifurcation Chaos* **1**(1), 13–26.

Lee, C.-K., and Moon, F. C. (1986). "An Optical Technique for Measuring Fractal Dimensions of Planar Poincaré Maps," *Phys. Lett. A* **114**(5), 222–226.

Leipnik, R. B., and Newton, T. A. (1981). "Double Strange Attractors in Rigid Body Motion with Linear Feedback Control," *Phys. Lett.* **86A**(2), 63–67.

Leong, C. W., and Ottino, J. M. (1989). "Experiments on Mixing Due to Chaotic Advection in a Cavity," *J. Fluid Mech.* **209,** 463–499.

Levi, B. G. (1990). "Are Fractures Fractal or Quakes Chaotic?" *Phys. Today,* 17–19.

Levin, P. W., and Koch, B. P. (1981). "Chaotic Behavior of a Parametrically Excited Damped Pendulum," *Phys. Lett. A* **86**(2), 71–74.

Lewis, T. J., and Guevara, M. R. (1990). "Chaotic Dynamics in an Ionic Model of the Propagated Cardiac Action Potential," *J. Theor. Biol.* **146,** 407–432.

Li, G.-X. (1984). *Chaotic Vibrations of a Nonlinear System with Five Equilibrium States,* M.S. Thesis, Cornell University, Ithaca, NY, August.

Li, G.-X. (1987). "Homoclinic Bifurcation and Chaos in a Two Degree-of-Freedom Magneto-Mechanical Nonlinear System," Doctoral Dissertation, Department of Theoretical and Applied Mechanics, Cornell University, Ithaca, NY.

Li, G.-X, and Moon, F. C. (1990a). "Criteria for Chaos of a Three-Well Potential Oscillator with Homoclinic and Heteroclinic Orbits," *J. Sound Vib.* **136,** 17–34.

Li, G.-X., and Moon, F. C. (1990b). "Fractal Basin Boundaries in a Two-Degree-of-Freedom Nonlinear System," *Nonlin. Dynam.* **1,** 209–219.

Li, G.-X., Rand, R. H., and Moon, F. C. (1990). "Bifurcations and Chaos in a Forced Zero-Stiffness Impact Oscillator," *Int. J. Non-Linear Mech.* **25**(4), 417–432.

Libchaber, A. (1982). "Convections and Turbulence in Liquid Helium I," *Physica* **109 & 110B,** 1583–1589.

Libchaber, A., Fauve, S., and Laroche, C. (1982). "Two-Parameter Study of the Routes to Chaos," in *Order in Chaos,* edited by D. K. Campbell and H. A. Rose, North–Holland, Amsterdam.

Libchaber, A., and Maurer (1978). *J. Phys. Lett.* **39,** 369.

Libchaber, A., and Maurer (1980). *J. Phys. Colloq.* **C3,** 41–45.

Lichtenberg, A. J., and Lieberman, M. A. (1983). *Regular and Stochastic Motion,* Springer-Verlag, New York.

Lieberman, M. A., and Tsang, K. Y. (1985). "Transient Chaos in Dissipatively Perturbed Near-Integrable Hamiltonian Systems," *Phys. Rev. Lett.* **55**(9), 908–911.

Liebovitch, L. S., and Czegledy, F. P. (1991). "Fractal, Chaotic, and Self-Organizing Critical System Descriptions of the Kinetics of Cell Membrane Ion Channels," in *Complexity, Chaos and Biological Evolution,* edited by E. Moskilde.

Liebovitch, L. S., and Toth, T. I. (1991). "A Model of Ion Channel Kinetics Using Deterministic Chaotic Rather than Stochastic Processes," *J. Theor. Biol.* **148,** 243–267.

Lin, Y. K. (1976). *Probabilistic Theory of Structural Dynamics,* Krieger, Huntington, NY.

Linsay, P. S. (1981). "Period Doubling and Chaotic Behavior in a Driven, Anharmonic Oscillator," *Phys. Rev. Lett.* **47**(19),1349–1352.

Linsay, P. S. (1985). The Structure of Chaotic Behavior in a PN Junction Oscillator, MIT, Department of Physics Report.

Linsay, P. S., and Cumming, A. W. (1989). "Three-Frequency Quasiperiodicity, Phase Locking, and the Onset of Chaos," *Physica D* **40,** 196–217.

Lipman, J. (1976). *Calder's Universe,* Viking Press, New York.

Lorenz, E. N. (1963). "Deterministic Non-periodic Flow," *J. Atmos. Sci.* **20,** 130–141.

Lorenz, E. N. (1984). "The Local Structure of a Chaotic Attractor in Four Dimensions," *Physica* **13D,** 90–104.

Louis, E., Guinea, F., and Flores, F. (1986). "The Fractal Nature of Fracture," in *Fractals in Physics,* edited by L. Pietronero and E. Tosatti, Elsevier Science Publishers, B.V., Amsterdam, pp. 177–180.

Love, A. E. H. (1922). *A Treatise on the Mathematical Theory of Elasticity,* 4th ed., Dover, New York.

Lung, C. W. (1986). "Fractals and the Fracture of Cracked Metals" in *Fractals in Physics,* edited by L. Pietronero and E. Tosatti, Elsevier Science Publishers, B.V, Amsterdam, 189–192.

L'vov, V. S., Predtechensky, A. A., and Chernykh, A. I. (1981). "Bifurcation and Chaos in a System of Taylor Vortices: A Natural and Numerical Experiment, *Soviet Phys. JETP* **53,** 562.

Maganza, C., Caussé, R. and Laloe, F. (1986). "Bifurcations, Period Doubling and Chaos in Clarinetlike Systems," *Europhys. Lett.* **1**(6), 295–302.

Mahaffey, R. A. (1976). "Anharmonic Oscillator Description of Plasma Oscillations," *Phys. Fluids* **19,** 1387–1391.

Måløy, K. J., Feder, J., and Jøssang T. (1985). "Viscous Fingering Fractals in Porous Media," *Phys. Rev. Lett.* **55**(24), 2688–2691.

Malraison, G., Atten, P., Bergé, P., and Dubois, M. (1983). "Dimension of Strange Attractors: An Experimental Determination of the Chaotic Regime of Two, Convective Systems," *J. Phys. Lett.* **44,** 897–902.

Mandelbrot, B. B. (1977). *Fractals, Form, Chance, and Dimension,* W. H. Freeman, San Francisco.

Mandelbrot, B. B. (1982). *The Fractal Geometry of Nature,* W. H. Freeman, San Francisco.

Mandelbrot, B. B., Passoja, D. E., and Paullay, A. J. (1984). "Fractal Character of Fracture Surfaces of Metals," *Nature* **308,** 721–722.

Manneville, P., and Pomeau, Y. (1980). "Different Ways to Turbulence in Dissipative Dynamical Systems," *Physica* **1D,** 219–226.

Marcus, P.S. (1988). "Numerical Simulation of Jupiter's Great Red Spot," *Nature* **331,** 693–696.

Markworth, A. J., McCoy, J. K., and Rollins, R. W. (1988). "Chaotic Dynamics in an Atomistic Model of Environmentally Assisted Fracture," *J. Mater. Res.* **3**(4), 675–686.

Martin, S., and Martienssen, W. (1986). "Circle Maps and Mode Locking in the Driven Electrical Conductivity of Barium Sodium Niobate Crystals," *Phys. Rev. Lett.* **56**(15), 1522–1525.

Marzec, C. J., and Spiegel, E. A. (1980). "Ordinary Differential Equations with Strange Attractors," *SIAM J. Appl. Math.* **38**(3), 403–421.

Matsumoto, T. (1984). "A Chaotic Attractor from Chua's Circuit," *IEEE. Trans Circuits Syst.* **CAS-31**(12), 1055–1058.

Matsumoto, T., Chua, L. O., and Tanaka, S. (1984). "Simplest Chaotic Nonautonomous Circuit," *Phys. Rev. A.* **30**(2),1155–1157.

Matsumoto, T., Chua, L. O., and Komuro, M. (1985). "The Double Scroll," *IEEE Trans. Circuits Syst.* **CAS-32**(8), 798–818.

May, R. M. (1976). "Simple Mathematical Models with Very Complicated Dynamics," *Nature* **261,** 459–467.

May, R. M. (1987). "Chaos and the Dynamics of Biological Populations," *Proceedings of the Royal Society,* **A413.**

Mayer-Kress, G. (1985). "Introductory Remarks," in *Dimensions and Entropies in Chaotic Systems,* edited by G. Mayer-Kress, Springer-Verlag, Berlin.

McCaughan, F. E. (1989). "Application of Bifurcation Theory to Axial Flow Compressor Instability," *Trans. ASME,* **111,** 426–433.

McDonald, S. W., Grebogi, C., Ott, E., and Yorke, J. A. (1985). "Fractal Basin Boundaries," *Physica* **17D,** 125–153.

McLaughlin, J. B. (1981). "Period-Doubling Bifurcations and Chaotic Motion for Parametrically Forced Pendulum," *J. Stat. Phys.* **24**(2), 375–388.

Meneveau, C., and Sreenivasan, K. R. (1987). "Simple Multifractal Cascade Model for Fully Developed Turbulence," *Phys. Rev. Lett.* **59**(13), 1424–1427.

Metropolis, N, Stein, N. L., and Stein, P. R. (1973). "On Finite Limit Sets for Transformations on the Unit Interval," *J. Combinatorial Theory* **15,** 25–44.

Meyers, S. D., Sommeria, J., and Swinney, H. L. (1989). "Laboratory Study of the Dynamics of Jovian-type Vortices," *Physica D* **37,** 515–530.

Mielke, A., and Holmes, P. (1988). "Spatially Complex Equilibria of Buckled Rods," *Arch. Rational Mech. Anal.* **101**(4), 319–348.

Miles, J. (1984a). "Resonant Motion of Spherical Pendulum," *Physica* **11D,** 309–323.

Miles, J. (1984b). "Resonantly Forced Motion of Two Quadratically Coupled Oscillators," *Physica* **13D,** 247–260.

Miles, J. (1984c). "Resonant, Nonplanar Motion of a Stretched String," *J. Acoust. Soc. Am.* **75**(5), 1505–1510.

Milonni, P. W., Shih, M.-L., and Ackerhalt, J. R. (1987). *Chaos in Laser-Matter Interactions,* World Scientific, Singapore.

Minorsky, N. (1962). *Nonlinear Oscillations,* Van Nostrand, Princeton, NJ.

Moon, F. C. (1980a). "Experiments on Chaotic Motions of a Forced Nonlinear Oscillator: Strange Attractors," *ASME J. Appl. Mech.* **47,** 638–644.

Moon, F. C. (1980b). "Experimental Models for Strange Attractor Vibration in Elastic Systems," in *New Approaches to Nonlinear Problems in Dynamics,* edited by P. J. Holmes, pp. 487–495.

Moon, F. C. (1984a). *Magneto-Solid Mechanics,* John Wiley & Sons, New York.

Moon, F. C. (1984b), "Fractal Boundary for Chaos in a Two State Mechanical Oscillator," *Phys. Rev. Lett.* **53**(60), 962–964.

Moon, F. C. (1986). "New Research Directions for Chaotic Phenomena in Solid Mechanics," in *Perspectives in Nonlinear Dynamics,* edited by R. Cawley and Shlesinger, World Scientific Publishers, Singapore, 230–258.

Moon, F. C. (1988). "Chaotic Vibrations of a Magnet Near a Superconductor," *Phys. Lett. A* **132**(5), 249–252.

Moon, F. C. (1990). "Spacial and Temporal Chaos in Elastic Continua," in *Nonlinear Dynamics in Engineering Systems,* IUTAM Symposium, Stuttgart, Germany, Springer-Verlag, Berlin, 194–204.

Moon, F. C. (1991). "Experimental Measurement of Chaotic Attractors in Solid Mechanics," *Chaos* **1**(1), 31–41.

Moon, F. C. (1993). *Superconducting Levitation and Bearings,* John Wiley & Sons, New York.

Moon, F. C., and Broschart, T. (1991). "Chaotic Sources of Noise in Machine Acoustics," *Arch. Appl. Mech.* **61**, 438–448.

Moon, F. C., and Holmes, P. J. (1979). "A Magnetoelastic Strange Attractor," *J. Sound Vib.* **65**(2), 275–296; "A Magnetoelastic Strange Attractor," *J. Sound Vib.* **69**(2), 339.

Moon, F. C., and Holmes, W. T. (1985). "Double Poincaré Sections of a Quasiperiodically Forced, Chaotic Attractor," *Phys. Lett. A* **111**(4), 157–160.

Moon, F. C., and Li, G.-X. (1985a). "The Fractal Dimension of the Two-Well Potential Strange Attractor," *Physica* **17D,** 99–108.

Moon, F. C., and Li, G.-X. (1985b). "Fractal Basin Boundaries and Homoclinic Orbits for Periodic Motion in a Two-Well Potential," *Phys. Rev. Lett.* **55**(14), 1439–1442.

Moon, F. C., and Shaw, S. W. (1983). "Chaotic Vibration of a Beam with Nonlinear Boundary Conditions," *J. Nonlinear Mech.* **18,** 465–477.

Moon, F. C., Cusumano, J., and Holmes, P. J. (1987). "Evidence for Homoclinic Orbits as a Precursor to Chaos in a Magnetic Pendulum," *Physica* **24D,** 383–390.

Moon, F. C., Holmes, W. T., and Khoury, P. (1991). "Symbol Dynamic Maps of Spatial-Temporal Chaotic Vibrations in a String of Impact Oscillators," *Chaos* **1**(1), 65–68.

Moore, D. W., and Spiegel, E. A. (1966). "A Thermally Excited Non-linear Oscillator," *Astrophys. J.* **143**(3), 871–887.

Murat, M., and Aharony, A. (1986). "Viscous Fingering and Diffusion-Limited Aggregates Near Percolation," *Phys. Rev. Lett.* **57**, 1875–1878.

Nayfeh, A. H., and Khdeir, A. A. (1986). "Nonlinear Rolling of Ships in Regular Beam Seas," *Int. Shipbuilding Prog.* **33**(379), 40–49.

Nayfeh, A. H., and Mook, D. T. (1979). *Nonlinear Oscillations,* John Wiley & Sons, New York.

Newhouse, S., Ruelle, D., and Takens, F. (1978). "Occurrence of Strange Axiom A Attractors Near Quasiperiodic Flows on T^m, $m \geq 3$," *Commun. Math. Phys.* **64**, 35–40.

Nussbaum, J., and Ruina, A. (1987). "A Two Degree-of-Freedom Earthquake Model with Static/Dynamic Friction," *Pure Appl. Geophys.* **125**(4), 629–656.

Olinger, D. J., and Sreenivasan, K. R. (1988). "Nonlinear Dynamics of the Wake of an Oscillating Cylinder," *Phys. Rev. Lett.* **60**(9), 797–800.

O'Reilly, O. M. (1991). "Global Bifurcations in the Forced Vibration of a Damped String," *Int. J. Non-Linear Mech..*

Ott, E. (1981). "Strange Attractors and Chaotic Motions of Dynamical Systems," *Rev. Mod. Phys.* **53**(4), Part 1, 655–671.

Ott, E., Grebogi, C., and Yorke, J. A. (1990). "Controlling Chaos," *Phys. Rev. Lett.* **64**, 1196–1199.

Ottino, J. M. (1989a). *The Kinematics of Mixing: Stretching, Chaos, and Transport.* Cambridge University Press, England.

Ottino, J. M. (1989b). "The Mixing of Fluids," *Sci. Am.* **260**, 56–67.

Packard, N. H., Crutchfield, J. P., Farmer, J. D., and Shaw, R. S. (1980). "Geometry from a Time Series," *Phys. Rev. Lett.* **45**, 712.

Païdoussis, M. P. (1980). "Flow-Induced Vibrations in Nuclear Reactors and Heat Exchangers," in *Practical Experiences with Flow-Induced Vibrations,* edited by E. Naudascher and D. Rockwell, IAHR/, Springer-Verlag, Berlin.

Païdoussis, M. P., Li, G.-X., and Moon, F. C. (1989). "Chaotic Oscillations of the Autonomous System of a Constrained Pipe Conveying Fluid," *J. Sound Vib.* **135**(1), 1–19.

Païdoussis, M. P., and Moon, F. C. (1988). "Nonlinear and Chaotic Fluidelastic Vibrations of a Flexible Pipe Conveying Fluid," *J. Fluids Struct.* **2**, 567–591.

Parker, T. S., and Chua, L. O. (1989). *Practical Numerical Algorithms for Chaotic Systems,* Springer-Verlag, New York.

Pecora, L., and Carroll, T. (1991). "Driving Nonlinear Systems with Chaotic Signals," in *Proceedings of the 1st Experimental Chaos Conference* (1–3 October 1991, Arlington, VA), World Scientific Publishers, Singapore.

Peitgen, H.-O. Jürgens, H., and Saupe D. (1992). *Fractals for the Classroom, Part One: Introduction to Fractals and Chaos,* Springer-Verlag, New York.

Peitgen, H.-O., and Richter, P. H. (1986). *The Beauty of Fractals,* Springer-Verlag, Berlin.

Pezeshki, C., and Dowell, E. H. (1987). "An Examination of Initial Condition Maps for the Sinusoidally Excited Buckled Beam Modeled by Duffing's Equation," *J. Sound Vib.* **117**, 219–232.

Pezeshki, C., and Dowell, E. H. (1988). "On Chaos and Fractal Behavior in a Generalized Duffing's System," *Physica D* **32**, 194–209.

Pezeshki, C., Miles, W. H., Yang, Y. M., and Troutt, T. R. (1992). "On the

Use of Wavelet Transforms for Nonlinear Structural Dynamic Systems,'' Department of Mechanical and Materials Engineering, Washington State University report, submitted to *J. Sound Vib.*

Pfeiffer, F. (1988). ''Seltsame Attraktoren in Zahnradgetrieben,'' *Ingenieur Archiv.* **58,** 113–125.

Pfeiffer, F., and Kunert A. (1990). ''Rattling Models from Deterministic to Stochastic Processes,'' *Nonlinear Dynam.* **1,** 63–74.

Poddar, B., Moon, F. C., and Mukherjee, S. (1988). ''Chaotic Motion of an Elastic–Plastic Beam,'' *J. Appl. Mech.* **55,** 185–189.

Poincaré, H. (1921). *The Foundation of Science: Science and Method,* English Translation, The Science Press, New York.

Pomeau, Y., and Manneville, P. (1980). ''Intermittent Transition to Turbulence in Dissipative Dynamical Systems,'' *Commun. Math. Phys.* **74,** 189–197.

Pool, R. (1989). ''Quantum Chaos: Enigma Wrapped in a Mystery,'' *Res. News,* 17 February, pp. 893–895.

Popp, K. and Stelter, P. (1990). ''Nonlinear Oscillations of Structures Induced by Dry Friction,'' in *Nonlinear Dynamics Engineering Systems,* edited by W. Schielen, Springer-Verlag, Berlin, 233–240.

Prigogine, I., and Stengers, I. (1984). *Order Out of Chaos,* Bantam Books, Toronto.

Purwins, H. G., Klempt, G., and Berkemeier, J. (1987). ''Temporal and Spatial Structures of Nonlinear Dynamical Systems,'' *Festkörperprobleme* **27,** 27–61.

Purwins, H. G., Radehaus, C., and Berkemeier, J. (1988). ''Experimental Investigation of Spatial Pattern Formation in Physical Systems of Activator Inhibitor Type,'' *Z. Naturforsch.* A **43,** 17–29.

Rand, D., Ostlund, S., Sethna, J., and Siggia, E. D. (1982). ''Universal Transition from Quasiperiodicity to Chaos in Dissipative Systems,'' *Phys. Rev. Lett.* **49**(2), 387–390.

Rasband, S.N. (1990). *Chaotic Dynamics of Nonlinear Systems,* John Wiley & Sons, New York.

Richter, P. H., and Scholz, H.-J. (1984). ''Chaos in Classical Mechanics: The Double Pendulum,'' in *Stochastic Phenomena and Chaotic Behavior in Complex Systems,* edited by P. Schuster, Springer-Verlag, Berlin, pp. 86–97.

Rigney, D. R., and Goldberger, A. L. (1989). ''Nonlinear Mechanics of the Heart's Swinging During Pericardial Effusion,'' *Am. J. Physiol.* **257,** H1292–H1305.

Robbins, K. A. (1977). ''A New Approach to Subcritical Instability and Turbulent Transitions in a Simple Dynamo,'' *Math. Proc. Camb. Philos. Soc.* **82,** 309–325.

Rollins, R. W., and Hunt, E. R. (1982). "Exactly Solvable Model of a Physical System Exhibition Universal Chaotic Behavior," *Phys. Rev. Lett.* **49**(18), 1295–1298.

Romeiras, F. J., Bondeson, A., Ott, E., Antonsen, T. M., Jr., and Grebogi, C. (1989). "Quasiperiodic Forcing and the Observability of Strange Nonchaotic Attractors," *Physica Scr.* **40**, 442–444.

Romeiras, F.J., Ott, E., Grebogi, C., and Dayawansa, W.P. (1991). "Controlling Chaotic Dynamical Systems," in Proceedings of the 1991 American Control Conference, American Council, IEEE Service Center, Piscataway, NJ.

Rössler, O. E. (1976a). "Chemical Turbulence: Chaos in a Small Reaction-Diffusion System," *Z. Naturforsch.* **31**, 1168–1172.

Rössler, O. E. (1976b). "An Equation for Continuous Chaos," *Phys. Lett. A* **57**, 397.

Rössler, O. E. (1979). "An Equation for Hyperchaos," *Phys. Lett.* **71A**(2,3), 155–157.

Rott, N. (1970). "A Multiple Pendulum for the Demonstration of Non-linear Coupling," *J. Appl. Math. Phys.* **21**, 570–582.

Roux, T. C., Simoyi, R. H., and Swinney, H. L. (1983). "Observation of a Strange Attractor," *Physica* **8D**, 257–266.

Rubio, M. A., Bigazzi, P., Albavetti, L., and Ciliberto, S. (1989). "Spatiotemporal Regimes in Rayleigh–Bénard Convection in a Small Rectangular Cell," *J. Fluid Mech.* **209**, 309–334.

Russel, D. A., Hanson, J. D., and Ott, E. (1980). "Dimension of Strange Attractors," *Phys. Rev. Lett.* **45**(14), 1175–1178.

Russell, S., Caroselli, R., Mason, J., and Shah, M. (1991). "A Conjectured Solder Fatigue Law Based on Non-linear Dynamics," presented at the ASME Winter Annual Meeting, December 1–6, 1991.

Sagdeev, R. Z., Usikov, D. A., and Zaslavsky, G. M. (1988). *Nonlinear Physics,* Harwood Academic Publishers, Chur, Switzerland.

Saltzman, B. (1962). "Finite Amplitude Free Convection as an Initial Value Problem—I," *J. Atmos. Sci.* **19**, 329–341.

Sanchez, N. E., and Nayfeh, A. H. (1990). "Nonlinear Rolling Motions of Ships in Longitudinal Waves," *Int. Shipbuild. Prog.* **37**(411), 247–272.

Schaefer, D. W., and Keefer, K. D. (1984a). "Fractal Geometry of Silica Condensation Polymers," *Phys. Rev. Lett.* **53**(14), 1383–1386.

Schaefer, D. W., and Keefer, K. D. (1984b). "Structure of Soluble Silicates," *Mater. Res. Soc. Symp. Proc.* **32**, 1–14.

Schaefer, D. W., and Keefer, K. D. (1986). "Structure of Random Silicates, Polymers, Colloids, and Porous Solids," in *Fractals in Physics,* edited by Pietronero, L., and Tosatti, E., Elsevier Science Publishers, B.V., Amsterdam, pp. 39–45.

Schreiber, I., Kubicek, M., and Marak, M. (1980). "On Coupled Cells," in *New Approaches to Nonlinear Problems in Dynamics,* edited by P. J. Holmes, SIAM, Philadelphia, pp. 496–508.

Schuster, H. G. (1984). *Deterministic Chaos,* Physik-Verlag GmbH, Weinheim (F.R.G.).

Schuster, H. G. (1988). *Deterministic Chaos,* 2nd ed. Physik-Verlag GmbH, Weinheim (F.R.G.).

Scott, P. J. (1989). "Nonlinear Dynamic Systems in Surface Metrology," *Surf. Topogr.* **2,** 345–366.

Sethna, P. R., and Shaw, S. W. (1987). "On Codimension-Three Bifurcations in the Motion of Articulated Tubes Conveying a Fluid," *Physica D* **24,** 305–327.

Shaw, J., and Shaw, S. W. (1989). "The Onset of Chaos in a Two-Degree-of-Freedom Impacting System," *Trans. ASME* **56,** 168–174.

Shaw, R. (1981). "Strange Attractors, Chaotic Behavior and Information Flow," *Z. Naturforsch. A* **36,** 80–112.

Shaw, R. (1984). *The Dripping Faucet as a Model Chaotic System,* Aerial Press, Santa Cruz, CA.

Shaw, S. W. (1985). "The Dynamics of a Harmonically Excited System Having Rigid Amplitude Constraints, Parts 1, 2," *J. Appl. Mech.* **52**(2), 453–464.

Shaw, S. W., and Holmes, P. J. (1983). "A Periodically Forced Piecewise Linear Oscillator," *J. Sound Vib.* **90**(1), 129–155.

Shaw, S. W., and Rand, R. H. (1989). "The Transition to Chaos in a Simple Mechanical System," *Int. J. Non-Linear Mech.* **24**(1), 41–56.

Shimada, I., and Nagashima, T. (1979). "A Numerical Approach to Ergodic Problem of Dissipative Dynamical Systems," *Prog. Theor. Phys.* **61**(6), 1605–1616.

Shinbrot, T., Ott, E., Grebogi, C., and Yorke, J. A. (1990). "Using Chaos to Direct Trajectories to Targets," *Phys. Rev. Lett.* **26,** 3215–3218.

Simiu, E., and Cook, G. R. (1991). "Chaotic Motions of Self-Excited Forced and Autonomous Square Prisms," *J. Eng. Mech.* **117**(2), 241–259.

Simoyi, R. H., Wolf, A., and Swinney, H. L. (1982). "One-Dimensional Dynamics in a Multi-component Chemical Reaction," *Phys. Rev. Lett.* **49,** 245.

Singh, A. and Joseph, D. D. (1989). "Autoregressive Methods for Chaos on Binary Sequences for the Lorenz Attractor," *Phys. Lett. A* **135,** 247.

Singh, R., Xie, H., and Comparin, R. J. (1989). "Analysis of Automotive Neutral Gear Rattle," *J. Sound Vib.* **131**(2), 177–196.

Soliman, M. S., and Thompson, J. M. T. (1989). "Integrity Measures Quantifying the Erosion of Smooth and Fractal Basins of Attraction," *J. Sound Vib.* **135**(3), 453–475.

Soong, T. T. (1973). *Random Differential Equations in Science and Engineering,* Academic Press, New York.

Spano, M. L., and Ditto, W. L. (1991). "Taming Chaos Experimentally: A Primer," in *Proceedings of the 1st Experimental Chaos Conference* (1–3 October 1991, Arlington, VA), World Scientific Publishers, Singapore.

Sparrow, C. T. (1981). "Chaos in a Three-Dimensional Single Loop Flow System with a Piecewise Linear Feedback Function," *J. Math. Anal. A* **83,** 275–291.

Sparrow, C .T. (1982). *The Lorenz Equations: Bifurcations, Chaos, and Strange Attractors,* Springer-Verlag, New York.

Sreenivasan, K. R. (1986). "Chaos in Open Flow Systems," in *Dimension and Entropies* in *Chaotic Systems,* edited by G. Mayer-Kress, Springer-Verlag, New York.

Sreenivasan, K. R. (1991). "Fractals and Multifractals in Fluid Turbulence," *Annu. Rev. Fluid Mech.* **23,** 539–600.

Stavans, J., Thomae, S., and Libchaber, A. (19xx). "Experimental Study of the Attractor of a Driven Rayleigh–Bénard System," in *Dimensions and Entropies in Chaotic Systems,* edited by G. Mayer-Kress, Springer-Verlag, New York.

Steindl, A., and Troger, H. (1988). "Flow Induced Bifurcations to Three-dimensional Motions of Tubes with an Elastic Support," in *Trends in Applications of Mathematics to Mechanics,* Springer-Verlag, New York.

Stoker, J. J. (1950). *Nonlinear Vibrations,* Interscience, New York.

Swanson, P. D., and Ottino, J. M. (1990). "A Comparative Computational and Experimental Study of Chaotic Mixing of Viscous Fluids," *J. Fluid Mech.* **213,** 227–249.

Swetits, J. J., and Buoncristiani, A. M. (1988). "Shilnikov Instabilities in Laser Systems," *Phys. Rev. A* **38**(10), 5430–5432.

Swinn, E. G., and Westervelt, R. M. (1987). "Scaling Structure of Attractors at the Transition from Quasiperiodicity to Chaos in Electronic Transport in Ge," *Phys. Rev. Lett.* **59**(2), 157–160.

Swinney, H. L. (1983). "Observations of Order and Chaos in Nonlinear Systems," in *Order and Chaos,* edited by D. K. Campbell and H. A. Rose, North-Holland, Amsterdam, pp. 3–15.

Swinney, H. L. (1985). "Observations of Complex Dynamics and Chaos," in *Fundamental Problems in Statistical Mechanics Vl,* edited by E. G. D. Cohen, Elsevier Science Publishers, New York, pp. 253–289.

Swinney, H. L., and Gollub, J. P. (1978). "The Transition of Turbulence," *Phys. Today,* **31**(8), 41.

Symonds, P. S., and Yu, T. X. (1985). "Counterintuitive Behavior in a Problem or Elastic–Plastic Beam Dynamics," *J. Appl. Mech.* **52,** 517–522.

Szczygielski, W., and Schweitzer, G. (1985). "Dynamics of a High-Speed

Rotor Touching a Boundary," presented at the *IUTAM/IFT0MM Symposium on the Dynamics of Multibody Systems,* CISM, Udine, Italy.

Szemplinska-Stupnicka, W. (1992). "Cross-Well Chaos and Escape Phenomena in Driven Oscillators," *Nonlin. Dyn.* **3,** 225–243.

Szemplinska-Stupnicka, W., and Bajkowski, J. (1986). "The ½ Subharmonic Resonance and Its Transition to Chaotic Motion in a Non-linear Oscillator," *Int. J. Non-Linear Mech.* **21,** 401–419.

Tabor, M. (1989). *Chaos and Integrability in Nonlinear Dynamics,* John Wiley & Sons, New York.

Tam, W. Y., and Swinney, H. L. (1990). "Spatiotemporal Patterns in a One-Dimensional Open Reaction-Diffusion System," *Physica D* **46,** 10–22.

Tang, D. M., and Dowell, E. H. (1988). "On the Threshold Force for Chaotic Motions for a Forced Buckled Beam," *J. Appl. Mech.*

Tasi, J. (1990). "Evolution of Shock Waves in a One-Dimensional Lattice," *J. Appl. Phys.* **51,** 5804–5815.

Tatsumi, T. (ed.) (1984). *Turbulence and Chaotic Phenomena in Fluids,* North-Holland, Amsterdam.

Tel, T. (1990). "Transient Chaos" in *Directions in Chaos,* edited by B.-L. Hao, World Sci. Co., Singapore.

Termonia, Y., and Alexandrowicz, Z. (1983). *Phys. Rev. Lett.* **51**(14), 1265.

Testa, J., Perez, J., and Jeffries, C. (1982). "Evidence for Universal Chaotic Behavior in a Driven Nonlinear Oscillator," *Phys. Rev. Lett.* **48,** 714.

Thompson, J. M. T. (1983). "Complex Dynamics of Compliant Off-Shore Structures," *Proc. R. Soc. London* **A387,** 407–427.

Thompson, J. M. T. (1989a). "Chaotic Dynamics and the Newtonian Legacy," *Appl. Mech. Rev.* **42**(1), 15–24.

Thompson, J. M. T. (1989b). "Chaotic Phenomena Triggering the Escape from a Potential Well," *Proc. R. Soc. London A* **421,** 195–225.

Thompson, J. M. T., and Ghaffari, R. (1982). "Chaos After Period-Doubling Bifurcations in the Resonance of an Impact Oscillator," *Phys. Lett. A* **91**(1), 5–8.

Thompson, J. M. T., and Stewart, H. B. (1986). *Nonlinear Dynamics and Chaos,* John Wiley & Sons, Chichester, England.

Thompson, J. M. T., and Virgin, L. N. (1988). "Spatial Chaos and Localization Phenomena in Nonlinear Elasticity," *Phys. Lett A* **126**(8,9), 491–496.

Thompson, J. M. T., Rainey, R. C. T., and Soliman, M. S. (1990). "Ship Stability Criteria Based on Chaotic Transients from Incursive Fractals," *Philos. Trans. R. Soc. London A* **332,** 149–167.

Thomson, W. T. (1965), *Vibration Theory,* Prentice–Hall, Englewood Cliffs, NJ.

Toda, M. (1981). *Theory of Nonlinear Lattices,* Springer Series on Solid State Sciences, Vol. 20, Springer-Verlag, Berlin.

Toda, M. (1989). *Theory of Nonlinear Lattices,* 2nd ed., Springer-Verlag, New York.

Tongue, B. H. (1987). "On Obtaining Global Nonlinear System Characteristics Through Interpolated Cell Mapping," *Physics D* **28**, 401–408.

Tousi, S., and Bajaj, A. K. (1985). "Period-Doubling Bifurcations and Modulated Motions in Forced Mechanical Systems," *J. Appl. Mech.* **52**(2), 446–452.

Tsang, K. Y., and Lieberman, M. A. (1984).

Tseng, W.-Y., and Dugundji, J. (1971). "Nonlinear Vibrations of a Buckled Beam Under Harmonic Excitation," *J. Appl. Mech.* **38**, 467–476.

Tufillaro, N. B., and Albano, A. M. (1986). "Chaotic Dynamics of a Bouncing Ball," *Am. J. Phys.* **54**(10), 939–944.

Tung, P. C. and Shaw, S. W. (1988). "A Method for the Improvement of Impact Printer Performance," *Trans. ASME* **110**, 528–532.

Turcotte, D. L. (1992). *Fractals and Chaos in Geology and Geophysics,* Cambridge University Press, Cambridge, England.

Ueda, Y. (1979). "Randomly Transitional Phenomena in the System Governed by Duffing's Equation," *J. Stat. Phys.* **20**, 181–196.

Ueda, Y. (1980). "Steady Motions Exhibited By Duffing's Equation: A Picture Book of Regular and Chaotic Motions," in *New Approaches to Nonlinear Problems in Dynamics,* edited by P. J. Holmes, SIAM, Philadelphia, 311–322.

Ueda, Y. (1985). "Random Phenomena Resulting from Non-linearity in the System Described by Duffing's Equation," *Int. J. Non-Linear Mech.* **20**(5/6), 481–491.

Ueda, Y. (1991). "Strange Attractors and the Origin of Chaos," presented at International Symposium, "The Impact of Chaos on Science and Society," United Nations University and the University of Tokyo, 15–17 April 1991.

Ueda, Y., and Akamatsu, N. (1981). "Chaotically Transitional Phenomcna in the Forced Negative Resistance Oscillator," *Proc. IEEE ISCA8 '80.* Also *IEEE Trans. Circuits Syst.* **CAS-28**(3).

Ueda, Y., and Yoshida, S. (1987). "Attractor-Basin Phase Portraits of the Forced Duffing's Oscillator," in *Proceedings of the 1987 European Conference on Circuit Theory Design,* Paris, September 1987.

Ueda, Y., Doumoto, H., and Nobumoto, K. (1978). "An Example of Random Oscillations in Three-Order Self-Restoring System," *Proceedings of the Electric and Electronic Communication Joint Meeting,* Kansai District, Japan, October.

Ueda, Y., Nakajima, H., Hikihara, T., and Stewart, H. B. (1988). "Forced Two-Well Potential Duffing Oscillator," in *Dynamical Systems Approaches to Nonlinear Problems in Circuits and Systems,* edited by F. M. A. Salam and M. L. Levi, SIAM, Philadelphia, 128–137.

Umberger, D. K., Grebogi, C., Ott, E., and Afeyan, B. (1989). "Spatiotemporal Dynamics in a Dispersively Coupled Chain of Nonlinear Oscillators," *Phys. Rev. A* **39**(9), 4835–4842.

Van Buskirk, R., and Jeffries, C. (1985). "Observation of Chaotic Dynamics of Coupled Nonlinear Oscillators," *Phys. Rev. J. A* **31**(5), 3332–3357.

Van der Pol, B., and Van der Mark, J. (1927). "Frequency Demultiplication," *Nature* **120** (3019), 363–364.

Van Dyke, M. (1982). *An Album of Fluid Motion,* Parabolic Press.

Vastano, J. A., Russo, T., and Swinney, H. L. (1990). "Bifurcation to Spatially Induced Chaos in a Reaction-Diffusion System," *Physica D* **46**, 23–42.

Viet, O., Westfreid, X., and Guyon, E. (1983). "Art cinetique et chaos méchanique, *Eur. J. Phys.* **4**, 74–76.

Virgin, L. N. (1986). "The Nonlinear Rolling Response of a Vessel Including Chaotic Motions Leading to Capsize in Regular Seas," *Applied Ocean Research* **9**, 89–95.

Widmann, P. J, Gorman, M., and Robbins, K. A. (1989). "Nonlinear Dynamics of a Convection Loop II. Chaos in Laminar and Turbulent Flows," *Physica D* **36**, 157–166.

Wiggins, S. (1988). *Global Bifurcations and Chaos,* Springer-Verlag, New York.

Wiggins, S. (1990). *Introduction to Applied Nonlinear Dynamical Systems and Chaos,* Springer-Verlag, New York.

Winful, H. G., Chen, Y. C., and Liu, J. M. (1986). "Frequency Locking, Quasiperiodicity, and Chaos in Modulated Self-Pulsing Semiconductor Lasers," *Appl. Phys. Lett.* **48**, 616–618.

Wisdom, J., Peale, S. J., and Mignard, F. (1984). "The Chaotic Rotation of Hyperion," *Icarus* **58**, 137–152.

Wolf, A. (1984). "Quantifying Chaos with Lyapunov Exponents," *Nonlinear Scientific Theory and Applications,* edited by A. V. Holden, Manchester University Press, Manchester, England.

Wolf, A., Swift, J. B., Swinney, H. L., and Vasano, J. A. (1985). "Determining Lyapunov Exponents from a Time Series," *Physica* **16D**, 285–317.

Wolfram, S. (1984). "Universality and Complexity in Cellular Automata," *Physica* **10D**, 1–35.

Wolfram, S. (1986). *Theory and Applications of Cellular Automata,* World Scientific Publications, Singapore.

Yorke, J. A., Yorke, E. D., and Maller-Paret, J. (1985). "Lorenz-like Chaos in Partial Differential Equations for a Heated Fluid Loop," University of Maryland Report.

Zaslavsky, G. M. (1978). "The Simplest Case of a Strange Attractor," *Phys. Lett. A* **69**(3), 145–147.

Zaslavsky, G. M., and Chirikov, B. V. (1972). "Stochastic Instability of Nonlinear Oscillations," *Sov. Phys. Usp.* **14**(5), 549–672.

Zaslavsky, G. M., Sagdeev, R. Z., Usikov, D. A., and Chernikov, A. A. (1991). *Weak Chaos and Quasi-Regular Patterns,* Cambridge University Press, Cambridge, England

Zhu, Z.-X. (1983). "Experiment on the Chaotic Phenomena of an Upside Down Pendulum," Report of Laboratory of General Mechanics, Beijing University.

AUTHOR INDEX

Page numbers in *italics* refer to References.

SUBJECT INDEX